The Geography of South America

A Scholarly Guide and Bibliography

Thomas A. Rumney

THE SCARECROW PRESS, INC.
Lanham • Toronto • Plymouth, UK
2013

Published by Scarecrow Press, Inc.
A wholly owned subsidiary of The Rowman & Littlefield Publishing Group, Inc.
4501 Forbes Boulevard, Suite 200, Lanham, Maryland 20706
www.rowman.com

10 Thornbury Road, Plymouth PL6 7PP, United Kingdom

British Library Cataloguing in Publication Information Available

Library of Congress Cataloging-in-Publication Data

Rumney, Thomas A.
The geography of South America : a scholarly guide and bibliography / Thomas A. Rumney.
Includes bibliographical references and index.
ISBN 978-0-8108-8634-6 (cloth : alk. paper) — ISBN 978-0-8108-8635-3 (ebook)
1. South America—Geography—Bibliography. I. Title.
Z6027.S68R85 2013 [F2211.5]
015.98—dc23
2012051744

♾™ The paper used in this publication meets the minimum requirements of American National Standard for Information Sciences Permanence of Paper for Printed Library Materials, ANSI/NISO Z39.48-1992.

Printed in the United States of America

Contents

Introduction

Lying far to the south and southeast of North America is the massive continent of South America. An area of significance, vast resources and environmental differences, the origin of many foods and products of the world, and huge populations of native peoples and people derived from around the world, South America has played many important roles for humankind for millennia. The lands and peoples of South America have encouraged a large body of research and publications that encompass the many subdivisions of the discipline of geography. It is the goal of this compendium to organize and present as many of these scholarly publications as possible to assist and stimulate efforts in the teaching, study, and continuing scholarship of the geography of this region. As a matter of clarity, the areas and countries covered here include Argentina, Bolivia, Brazil, Chile, Colombia, Ecuador, French Guiana, Guyana, Paraguay, Peru, Surinam, Uruguay, and Venezuela, as well as the continent as a whole.

The organization begins with a chapter on the region as a whole, followed by chapters on the nations of the continent. Within each chapter, the organization is arranged to identify and record works about the subfields of geography. These begin with a General Works section, followed by sections on cultural and social geography, economic geography, historical geography, physical and environmental geography, political geography, and urban geography. Each section is then subdivided into topics specific to those areas of geographical specialization. The types of entries include atlases, books, monographs, textbooks, book chapters, scholarly articles, doctoral dissertations, and master's theses. Most of these entries are written in English, Spanish, and Portuguese, but others are written in German, Dutch, French, and other languages. Where possible, English titles translated from other languages are provided. The inclusions for each section and subtopic are arranged alphabetically by author's last name.

While the totality of entries recorded here represents a majority of research and published works done by geographers on the various aspects of the geographical study of South America and from a very early date, it is probable that some works have been missed. This is my responsibility, and I hope that at least two types of services are provided by this volume. The first is that this is a convenient and useful collection of existing published references on the geography of South America that can assist teachers, students, and scholars in the study and further research of the region. The second is to offer a view of what has not been studied and written about on the various aspects of the region's geographies. Such knowledge could encourage more efforts by geographers to further our knowledge and understanding of South America and its interesting geography.

List of Journals Used

The following scholarly journals served as sources of entries for the inclusions in this collection.

Acta Universitatis Carolinae, Geographica
Agricultural and Forest Meteorology
Agro-Ecosystems
Amazonia
Ambio
American Antiquity
American Journal of Economics and Sociology
Americas
Anales de geografía de la Universidad Complutense
Annales de Geographie
Annals, Japan Association of Economic Geographers
Annals of the American Association of Geographers
Annals of Regional Science
Annals of Tourism Research
Anthropologia
Anthropos
Antipode
Applied Geography
Applied Geography and Development
Applied Sciences and Development
Arctic and Alpine Research
Area
Australian Geographer
Australian Geographical Studies
BC Geographical Series/Western Geography

Bioscience
Bulletin de l'Association de Geographes Francais
Bulletin, Illinois Geographical Society
Bulletin of Latin American Studies
Bulletin of the American Geographical Society
Bulletin of the Geographical Society of Philadelphia
Bulletin of the North Dakota Geographical Society
Bulletin of the Pan American Union
Cahiers de Geographie du Quebec
California Geographer
Canadian Geographer
Canadian Journal of Development Studies
Canadian Journal of Latin American and Caribbean Studies
Canadian Journal of Remote Sensing
Caribbean Geography
Caribbean Quarterly
Cartographica
Catena
Chiri Kagaku/Geographical Sciences
Cities
Climatic Change
Cold Regions Science and Technology
Condor
Contemporary Review
Culture and Agriculture
Current Anthropology
Demography
Development and Change
Earth Surface Processes and Landforms
Earth Surfaces Reviews
East Lakes Geographer
Ecology
Economic Botany
Economic Development and Cultural Change
Economic Geography
Ecumene/Cultural Geographies
Educational Bi-monthly
Ekistics
Environment and Planning A
Environment and Planning D
Environmental Pollution
Die Erde
Erdkunde

L'Espace Geographique
Espace, Populations, Societe
Espacio y Dessarrollo
Ethnographical Studies
Explorers Journal
Fennia
Forest Ecology and Management
Gender, Place, and Culture
Geoforum
Geografisch Tijdshrift, Netherlands
Geografiska Annaler A
Geografiska Annaler B
Geographia Polonica
Geographica Helvetica
Geographical Analysis
Geographical Bulletin
Geographical Journal
Geographical Perspectives
Geographical Review
Geographical Review of Japan
Geographical Survey
Geographie et Cultures
Geography
Geography Compass
Geography Research Forum
Geography Review
Geography Review, Australia
GeoJournal
Geomorphology
Geopolitics
Geoscope
Great Plains–Rocky Mountain Geographical Journal
Growth and Change
Health and Place
Hispanic American Historical Review
Historical Geography
The Holocene
Housing Studies
Human Organization
International Journal of Administrative Sciences
International Journal of Remote Sensing
Jimbun Chiri/Human Geography
Journal of Biogeography

Journal of Climate
Journal of Coastal Research
Journal of Cultural Geography
Journal of Developing Areas
Journal of Ecology
Journal of Economic Geography
Journal of Environmental Management
Journal of Geography
Journal of Glaciology
Journal of Historical Geography
Journal of Hydrology
Journal of Latin American Geography
Journal of Latin American Studies
Journal of Quaternary Science
Journal of Transport Geography
Journal of Urban Economics
Journal of Wine Research
Latin American Studies, Japan
Luso-Brazilian Review
Malaysian Journal of Tropical Geography
Middle States Geographer
Names
Natural History
Nature
Norsk Geografisk Tidsskrift
Ohio Geographer
Ontario Geographer
Papers of the Applied Geography Conferences
Pennsylvania Geographer
Permafrost and Periglacial Processes
Physical Geography
Polar Geography
Political Geography
Proceedings of the Association of American Geographers
Professional Geographer
Progress in Human Geography
Quaternary Research
Quaternary Science Reviews
Quaternary Studies
Regional Science Review
Regional Studies
Remote Sensing of the Environment
Resource Management and Optimization

Revista Geografica
Revista Geografica Academica
Revue Belge de Geographie
Revue de Geographie Alpine
Rocky Mountain Social Science Journal
Rural Systems
Science
Science Reports, Tohoku University, Series 7, Geography
Scientific American
Scientific Monthly
Scottish Geographical Journal/Magazine
Sedimentology
Singapore Journal of Tropical Geography
Smithsonian Institute Report
Social and Cultural Geography
Social and Economic Studies
Social Science Journal
Social Science and Medicine
Society and Natural Resources
Southeastern Geographer
Space and Polity
Technology and Culture
Tellus
Terra Incognita
Third World Planning Review
Tijdschrift voor Economische en Sociale Geographie
Tools and Tillage
Tourism Geographies
Town Planning Review
Transactions, Institute of British Geographers
Transactions, New York Academy of Science
Travaux et Documents de Geographie Tropicale, France
Tropical Ecology
Tsukuba Studies in Human Geography, Japan
Urban Ecosystems
Urban Geography
Urban Studies
Vegetatio
Yearbook, Association of Pacific Coast Geographers
Yearbook, Conference of Latin Americanist Geographers
Zeitschrift fur Geomorphologie
Zeitschrift fur Wirtschaftsgeographie

Chapter One

The Region as a Whole

GENERAL WORKS

Atlases and Graphic Presentations

Brawer, Moshe. *Atlas of South America*. New York: Simon & Schuster, 1991.

Herlihi, Peter H. "Indigenous Mapmaking in the Americas: A Typology." In *Cultural and Physical Exposition: Geographic Studies in the Southern United States and Latin America*, edited by Michael K. Steinberg and Paul F. Hudson, 133–50. Baton Rouge: Louisiana State University, Geoscience and Man 36, 2002.

Hesse, Ralf. "Using SRTM to Quantify Size Parameters and Spatial Distribution of Endorheic Basic in Southern South America." *Revista Geografica Academica* 2, no. 2 (2008): 5–13.

Matthews, R. P. *Regional Atlas of Latin America*. Portsmouth, England: Portsmouth Polytechnic Institute, Department of Geography, 1984.

Platt, Raye R. "Catalogues of Maps of Hispanic America." *Geographical Review* 23, no. 4 (1933): 660–63.

———. "The Millionth Map of Hispanic America." *Geographical Review* 17 (1927): 301–8.

———. "Surveys in Hispanic America." *Geographical Review* 20 (1930): 138–42.

Rich, John L. *The Face of South America: An Aerial Traverse*. New York: American Geographical Society, 1942.

Schneider, Ronald M., and Robert C. Kingsbury. *An Atlas of Latin American Affairs*. New York: Praeger, 1965.

Walker, Robert. "Mapping Process to Pattern in the Landscape Change of the Amazonian Frontier." *Annals of the Association of American Geographers* 93, no. 2 (2003): 376–98.

Whitehead, Neil L. "Indigenous Cartography in Lowland South American and the Caribbean." In *The History of Cartography*, edited by David Woodward and M. Lewis, 301–26. Chicago: University of Chicago Press, 1998.

Books, Monographs, and Texts

Bebbington, Anthony J. *Geographies of Development in Latin America*. New York: Routledge, 2006.

Black, Jan K., ed. *Latin America: Its Problems and Its Promise: A Multidisciplinary Introduction*. Boulder, CO: Westview, 1984.

Blakemore, Harold, and Clifford T. Smith, eds. *Latin America: Geographical Perspectives.* 2nd ed. London: Methuen, 1983.
Blouet, Brian, and Olwyn Blouet. *Latin America and the Caribbean: A Systematic and Regional Survey.* New York: Wiley, 1993, 1997, and 2005.
———, eds. *Latin America: An Introductory Survey.* New York: Wiley, 1982.
Boehm, Richard G., and Sent Visser, eds. *Latin America: Case Studies.* Dubuque, IA: Kendall/ Hunt, 1984.
Borsdorf, Axel, ed. *Lateinamerika-Krise Ohne Ende?* Innsbruck, Austria: Innsbrucker Geographische Studien, Band 21, 1994.
Bowman, Isaiah. *Desert Trails of Atacama.* New York: American Geographical Society, 1924.
———. *South America: A Geography Reader.* Chicago: Rand McNally, 1915.
Butland, Gilbert J. *Latin America: A Regional Geography.* New York: Wiley, 1972.
Carslon, Fred A. *Geography of Latin America.* New York: Prentice Hall, 1952.
Caviedes, Cesar N., and Gregory W. Knapp. *South America.* Englewood Cliffs, NJ: Prentice Hall, 1995.
Clawson, David L. *Latin America and the Caribbean: Lands and Peoples.* Dubuque, IA: Brown, 1996.
Cole, John P. *Latin America: An Economic and Social Geography.* Totowa, NJ: Rowman & Littlefield, 1975.
Crist, Raymond E., Robert Harper, and Clarence Sorenson. *Learning about Latin America.* Morristown, NJ: Silver Burdett, 1961.
Crist, Raymond E., and Charles M. Nissly. *East from the Andes.* Gainesville: University of Florida, Social Science Monograph 50, 1973.
Devenan, William M., ed. *Hispanic Lands and Peoples: Selected Writings of James J. Parsons.* Boulder, CO: Westview, Dellplain Latin American Studies 23, 1989.
Dym, Jordana, and Karl Oflen, eds. *Mapping Latin America: A Cartographic Reader.* Chicago: University of Chicago Press, 2011.
Gauhar, Atlof, ed. *Regional Integration: The Latin American Experience.* Boulder, CO: Westview, 1986.
Gavina, Alejandro, Eduardo Lara, and John Gallup. *Is Geography Resting? Lessons from Latin America.* Palo Alto, CA: Stanford University Press, 2003.
Gwynne, Robert N., and Cristobal Kay. *Latin America Transformed: Globalization and Modernity.* London: Arnold, 1999.
Jackiewicz, Edward, and Fernando J. Bosco, eds. *Placing Latin America: Contemporary Themes in Latin America.* Lanham, MD: Rowman & Littlefield, 2008.
James, Preston E., and Clarence W. Minkel. *Latin America.* New York: Wiley, 1986.
Jones, Clarence F. *South America.* New York: Holt, 1930, 1942.
Kent, Robert B. *Latin America: Regions and People.* New York: Guilford, 2005.
Lee, Jeon. *Geography of Latin America.* Seoul, South Korea: Minum-sa, 1994.
Morris, Arthur S. *South America.* Totowa, NJ: Barnes and Noble, 1980, 1987.
Odell, Peter R., and David A. Preston. *Economies and Societies in Latin America.* Chichester, England: Wiley, 1973, 1978.
Ogilvie, Alan G. *Geography of the Central Andes.* New York: American Geographical Society, 1922.
Place, Susan E., ed. *Tropical Rainforests: Latin American Nature and Society in Transition.* Wilmington, DE: Scholarly Resources, 1993.
Platt, Robert S. *Latin America: Countryside and United Regions.* New York: McGraw-Hill, 1942.
Preston, David A., ed. *Latin American Development: Geographical Perspectives.* Essex, England: Longman, 1996.
Renner, John M. *Source Book on South American Geography.* Wellington, New Zealand: Hicks-Smith, 1972.
Rudolph, William E. *Vanishing Trails of Atacama.* New York: American Geographical Society, 1963.
Schmeider, Oscar. *Geografia de America Latino.* Mexico City, Mexico: Fondo de Cultura Economica, 1965.

Schmink, Marianne, and Charles Wood, eds. *Frontier Expansion in Amazonia*. Gainesville: University of Florida Press, 1984.

Shanahan, E. W. *South America: An Economic and Regional Geography*. London: Methuen, 1963.

Smith, Nigel J. H. *Amazonia: Resiliency and Dynamism of the Land and Its People*. New York: UN University Press, 1995.

———. *The Enchanted Amazon Rain Forest: Stories from a Vanishing World*. Gainesville: University of Florida Press, 1996.

———. *Man, Fishes, and the Amazon*. New York: Columbia University Press, 1981.

Stanford, Edward. *Notes on the Geography of South America, Physical and Political*. London: E. Stanford, 1873.

Stohr, Walter B. *Regional Development Experiences and Prospects in Latin America*. The Hague, Netherlands: Mouton, 1975.

Taylor, Alice, ed. *Focus on South America*. New York: Praeger and the American Geographical Society, 1973.

Timmons-Roberts, J., and N. D. Thomas. *Trouble in Paradise: Globalization and Environmental Crisis in Latin America*. New York: Routledge, 2003.

Vadjunec, Jacqueline, and Marianne Schmink, eds. *Amazonian Geographies*. New York: Routledge, 2011.

Wagley, Charles, ed. *Man in the Amazon*. Gainesville: University of Florida Press, 1974.

West, Robert C., ed. *Andean Reflections: Letters from Carl O. Sauer While on a South American Trip under a Grant from the Rockefeller Foundation, 1942*. Boulder, CO: Westview, Dellplain Latin American Studies 11, 1982.

Wilkie, Richard W. *Latin American Population and Urbanization Analysis: Maps and Statistics, 1950–1982*. Westwood: University of California, Los Angeles Latin American Center, 1984.

Articles and Book Chapters

Amilhat Szary, A. "L'Integration continentale aux marges du mercosur. Les echelles d'un processes trans-frontalier et transordin." *Revue de Geographie Alpine* 91, no. 3 (2003): 47–56.

Barker, Maude E. "An Avenue of Approach to the Study of South America." *Journal of Geography* 23 (1924): 238–44.

Barrett, John. "South America: Its General Geographical Features and Opportunities." *Bulletin of the Geographical Society of Philadelphia* 7, no. 2 (1909): 1–6.

Bebbington, Anthony J. "Globalized Andes? Livelihoods, Landscapes, and Development." *Ecumene* 8, no. 4 (2001): 414–36.

———. "Latin America: Contesting Extraction, Producing Geographies." *Singapore Journal of Tropical Geography* 30, no. 1 (2009): 7–12.

———. "Reencountering Development: Livelihood Transitions and Place Transformations in the Andes." *Annals of the Association of American Geographers* 90, no. 3 (2000): 495–520.

Bebbington, Anthony J., and D. H. Bebbington. "Development Alternatives: Practice, Dilemmas and Theory." *Area* 33, no. 1 (2001): 7–17.

Bowman, Isaiah. "Results of an Expedition to the Central Andes." *Bulletin of the American Geographical Society* 46, no. 3 (1914): 161–84.

Capitanelli, Ricardo G. "Patagonia." *Revista Geografica* 95 (1982): 30–45.

Carlson, Fred A. "Notes on the Amazon Valley." *Journal of Geography* 55, no. 6 (1954): 277–85.

Caviedes, Cesar N. "Globalization in Geography: A Latin American Perspective." *GeoJournal* 45, nos. 1–2 (1998): 97–100.

———. "Tangible and Mythical Places in Jose M. Arguedas, Gabriel Garcia Marques, and Pablo Neruda." *GeoJournal* 38, no. 1 (1996): 99–107.

Crist, Raymond E. "The Land and People of the Guajira Peninsula." In *Smithsonian Institute Report for 1957*, 339–55. Washington, DC: GPO, Publication 4324, 1958.

Crist, Raymond E., and Edward P. Leahy. "Four Profiles." In *Focus on South America*, edited by Alice Taylor, 15–21. New York: Praeger, 1973.

Cunningham, Allan. "Explorations in the Northernmost Andes." *Geographical Journal* 123, no. 3 (1957): 344–55.

Denevan, William M. "Latin America." In *World Systems of Traditional Management*, edited by G. A. Klee, 217–44. London: Arnold, 1980.

Driever, Steven L. "Geographic Narratives in the South American 53 Travelogues of Harry A. Franck: 1917–1943." *Journal of Latin American Geography* 10, no. 1 (2011): 53–70.

Farnbee, William C. "A Pioneer in Amazonia: The Narrative of a Journey from Manaus to Georgetown." *Bulletin of the Geographical Society of Philadelphia* 15, no. 2 (1917): 51–103.

Fifer, J. Valerie. "Andes Crossing: Old Tracks and New Opportunities at the Uspallata Pass." *Yearbook, Conference of Latin Americanist Geographers* 20 (1994): 35–48.

Furley, Peter A. "Environmental Issues and the Impact of Development." In *Latin American Development: Geographical Perspectives*, edited by David A. Preston, 70–115. London: Longman, 1996.

Hale, Albert. "The River Plate Region: Its Possibilities." *Bulletin of the Geographical Society of Philadelphia* 7 (1909): 21–27, 67–73.

Higbee, Edward C. "Of Man and the Amazon." *Geographical Review* 41, no. 3 (1951): 401–20.

Hitchcock, Charles B. "Fourth General Assembly of the Pan American Institute of Geography and History, with Notes on a Trip from Caracas to Bogota." *Geographical Review* 37 (1947): 121–36.

James, Preston E. "Studies of Latin America." In *On Geography: Selected Writings of Preston E. James*, edited by Donald W. Meinig, 197–211. Syracuse, NY: Syracuse University Press, 1970.

Johnson, Emory R. "Some Impressions of South America." *Bulletin of the Geographical Society of Philadelphia* 21, no. 2 (1923): 16–23.

Keeling, David J. "Latin American Development and the Globalization Imperative: New Directions, Familiar Cases." *Journal of Latin American Geography* 3, no. 1 (2004): 1–21.

Kohlhepp, Gerd. "A Challenge to Science and Regional Development Policy: Reflections on the Future Development of Amazonia." *Applied Geography and Development* 33 (1989): 52–67.

Lawson, Victoria A. "Institutional Research and Philosophical Domains of Concern." In *Future Directions in Latin Americanist Geography: Research Agendas for the Nineties and Beyond*, edited by Gary S. Elbow, 13–24. Auburn, AL: Conference of Latin Americanist Geographers, Special Publication 3, 1992.

LeCointe, P. "Le bas Amazone." *Annales de Geographie* 12 (1903): 54–66.

Light, Mary, and Richard Light. "Atacama Revisited: Desert Trails Seen from the Air." *Geographical Review* 36, no. 4 (1946): 525–45.

McClintock, John C. "The Battle of the Amazon." *Journal of Geography* 43 (1944): 27–33.

Miller, Leo E. "The Descent of the Rio Gy-Parana." *Geographical Review* 1 (1916): 169–91.

Morong, Thomas. "The Rio de la Plata: Its Basin, Geography, and Inhabitants." *Journal of the American Geographical Society of New York* 24 (1892): 479–509.

Murphy, Robert C. "The Littoral of Pacific Colombia and Ecuador." *Geographical Review* 29, no. 1 (1939): 1–33.

Naylor, Simon, and Gareth A. Jones. "Writing Orderly Geographies of Distant Places: The Regional Survey Movement and Latin America." *Ecumene* 4, no. 3 (1997): 273–99.

Parsons, James J. "Geography." In *Latin American and Caribbean Studies: A Critical Guide*, edited by P. H. Covington, 267–77. Westport, CT: Greenwood, 1992.

Platt, Robert S. "Reconnaissance in Dynamic Regional Geography: Tierra del Fuego." *Revista Geografica*, 13–24 (1949): 3–22.

Posey, C. J. "Some Points in the Geography of South America." *Journal of Geography* 12 (1913): 74–79.

Preston, David A. "Contemporary Issues in Rural Latin America." *Progress in Human Geography* 7 (1983): 276–92.

——. "Themes in Contemporary Latin America." In *Latin American Development: Geographical Perspectives*, edited by David A. Preston. Essex, England: Longman, 1996.

Radley, Ian. "The South American Dream." *Geography Review* 20, no. 3 (2007): 21–25.

Rice, A. Hamilton. "Plans for Exploration at the Headwaters of the Branco and Orinoco." *Geographical Review* 15 (1925): 115–22.

——. "The Rio Negro, the Casiquiane Canal and the Upper Orinoco." *Bulletin of the Geographical Society of Philadelphia* 20, no. 1 (1922): 1–27.

Scherm, Georg. "The Guayanan Countries." *Applied Geography and Development* 29 (1987): 27–43.

Smith, Nigel J. H. "Amazonia." In *Regions at Risk: Comparisons of Threatened Environments*, edited by Jeanne V. Kasperson, Roger E. Kasperson, and B. L. Turner II, 41–91. Tokyo, Japan: UN University Press, 1995.

Stears, J. A., and L. Dudley Stamp. "South American Prospect." *Geographical Journal* 123, no. 3 (1957): 329–43.

Sternberg, Rolf. "Environment, Energy, and Population Shifts: The View from Asia and South America." *Revista Geografica* 137 (2005): 123–56.

Sundberg, Juanita. "Looking for the Critical Geographer, or Why Bodies and Geographies Matter to the Emergence of Critical Geographies of Latin America." *Geoforum* 36, no. 1 (2005): 17–28.

Tower, Walter S. "The Andes as a Factor in South American Geography." *Journal of Geography* 15 (1916): 1–8.

Townsend, Janet G. "Rural Change: Progress for Whom?" In *Latin American Development: Geographical Perspectives*, edited by David A. Preston, 199–228. Harlow, England: Longman, 1987.

Troll, Carl. "An Expedition to the Central Andes." *Geographical Review* 19 (1929): 234–47.

Valverde, Orlando. "L'Amazonie a la fin du XXe siècle." *Les Cahiers d'Outre Mer* 49, no. 1 (1996): 53–94.

Van Paasen, C. "The General Situation of Latin and Especially Tropical Spanish America at Mid-century." *Tijdschrift voor Economische en Sociale Geographie* 56, no. 1 (1965): 161–70.

White, C. Langdon. "Is the Twentieth Century South America's?" *Economic Geography* 21, no. 3 (1945): 79–87.

——. "Sleepy Orinoco Valley Comes to Life." *Journal of Geography* 55 (1956): 111–21.

Williams, Frank E. "Crossing the Andes at 41 Degrees South." *Bulletin of the Geographical Society of Philadelphia* 32 (1934): 1–9.

Wilson, Huntington. "From Mollendo to La Paz Via Cuzco." *Bulletin of the Geographical Society of Philadelphia* 16 (1918): 1–22.

Zimmerer, Karl S. "Humbolt's Nodes and Modes of Interdisciplinary Environmental Science in the Andean World." *Geographical Review* 96, no. 3 (2006): 335–60.

Theses and Dissertations

Bowman, Isaiah. "The Geography of the Central Andes." PhD diss., Yale University, New Haven, CT, 1909.

Hernandez, Jose L. "Latin America: A Regional Overview." Master's thesis, State University of New York, Buffalo, 1990.

CULTURAL AND SOCIAL GEOGRAPHY

General Works

Butland, Gilbert J. "Southern Affinities and Contrasts: A Comparative Review of the Cultural Landscapes of South America and Australia." *Revista Geografica* 63 (1965): 117–30.

Crist, Raymond E. "Acculturation in the Guajira." In *Smithsonian Institute Report for 1958*, 481–99. Washington, DC: Smithsonian Institute, Publication 4370, 1959.

———. "The Latin American Way of Life, I: A Culturally Diverse Continent-Balkanized." *American Journal of Economics and Sociology* 27, no. 2 (1968): 63–7.

———. "Some Aspects of Human Geography in Latin American Literature." *American Journal of Economics and Sociology* 21 (1962): 407–12.

Driever, Steven L. "The Conquest Considered in Pablo Neruda's Canto general." *Journal of Cultural Geography* 13, no. 1 (1992): 69–84.

Edwards, Clinton R. *Aboriginal Watercraft on the Pacific Coast of South America*. Berkeley: University of California Press, Ibero-Americana 42, 1965.

Forman, Sylvia H. "The Future Value of the 'Verticality' Concept: Implications and Possible Applications in the Andes." In *Human Impact on Mountains*, edited by Nigel Allan et al., 133–53. Lanham, MD: Rowman & Littlefield, 1988.

Gade, Daniel. "Carl Troll on Nature and Culture in the Andes." *Erdkunde* 50, no. 4 (1996): 301–16.

———. "The Guinea Pig in Andean Folk Culture." *Geographical Review* 57, no. 2 (1967): 213–24.

———. "Landscape System and Identity in the Post-conquest Andes." *Annals of the Association of American Geographers* 82, no. 3 (1992): 461–77.

———. "Lightning in the Folk Life and Religion of the Central Andes." *Anthropos* 78 (1983): 778–88.

———. *Nature and Culture in the Andes*. Madison: University of Wisconsin Press, 1999.

Hills, Theo L. "The Ecology of Hazardousness, the Experience of South America." *GeoJournal* 6, no. 2 (1982): 151–56.

Kleinpenning, J. M. G., ed. *Competition for Rural and Urban Spaces in Latin America: Its Consequences for Low Income Groups*. Nijmejen, Netherlands: Knag/Geografisch en Planologisch Institut Katholieke Universiteit Nijmegen, 1986.

Knapp, Gregory W. "The Andes: Personal Reflections on Cultural Change, 1977–2010." *Journal of Cultural Geography* 27, no. 3 (2010): 307–16.

———. "Cultural and Historical Geography of the Andes." *Yearbook, Conference of Latin Americanist Geographers*, 17–18 (1990): 165–76.

———. "Potential Ethnic Territories: Mapping Linguistic Data from Modern Andean Censuses." Texas Papers on Latin America 89–13. Austin: University of Texas, Institute of Latin American Studies, 1989.

Lauer, Wilhelm. "Human Development and Environment in the Andes: A Geoecological Overview." *Mountain Research and Development* 13, no. 2 (1993): 157–66.

Mathewson, Kent. "Human Geography of the American Tropics: A Forty-Year Review." *Singapore Journal of Tropical Geography* 14, no. 2 (1993): 123–56.

Mathewson, Kent, and Michael Yoder. "Human Ecology." In *Future Directions in Latin Americanist Geography: Research Aspects for the Nineties and Beyond*, edited by Gary S. Elbow, 38–44. Auburn, AL: Conference of Latin Americanist Geographers, Special Publication 3, 1992.

Nelson, Burton, and Robert Aron. "The Bolas: Its Origin and Distribution." *Geographical Perspectives* 59 (1987): 111–15.

Preston, David A. "Geographers among the Peasants: Research on Rural Societies in Latin America." *Progress in Geography* 6 (1975): 143–78.

Reboratti, Carlos E. "Human Geography in Latin America." *Progress in Human Geography* 6 (1982): 397–407.

Tavernor, Jean. "Musical Themes in Latin American Culture: A Geographical Appraisal." *Bloomsbury Geographer* 3 (1970): 60–66.

Thomas, R. Brooke. "Simulation Models of Andean Adaptability and Change." In *Human Impact on Mountains*, edited by Nigel Allen et al., 165–84. Lanham, MD: Rowman & Littlefield, 1983.

Valjunec, Jacqueline, et al. "New Amazonia Geographies: Emerging Identities and Landscapes." *Journal of Cultural Geography* 28, no. 1 (2011): 1–20.

Young, Kenneth R. "Wildlife Conservation in the Cultural Landscape of the Central Andes." *Landscape and Urban Planning* 38 (1997): 137–47.

The Built Environment

Gade, Daniel. "Bridge Types in the Central Andes." *Annals of the Association of American Geographers* 62 (1972): 94–109.

———. "Grist Milling with the Horizontal Waterwheel in the Central Andes." *Technology and Culture* 12, no. 1 (1971): 43–51.

Gilbert, Alan. "Housing during Recession: Illustrations from Latin America." *Housing Studies* 4 (1989): 155–66.

———. *In Search of a Home: Rental and Shared Housing in Latin America*. London: University College of London Press, 1992.

Jett, Stephen C. "Cairn Trail Shrines in Middle and South America." *Yearbook, Conference of Latin Americanist Geographers* 20 (1994): 1–8.

Parsons, James J. "Giant American Bamboo in the Vernacular Architecture of Colombia and Ecuador." *Geographical Review* 81, no. 2 (1991): 131–52.

Medical Geography

Betancourt, Jose. "Different Roads to a Common Goal: The Lowering of Infant Mortality Rates in Latin America." *Revista Geografica* 107 (1988): 49–66.

Curton de Casas, Susana I. "Geographical Inequalities in Mortality in Latin America." *Social Science and Medicine* 36, no. 10 (1993): 1349–56.

Gagnan, A. S., et al. "The El Nino Southern Oscillation and Malaria Epidemics in South America." *International Journal of Biomedicine* 46 (2002): 81–89.

Hunter, John M. "Bot-Fly Maggot Infestation in Latin America." *Geographical Review* 80, no. 4 (1990): 382–98.

May, Jacques M., and D. L. McLellan. *The Ecology of Malnutrition in Eastern South America*. New York: Hafner/Macmillan, 1974.

———. *The Ecology of Malnutrition in Western South America*. New York: Hafner/Macmillan, 1974.

Prothero, R. M. "Malaria in Latin America: Environmental and Human Factors." *Bulletin of Latin American Research* 14, no. 3 (1995): 357–65.

Scarpaci, Joseph L. "Medical Care, Welfare State, and Deindustrialization in the Southern Cone." *Environment and Planning D* 8, no. 2 (1990): 191–210.

———. "Physician Proletarianization and Medical Care Restructuring in Argentina and Uruguay." *Economic Geography* 66, no. 4 (1990): 362–77.

———. "Primary-Care Decentralization in the Southern Cone: Shantytown Health Care as Urban Social Movement." *Annals of the Association of American Geographers* 81, no. 1 (1991): 103–26.

Scarpaci, Joseph L., and Ignacio Irrarazabol. "Caring for People: Health Care and Education Provision." In *Latin American Development: Geographical Perspectives*, edited by David A. Preston, 188–215. London: Longman, 1994.

Weil, Connie, ed. *Medical Geographic Research in Latin America*. Elmsford, NY: Pergamon, 1982.

Ethnic, Social, and Population Geography

Aragon, Luis. "El problema migratorio en la pan-amazonia: Una aproximacion teorica y una alternative metodologica." *Revista Geografica* 97 (1983): 44–55.

Aschmann, Homer. "The Persistent Guajiro." *Natural History* 84, no. 3 (1975): 28–37.

Birraux-Ziegler, Pierrette. "Les Yamomani, la foret et les Blancs." *Geographie et Cultures* 4 (1992): 25–34.

Bowman, Isaiah. "Regional Population Groups of Atacama, Part 1." *Bulletin of the American Geographical Society* 41 (1909): 142–54.

———. "Regional Population Groups of Atacama, Part II." *Bulletin of the American Geographical Society* 41 (1909): 193–211.

Crist, Raymond E. "Andean America: Some Aspects of Human Migration and Settlement." *Annals of the Southwestern Conference on Latin American Studies* 1 (1970): 77–92.

———. "Andean America: Some Aspects of Human Migration and Settlement." Occasional Paper 3, Vanderbilt University, Graduate Center for Latin American Studies, Nashville, TN, 1964.

———. "The Indians in Andean America, I." *American Journal of Economics and Sociology* 23, no. 2 (1964): 131–43.

———. "The Indians of Andean America, II." *American Journal of Economics and Sociology* 23, no. 3 (1964): 303–13.

Crowley, William K. "The Levantine Arabs: Diaspora in the New World." *Proceedings of the Association of American Geographers* 6 (1974): 137–42.

Denevan, William M. "The Aboriginal Population of Tropical America: Problems and Methods of Estimation." In *Population and Economics*, edited by P. Deprez, 251–69. Winnipeg, Manitoba: University of Manitoba Press, 1970.

———. "The Aboriginal Population of Western Amazonia in Relation to Habitat and Subsistence." *Revista Geografica* 72 (1970): 61–86.

Denevan, William M., et al. "Feeding a Growing Population on an Increasingly Fragile Planet." In *Latin America in the Twenty-First Century: Challenges and Solutions*, edited by Gregory W. Knapp, 45–76. Austin: University of Texas Press and Conference of Latin Americanist Geographers, 2001.

Fantin, Maria A. "Analisis demografico de la poblacion en la franja fronteriza entre Argentina y Paraguay, anos 2001 y 2002." *Revista Geografica* 145 (2009): 149–68.

Furlong, Charles W. "Some Effects of Environment on the Fuegian Tribes." *Geographical Review* 3 (1917): 1–15.

Gade, Daniel. "Names for *Manihot esculenta*: Geographical Variation and Lexical Clarification." *Journal of Latin American Geography* 1, no. 1 (2002): 43–57.

Geiger, Pedro P. "Interregional Migrations in Latin America." In *People on the Move: Studies on Internal Migration*, edited by Leszek Kosinski and R. M. Prothero, 165–80. London: Methuen, 1975.

Gonzalez, Alfonso. "Latin America: Recent Trends in Population, Agriculture, and Food Supply." *Canadian Journal of Latin American and Caribbean Studies* 10, no. 20 (1985): 3–14.

———. "Population Geography of Mainland Hispanic America: Inventory of the 1980s." *Yearbook, Conference of Latin Americanist Geographers* 17–18 (1990): 99–108.

Greenow, Linda. "Geographic Perspectives on Latin American Women." *Yearbook, Conference of Latin Americanist Geographers*, 17–18 (1990): 231–38.

Guyot, Sylvain. "The Instrumentalization of Participatory Management in Protected Areas: The Ethnicization of Participation in the Kolla-Atacamena Region of the Central Andes of Argentina and Chile." *Journal of Latin American Geography* 10, no. 2 (2011): 9–36.

Haas, William H. "The Plateau Indians of South America." *Journal of Geography* 45, no. 6 (1946): 243–53.

Hanson, Earl. "Social Regression in the Orinoco and Amazon Basins." *Geographical Review* 23, no. 4 (1933): 578–98.

Heinenberg, H. *Investigaciones alemanos de geografia en America Latina*. Tubingen, Germany: Institut fur Wissentliche Zusammenarbeit, 1995.

Hiraoka, Mario. "Mestizo Subsistence in Riparian Amazonia." *National Geographic Research* 1 (1985): 236–46.

James, Preston E. "The Distribution of People in South America." In *Geographic Aspects of International Relations*, edited by C. C. Colby, 217–42. Chicago: University of Chicago Press, 1938.

Jefferson, Mark. "The Distribution of People in South America." *Bulletin of the Geographical Society of Philadelphia* 5 (1907): 1–11, 42–52, 182–192.

Jones, Richard C. "International Migration." In *Future Directions in Latin Americanist Geography: Research Agendas for the Nineties and Beyond*, edited by Gary S. Elbow, 31–37. Auburn, AL: Conference of Latin Americanist Geographers, Special Publication 3, 1992.

Kent, Robert B. "A Diaspora of Chinese Settlement in Latin America and the Caribbean." In *The Chinese Diaspora: Space, Place, Mobility, and Identity*, edited by Laurence J. C. Ma and Carolyn Cartier, 117–40. Lanham, MD: Rowman & Littlefield, 2003.

Lawson, Victoria A. "Hierarchical Households and Gendered Migration in Latin America: Feminist Extensions to Migration Research." *Progress in Human Geography* 22, no. 1 (1998): 39–53.

Lee, Jeon. "The Indian Culture and Aboriginal Heritage in Latin America." *Journal of Cultural and Historical Geography, South Korea* 4 (1992): 305–20.

Lowenthal, David. "Population Contrasts in the Guianas." *Geographical Review* 50, no. 1 (1960): 41–58.

Nolan, M. L., and S. D. Nolan. "The European Roots of Latin American Pilgrimage." In *Pilgrimage in Latin America*, edited by N. R. Crumrine and A. Marinas, 19–49. New York: Greenwood Press, 1991.

Nordenskiold, Erland. *An Ethno-geographical Analysis of the Material Culture of Two Indian Tribes in the Gran Chaco*. Goteburg, Sweden: Ethnological Studies 1, 1919.

Parsons, James J. "The Africanization of the New World Tropical Grasslands." *Tubinger Geographische Studien* 34 (1970): 141–53.

Preston, David A. *Environment, Society, and Rural Change in Latin America: Past, Present, and Future in the Countryside*. New York: Wiley Halsted, 1980.

———. "Negro, Mestizo, and Indian in an Andean Environment." *Geographical Journal* 131, no. 2 (1965): 220–35.

———. "People on the Move: Migrations Past and Present." In *Latin American Development: Geographical Perspectives*, edited by David A. Preston, 165–187. London: Longman, 1994.

———. "Rural Emigration in Andean America." *Human Organization* 28 (1969): 279–86.

Radcliffe, Sarah A. "Latin American Indigenous Geographies of Fear: Living in the Shadow of Racism, Lack of Development, and Antiterror Measures." *Annals of the Association of American Geographers* 97, no. 2 (2007): 385–97.

———. "Race, Gender, and Generation: Cultural Geographies." In *Latin American Development: Geographical Perspectives*, edited by David A. Preston, 146–64. London: Longman, 1994.

Radcliffe, Sarah A., and Nina Laurie. "Culture and Development: Taking Culture Seriously in Development for Andean Indigenous People." *Environment and Planning D* 24, no. 2 (2006): 231–48.

Robinson, David J. "The Language and Significance of Place in Latin America." In *The Power of Place: Bringing Together Geographical and Sociological Imaginations*, edited by John A. Agnew and James S. Duncan, 157–84. Boston: Unwin Hyman, 1989.

———. *Studies in Spanish American Population*. Boulder, CO: Westview, 1981.

Roorback, G. B. "The Food Surplus of South America in Time of War." *Journal of Geography* 16 (1917): 129–34.

Scarpaci, Joseph L. "Social Theory." In *Future Directions in Latin Americanist Geography: Research Agendas for the Nineties and Beyond*, edited by Gary S. Elbow, 7–12. Auburn, AL: Conference of Latin Americanist Geographers, Special Publication 2, 1992.

Shrestha, N. R., and J. G. Patterson. "Population and Poverty in Dependent States: Latin America Considered." *Antipode* 22, no. 2 (1990): 121–55.

Smole, William J. *The Yanomama Indians: A Cultural Geography*. Austin: University of Texas Press, 1976.

Stewart, Norman R. "On Population Pressures and Population Movements in the Central Andes." *Rocky Mountain Social Science Journal* 4, no. 1 (1967): 62–67.

Takasaki, Yoshito, et al. "Amazonian Peasants, Rain Forest Use, and Income Generation: The Role of Wealth and Geographical Factors." *Society and Natural Resources* 14, no. 4 (2001): 291–308.

Thomas, L. "Des peoples en voie de disparition: les Fuegiens." *Les Cahiers d'Outre Mer* 6 (1953): 379–98.

Timmons-Roberts, J., and F. Nii-Amoo Dodoo. "Population Growth, Sex Ratios, and Women's Work on the Contemporary Amazon Frontier." *Yearbook, Conference of Latin Americanist Geographers* 21 (1995): 91–106.
White, C. Landon. "Whither South America: Population and Natural Resources." *Journal of Geography* 60, no. 3 (1961): 103–11.

Community and Settlement Studies

Bromley, Raymond J. "Agricultural Colonization in the Upper Amazon Basin: The Impact of Oil Discoveries." *Tijdschrift voor Economische en Sociale Geographie* 63, no. 4 (1972): 278–94.
Butland, Gilbert J. "Frontiers of Settlement in South America." *Revista Geográfica* 65 (1966): 93–108.
Caviedes, Cesar N. "The Latin-American Boom Town in the Literary View of Jose Maria Arguedas." In *Geography and Literature: A Meeting of the Disciplines*, edited by William E. Mallory and P. Simpson-Housely, 57–80. Syracuse, NY: Syracuse University Press, 1987.
Charbonneau, M. "Scattered Development of Settlement and Grouping in the Andean Pastoral Systems [in French]." *Annales de Geographie* 118, no. 6 (2009): 637–58.
Chevalier, Jacques, and Cristina Carballo. "Residential Enclosure and the Quest for Togetherness in North and South America [in French]." *L'Information Geographique* 33, no. 4 (2004): 325–35.
Clawson, David L. "Obstacles to Successful Highland Colonization of the Amazon and Orinoco Basins." *American Journal of Economics and Sociology* 41, no. 4 (1982): 351–62.
Collins, Jane L. "Smallholder Settlement of Tropical South America: The Social Causes of Ecological Destruction." *Human Organization* 45, no. 1 (1986): 1–10.
Crist, Raymond E., and Charles M. Nissly. *East from the Andes: Pioneer Settlement in the South American Heartland*. Gainesville: University of Florida Press, Social Science Monograph 50, 1973.
Denevan, William M. "A Cultural-Ecological View of the Former Aboriginal Settlement in the Amazon Basin." *Professional Geographer* 18, no. 4 (1966): 346–51.
Eidt, Robert C. "Comparative Problems and Techniques in Tropical and Semi-tropical Settlements: Colombia, Peru, and Argentina." *Yearbook, Association of Pacific Coast Geographers* 26 (1964): 37–42.
Furlong, Charles W. "Tribal Distribution and Settlements of the Fuegians." *Geographical Review* 3 (1917): 169–87.
Greinacher, N. "Bekehrung durch Eroberung-Kritische Reflexion auf die Kolonisations-und missionsgeschichte in Lateinamerika." In *Lateinamerika-krise ohne Ende?* edited by Axel Borsdorf, 131–45. Innsbruck, Austria: Innsbrucker Geographische Studien, Band 21, 1994.
James, Preston E. "Expanding Frontiers of Settlement in Latin America: A Project for Future Study." *Hispanic American Historical Review* 21 (1941): 183–95.
Maos, Jacob O. *The Spatial Organization of New Land Settlement in Eastern South America*. Boulder, CO: Westview, 1984.
Renner, John M. "Indian Settlements of the Altiplano: Indications of Change." *New Zealand Journal of Geography* 51 (1971): 22–26.
Schurman, F. J. "From Resource Frontier to Periphery: Agricultural Colonization East of the Andes." *Tijdschrift voor Economische en Sociale Geographie* 69, nos. 1–2 (1978): 95–104.
Sjoholt, Peter. "Tropical Colonization: Some Important Research Issues and Their Wider Implications." *Norsk Geografisk Tidsskrift* 43, no. 2 (1989): 77–84.
Smith, Nigel J. H. *Rainforest Corridors: The Transmission Colonization Scheme*. Berkeley: University of California Press, 1982.
Stewart, Norman R. "Some Problems in the Development of Agricultural Colonization in the Andean Oriente." *Professional Geographer* 20, no. 1 (1968): 33–38.

Tourism and Recreation

Eyre, L. A. "The Tropical National Parks of Latin America and the Caribbean: Present Problems and Future Potential?" *Yearbook, Conference of Latin Americanist Geographers* 16 (1990): 15–33.

Gormsen, Erdmann. "Tourism in Latin America: Spatial Distribution and Impact on Regional Change." *Applied Geography and Development* 32 (1988): 65–80.

Meyer-Arendt, Klaus J. "Geographic Research on Tourism in Latin America, 1980–1990." *Yearbook, Conference of Latin Americanist Geographers*, 17–18 (1990): 199–208.

Scarpaci, Joseph L. *Plazas and Barrios: Heritage Tourism and Globalization in the Latin American Centro Historico*. Tucson: University of Arizona Press, 2005.

Place Names

Barr, Glenn. "South American Place Names." *Journal of Geography* 30 (1931): 79–81.

Columbus, Claudette K. "Map, Metaphor, Topos, and Toponym: Some Andean Instances." *Yearbook, Conference of Latin Americanist Geographers* 20 (1994): 9–20.

Theses and Dissertations

Angelback, Wilma M. "Population Limitations of the Gran Chaco." Master's thesis, Columbia University, New York, 1936.

Bebbington, Anthony J. "Construction and Change of Indigenous Technical Knowledge: Cases from Ecuador and Peru." PhD diss., Clark University, Worcester, Massachusetts, 1900.

Bergman, Roland. "Shipibo Subsistence in the Upper Amazon Rainforest." PhD diss., University of Wisconsin, Madison, 1974.

Edwards, Clinton R. "Aboriginal Water-Craft of Western South America: Distribution, History, and Problems of Origins." PhD diss., University of California, Berkeley, 1962.

Gagon, Alexandre. "The Relationship between the El Nino Southern Oscillation and the Incidence of Dengue/DHF and Malaria in South America and Southeast Asia." Master's thesis, University of Alberta, Edmonton, 2000.

McCall, John. "Local Resistance to Globalization in the Andes: A Cultural Continuum in Andean Time and Space." Master's thesis, Appalachian State University, Boone, North Carolina, 2004.

Pruitt, Amy M. "The Influence of Geographic Factors on the Distribution of Dysentary in South America." Master's thesis, University of California, Los Angeles, 1949.

Racovitaeaunce, Adina. "Sacred Mountains and Glacial Archeology in the Andes." Master's thesis, University of Colorado, Boulder, 2004.

Wesche, Rolf. "The Settler Wedge of the Upper Putumayo River." PhD diss., University of Florida, Gainesville, 1967.

Whitesell, Edward A. "Changing Courses: The Jurua River, Its People, and Amazonian Extractive Reserves." PhD diss., University of California, Berkeley, 1993.

ECONOMIC GEOGRAPHY

General Works

Barton, Jonathan R. "Eco-dependency in Latin America." *Singapore Journal of Tropical Geography* 27, no. 2 (2006): 134–49.

Bromley, Raymond J., and Rosemary D. F. Bromley. *South American Development: A Geographical Introduction*. New York: Cambridge University Press, 1982.

Caviedes, Cesar N. "Natural Resource Exploitation in Latin America: Espoliation or Tool for Development." *GeoJournal* 11, no. 1 (1985): 111–19.

Craig, Alan K. "Depletion of Natural Resources and the Status of Conservation in Latin America." *Yearbook, Conference of Latin Americanist Geographers* 17–18 (1990): 61–66.

Cunningham, Susan M. "Multinationals and Restructuring in Latin America." In *Multination Corporations and the Third World*, edited by Chris Dixon et al., 39–65. Boulder, CO: Westview, 1986.

Denevan, William M. "Development and the Imminent Demise of the Amazon Rain Forest." *Professional Geographer* 25, no. 2 (1973): 130–35.

———. "Latin America." In *World Systems of Traditional Resource Management*, edited by Gary Klee, 217–44. New York: Wiley, 1980.

Faissol, Speridiao. "As Crises de Energia da Divida Externa na America Latina e seu Impacto no Processo de Desenvolvimento." *Revista Geografica* 111 (1990): 207–21.

Gilbert, Alan, ed. *Latin American Development: A Geographical Perspective*. Baltimore: Penguin, 1974.

Gow, David. "Development of Fragile Lands: An Integrated Approach Reconsidered." In *Fragile Lands of Latin America: Strategies for Sustainable Development*, edited by John O. Browder, 25–43. Boulder, CO: Westview, 1989.

Gwynne, Robert N. "Multinational Corporations and the Triple Alliance in Latin America." In *Multinational Corporations and the Third World*, edited by Chris Dixon et al., 118–36. Boulder, CO: Westview, 1986.

Hays-Mitchell, Maureen. "Development Versus Empowerment: The Gendered Legacy of Economic Restructuring in Latin America." *Yearbook, Conference of Latin Americanist Geographers* 23 (1997): 119–32.

Johnson, Emory R. "The Trade and Industries of Western South America." *Journal of Geography* 1 (1902): 25–35, 51–62, 109–122.

Keeling, David J. "Developing MERCOSUR: Can Argentina and Brazil Replicate the European Common Market?" *Proceedings, Kentucky Academy of Science* (1993): 59–72.

———. "A Graphic Representation of Economic Integration Trends in the Southern Cone of South America." *Proceedings, Kentucky Academy of Science* (1999): 18–21.

Morris, Arthur S. *Latin America: Economic Development and Regional Differentiation*. London: Hutchinson, 1981.

Pedersen, P., and Walter B. Stohr. "Economic Integration and Spatial Development of South America." *American Behavioral Scientist* 16 (1926): 82–97.

Place, Susan E. "Natural Resources and Environmental Issues." In *Future Directions in Latin Americanist Geography: Research Agendas for the Nineties and Beyond*, edited by Gary S. Elbow, 53–58. Auburn, AL: Conference of Latin Americanist Geographers, Special Publication 3, 1992.

Saludjian, Alexis. "Critiques du Regionalisme Ouvert a partier de l'economie geographique appliqué au Merco sur." *Journal of Latin American Geography* 4, no. 2 (2005): 77–96.

Smith, Nigel J. H. "Destructive Exploitation of the South American River Turtle." *Yearbook, Association of Pacific Coast Geographers* 36 (1974): 85–102.

Stadel, Christoph. "Environmental Stress and Sustainable Rural Development in the Tropical Andes." *Mountain Research and Development* 11, no. 3 (1991): 213–23.

Werlhof, C. "Die Zukundt der Entwicklung und die Zukunft der Subsistenz, oder 'Subsistanz statt Entwicklung." In *Lateinamerika-krise ohne Ende?* edited by Axel Borsdorf, 161–70. Innsbruck, Austria: Innsbrucker Geographische Studien, Band 21, 1994.

Wilhelmy, Herbert. "Amazonia as a Living Area and an Economic Area." *Applied Sciences and Development* 1 (1973): 115–35.

Wilkens, Gene C. "Transferring Traditional Technology: A Bottom-Up Approach to Fragile Lands." In *Fragile Lands of Latin America: Strategies for Sustainable Development*, edited by John O. Browder, 44–60. Boulder, CO: Westview, 1989.

Winberry, John. "Revional Economic Integration in Latin America: Strategies and the Role of Gateway Cities." *Southeastern Latin Americanist* 29 (1998): 45–56.

Agriculture and Land Use

Aldrich, Stephen P., et al. "Contentious Land Change in the Amazon's Arc of Deforestation." *Annals of the Association of American Geographers* 102, no. 1 (2012): 103–28.

Baied, Carlos A. "Transhumance and Land Use in the Northern Patagonian Andes." *Mountain Research and Development* 9, no. 4 (1989): 365–80.

Boorstein, Margaret. "Three Approaches along the Rio de la Plata: Landfill, Development, and Monumentalizing the Past." *Middle States Geographer* 40 (2007): 171–75.

Brannstrom, Christian. "South America's Neoliberal Agricultural Frontiers: Places of Environmental Sacrifice or Conservation Opportunity?" *Ambio* 38, no. 3 (2009): 141–49.

Caldas, Marcellus, et al. "Theorizing Land Cover and Land Use Change: The Peasant Economy of Amazonian Deforestation." *Annals of the Association of American Geographers* 97, no. 1 (2007): 86–110.

Clawson, David L. "Peasant Agriculture and Farming Systems." In *Future Directions in Latin Americanist Geography: Research Agendas for the Nineties and Beyond*, edited by Gary S. Elbow, 45–52. Auburn, AL: Conference of Latin Americanist Geographers, Special Publication 3, 1992.

Coomes, Oliver T. "Of Stakes, Stems, and Cuttings: The Importance of Local Seed Systems in Traditional Amazonian Societies." *Professional Geographer* 62, no. 3 (2010): 323–34.

Coomes, Oliver T., et al. "Targeting Conservation-Development Initiatives in Tropical Forests: Insights from Analyses of Rain Forest Use and Economic Reliance among Amazonian Peasants." *Ecological Economics* 51, nos. 1–2 (2004): 47–64.

Crowley, William K. "Order and Disorder: A Model of Latin American Land Use." *Yearbook, Association of Pacific Coast Geographers* 57 (1995): 9–31.

Deffontaines, Pierre. "La vie pastorale dans les Andes du Nord et du Centre." *Les Cahiers d'Outre Mer* 26 (1973): 5–38.

Denevan, William M. "Aboriginal Drained-Field Cultivation in the Americas." *Science* 169 (1970): 647–54.

———. *Cultivated Landscapes of Native Amazonia and the Andes*. New York: Oxford University Press, 2003.

———. "Swiddens and Cattle Versus Forest: The Imminent Demise of the Amazon Rainforest Reexamined." *Studies in Third World Societies* 13 (1981): 25–44.

Dorch, J. P. "Drained Field Agriculture in Tropical Latin America: Parallels from Past to Present." *Journal of Biogeography* 15, no. 1 (1988): 87–96.

Dozier, Craig L. *Land Development and Colonization in Latin America: Case Studies of Peru, Bolivia, and Mexico*. New York: Praeger, 1969.

Eden, Michael J. "Ecology and Land Development: The Case of Amazonian Rainforest." *Transactions, Institute of British Geographers* 3, no. 4 (1978): 444–63.

Eiselen, Elizabeth. "Quinoa, a Potentially Important Food Crop of the Andes." *Journal of Geography* 55 (1956): 330–33.

Enjalbert, H. "L'agriculture europeenne en Americque du Sud." *Les Cahiers d'Outre Mer* 1 (1948): 201–28.

Foreign Agriculture Service. *Agricultural Geography of Latin America*. Washington, DC: Department of Agriculture, Miscellaneous Publication 743, 1958.

Gade, Daniel. "Crops and Boundaries: Manioc at Its Meridional Limits in South America." *Revista Geografica* 133 (2003): 103–26.

———. "Ethnobotany of Canihau (Chenopodium pallidicaule), Rustic Seed Crop of the Altiplano." *Economic Botany* 24, no. 1 (1974): 55–61.

———. "The Iberian Pig in the Central Andes." *Journal of Cultural Geography* 7, no. 2 (1987): 35–50.

———. "The Llama, Alpaca, and Vicuna: A Fact versus Fiction." *Journal of Geography* 68 (1969): 339–43.

———. "Setting the Stage for Domestication: Brassica Weeds in Andean Ecology." *Proceedings of the Association of American Geographers* 4 (1972): 38–40.

———. "Vanishing Crops of Traditional Agriculture: The Case of Tarwi in the Andes." *Proceedings of the Association of American Geographers* 1 (1969): 47–51.

Gade, Daniel, and Roberto Rios. "Chaquittaella: The Native Footplough and Its Persistence in Central Andean Agriculture." *Tools and Tillage* 2 (1972): 3–15.

Godoy, Ricardo A. "Ecological Degradation and Agricultural Intensification in the Andean Highlands." *Human Ecology* 12, no. 4 (1984): 359–83.

Gomes, Carlos V. A., et al. "Convergence and Contrasts in the Adaption of Cattle Ranchign: Comparisons of Small Holder Agriculturalists and Forest Extractivists in the Amazon." *Journal of Latin American Geography* 11, no. 1 (2012): 99–120.

Hecht, Susanna B. "Indigenous Soil Management in the Amazon Basin: Some Implications for Development." In *Fragile Lands of Latin America: Strategies for Sustainable Development,* edited by John O. Browder, 166–81. Boulder, CO: Westview, 1989.

Jonasson, Olo F. "Potential Areas of Cacao Cultivation in South America: A Review." *Economic Geography* 27, no. 1 (1951): 90–93.

Jones, Clarence F. "Agricultural Regions of South America: I–VI." *Economic Geography* 4 (1928): 1–30; 4 (1928): 159–86; 4 (1928): 267–94; 5 (1929): 109–40; 5 (1929): 277–307; 5 (1929): 390–421; 6 (1930): 1–36.

Kent, Robert B. "The Introduction and Diffusion of the African Honeybee in South America." *Yearbook, Association of Pacific Coast Geographers* 50 (1988): 21–44.

Kirby, John M. "Agricultural Land Use and the Settlement of Amazonia." *Pacific Viewpoint* 17, no. 2 (1976): 105–32.

Knapp, Gregory W. "Una perspective de la irrigacion en los Andes del Norte." *America Indigena* 46 (1986): 349–56.

Kohlhepp, Gerd. "Problems of Agriculture in Latin America: Production of Food-Crops versus Production of Energy-Plants and Export." *Applied Geography and Development* 27 (1986): 60–92.

Martinson, Tom L. "Geographers' Contributions to Agricultural Planning in Latin America." *Proceedings of the Indiana Academy of the Social Sciences* 11 (1976): 46–55.

McBride, G. M. "Cotton Growing in South America." *Geographical Review* 9 (1920): 35–50.

Messerli, Bruno, et al. "Water Availability, Protected Areas, and Natural Resources in the Andean Desert Altiplano." *Mountain Research and Development* 17, no. 3 (1997): 229–38.

Morales, Ignacio. "Agrarian Reform in Latin America: Dignity for Empty Stomachs." *Kansas Geographer* 11 (1976): 13–26.

Nilsen, A. F., and Peter Sjoholt. "Colonization and Adaptation of Agricultural Systems: Some Critical Issues in the Andean Montana." *Norsk Geografisk Tidsskrift* 47, no. 4 (1993): 229–40.

Pacheco, Pablo, et al. "Landscape Transformation in Tropical Latin America: Assessing Trends and Policy Implications for REDD+." *Forests* 2, no. 1 (2011): 1–29.

Parsons, James J. "Southern Blooms: Latin America and the World of Flowers." *Queen's Quarterly* 99 (1992): 542–61.

Platt, Robert S. "Six Farms in the Central Andes." *Geographical Review* 22, no. 2 (1932): 245–59.

Posner, Joshua, et al. "Land Systems of Hill and Highland Tropical America." *Revista Geografica* 98 (1983): 5–22.

Rama, R. "Latin America and the Geographical Priorities of Multinational Agro-industries." *Geoforum* 27, no. 1 (1996): 39–52.

Riffo Rosas, Margarita. "Impactos espaciales y socioeconomicos de la vitivinicultura en Chile y Argentina." *Revista Geografica* 143 (2008): 163–210.

Ruddle, Kenneth R. *The Yukpa Cultivation System: A Study of Shifting Cultivation in Colombia and Venezuela.* Berkeley: University of California Press, Ibero-Americana 52, 1974.

Salisbury, David S., and Marianne Schmink. "Cows versus Rubber: Changing Livelihoods among Amazonian Extractivists." *Geoforum* 38, no. 6 (2007): 1233–49.

Sasaki, Komei. "Primitive Agriculture in Native South America [in Japanese]." *Jimbun Chiri/ Human Geography* 9, no. 3 (1957): 1–22.

Sauer, Carl O. "Cultivated Plants of South and Central America." In *Handbook of South American Indians,* edited by J. H. Steward, 487–543. Washington, DC: Smithsonian Insitute, Bureau of American Ethnology, Bulletin 143, 1950.

Simmons, Cynthia S., et al. "Wildfires in Amazonia: A Pilot Study Examining the Role of Farming Systems, Social Capital, and Fire Contagion." *Journal of Latin American Geography* 3, no. 1 (2004): 81–95.

Siren, Anders H., and Eduardo S. Brandizio. "Detecting Subtle Land Use Change in Tropical Forests." *Applied Geography* 28, no. 2 (2009): 201–11.

Smith, Nigel J. H. "Home Gardens as a Springboard for Agroforestry Developments in Amazonia." *International Tree Crops Journal* 9 (1996): 11–30.

———. "Human-Induced Landscape Changes in Amazonia and Implications for Development." In *Global Land Use Change: A Perspective from the Columbia Encounter*, edited by B. L. Turner II, 221–51. Madrid, Spain: Consejo Superio de Investigaciones Cientificos, 1995.

Stewart, Norman R., et al. "Transhumance in the Central Andes." *Annals of the Association of American Geographers* 66, no. 3 (1976): 377–97.

Tulet, Jean-Christian. "La jeunesse de la cafeculture latino-americaine." *Les Cahiers d'Outre Mer* 53, nos. 209–210 (2000): 31–54.

Une, R. Y. "An Analysis of the Effects of Frosts on the Principal Coffee Areas." *GeoJournal* 6, no. 2 (1982): 129–40.

Van Der Glas, M. *Gaining Ground: Land Use and Soil Conservation in Areas of Agricultural Colonisation in South Brazil and East Paraguay.* Utrecht, Netherlands: Netherlands Geographical Studies 248, 1998.

Walker, Robert, et al. "Ranching and the New Global Range: Amazonia in the Twenty-First Centry." *Geoforum* 40, no. 5 (2009): 732–45.

Watts, Michael. "Latin America: Implications of Land Use and Abuse." *Land Use Policy* 1, no. 3 (1984): 243–51.

Wesche, Rolf. "Land Use Alternatives for Small Producers in the Amazon: A Review Article." *Canadian Journal of Development Studies* 13, no. 2 (1992): 277–85.

White, C. Langdon, and John Thompson. "The Llanos: A Neglected Grazing Resource." *Journal of Range Management* 8 (1955): 11–16.

Williams, Lyndon S. "Agricultural Terrace Evolution in Latin America." *Yearbook, Conference of Latin Americanist Geographers* 16 (1990): 82–93.

Works, Martha A. "Continuity and Conversion of House Gardens in Western Amazonia." *Yearbook, Conference of Pacific Coast Geographers* 52 (1991): 31–64.

Zermonski, A. M. "Systems of Agriculture in Andean Countries." *Geographia Polonica* 35 (1977): 127–42.

Zimmerer, Karl S. "Ecogeography of the Cultivated Andean Potatoes." *Bioscience* 48, no. 6 (1998): 445–54.

———. "The Origins of Andean Irrigation." *Nature* 378 (1995): 481–83.

———. "Overlapping Patchworks of Mountain Agriculture in Peru and Bolivia: Toward a Regional-Global Landscape Model." *Human Ecology* 27, no. 1 (1999): 135–66.

———. "Rescaling Irrigation in Latin America: The Cultural Images and Political Ecology of Water Resources." *Ecumene* 7, no. 2 (2000): 150–76.

———. "Transformating Colquepata Wetlands: Landscapes of Knowledge and Practice in Andean Agriculture." In *Irrigation at High Altitudes: The Social Organization of Water Control Systems in the Andes*, edited by W. P. Mitchell and D. Guillet, 115–40. Washington, DC: Society of Latin American Anthropology, 1994.

Commerce and Trade

Behrens, Alfredo. "Regional Energy Trade: Its Role in South America." *Energy Policy* 18, no. 2 (1990): 175–85.

Bromley, Raymond J., and Richard Symanski. "Market-Place Trade in Latin America." *Latin American Research Review* 9, no. 3 (1974): 3–38.

Elbow, Gary S. "Marketing in Latin America: A Photo Essay." *Journal of Cultural Geography* 15, no. 2 (1995): 55–78.

Jones, Clarence F. "The Character and Distribution of South American Trade." *Economic Geography* 2 (1926): 358–93.

————. *Commerce of South America*. Boston: Ginn, 1926.

Kearns, Kevin C. "The Andean Common Market: A New Thrust at Economic Integration in Latin America." *Journal of Inter-American Studies and World Affairs* 14 (1972): 225–49.

————. "International Cooperation for Development: The Andean Common Market." In *Focus on South America*, edited by Alice Taylor, 47–66. New York: Praeger, 1973.

Tower, Walter S. "Notes on the Commercial Geography of South America." *Bulletin of the American Geographical Society* 45, no. 12 (1913): 881–901.

Wales, Michael L., and David A. Preston. "Peasants and Smugglers: Frontier Trade between Peru and Bolivia." *Inter-American Economic Affairs* 26, no. 2 (1972): 35–50.

Wiley, James. "The European Union's Single Market and Latin America's Banana Exporting Countries." *Yearbook, Conference of Latin Americanist Geographers* 22 (1996): 31–40.

Wilson, Leonard S. "Latin American Foreign Trade." *Geographical Review* 31, no. 2 (1941): 135–41.

Wrigley, Gladys M. "Fairs of the Central Andes." *Geographical Review* 7 (1919): 65–80.

Manufacturing and Labor Studies

Dickenson, John P. "The Rise of Industry in a World Periphery." In *Latin American Development: Geographical Perspectives*, edited by David A. Preston, 41–94. London: Longman, 1994.

Lawson, Victoria A. "Industrial Subcontracting and Employment Form in Latin America: A Framework for Contextual Analysis." *Progress in Human Geography* 16, no. 1 (1992): 1–23.

Mertins, Gunter. "Recent Industrial Development in the Andean Nations as Demonstrated by the Examples of Peru and Colombia." *Applied Geography and Development* 31 (1988): 54–71.

Mikus, Werner. "International Industrial Cooperation: Evaluation of Andean Pact Countries." *Tijdscrift voor Economische en Sociale Geographie* 86, no. 4 (1995): 357–67.

Morse, David A. "Unemployment: Bitter Burden of Millions." In *Focus on South America*, edited by Alice Taylor, 22–31. New York: Praeger, 1973.

Shaw, Robert. "A Rural Employment Strategy for South America." In *Focus on South America*, edited by Alice Taylor, 32–46. New York: Praeger, 1973.

Energy, Mining, Fishing, and Lumbering

Arima, Eugenio, et al. "Loggers and Forest Fragmentation: Behavioral Models of Road Building in the Amazon Basin." *Annals of the Association of American Geographers* 95, no. 3 (2005): 525–41.

Auty, R. M. "Contrasts in Mining Sector Resilience." In *Sustaining Development in Mineral Economies: The Resource Curse Thesis*, edited by R. M. Auty, 143–74. London: Routledge, 1993.

Caviedes, Cesar N., and Timothy J. Fisk. "Modelling Change in the Peruvian-Chilean Eastern Pacific Fisheries." *GeoJournal* 30, no. 4 (1993): 369–80.

————. "The Peruvian-Chile Eastern Pacific Fisheries and Climatic Oscillation." In *Climate Variability, Climate Change, and Fisheries*, edited by M. H. Glantz, 355–72. Cambridge, England: Cambridge University Press, 1992.

Coomes, Oliver T. "A Century of Rain Forest Use in Western Amazonia: Lessons for Extraction-Based Conservation of Tropical Forest Resources." *Forest and Conservation History* 38, no. 3 (1995): 108–20.

Coomes, Oliver T., and Bradford L. Barham. "Rain Forest Extraction and Conservation in Amazonia." *Geographical Journal* 163, no. 2 (1997): 180–88.

Coomes, Oliver T., and G. J. Burt. "Indigenous Market-Oriented Agroforestry: Dissecting Local Diversity in Western Amazonia." *Agroforestry Systems* 37, no. 1 (1997): 27–44.

Hecht, Susanna B., and A. Cockburn. *The Fate of the Forest: Developers, Destroyers, and Defenders of the Amazon*. London: Verso, 1989.

Lorena, Rodrigo B., and Eric F. Lambin. "The Spatial Dynamics of Deforestation and Agent Use in the Amazon." *Applied Geography* 29, no. 2 (2009): 171–81.

Pinard, J. "La centrale hydro-electrique d'Itaipu (Bresil-Paraguay)." *Les Cahiers d'Outre Mer* 36 (1983): 293–98.

Platt, Robert S. "Mining Patterns of Occupance in Five South American Districts." *Economic Geography* 12, no. 4 (1936): 340–50.

Polorny, Benno, et al. "Market-Based Conservation of the Amazonian Forests: Revisiting Win-Win Expectations." *Geoforum* 43, no. 3 (2012): 387–501.

Schwartznau, Stephen. "Extractive Reserves: The Rubber Tappers' Strategy for Sustainable Use of the Amazon Rainforest." In *Fragile Lands of Latin America: Strategies for Sustainable Development*, edited by John O. Browder, 150–65. Boulder, CO: Westview, 1989.

Smith, Nigel J. H. *The Amazon River Forest.* New York: Oxford University Press, 1999.

———. "The Impact of Cultural and Ecological Change on Amazonian Fisheries." *Biological Conservation* 32 (1985): 355–73.

———. *Man, Fishes, and the Amazon.* New York: Columbia University Press, 1981.

Smith, Nigel J. H., et al. "Environmental Impacts of Resource Exploitation in Amazonia." *Global Environmental Change A* 1, no. 4 (1991): 313–20.

Sternberg, Rolf. "Hydroelectric Energy, Repressed Demand, and Economic Change in Amazonia." *Acta Amazonia* 13 (1983): 371–91.

———. "The Latinamericanization of Electricity Systems in South America." *Revista Geografica* 131 (2002): 21–40.

Vellard, G. "Peuples pecheurs du Titicaca: Les Urus et leurs voisins." *Les Cahiers d'Outre Mer* 5 (1952): 135–48.

Transportation and Communications

Howenstine, Erick. "Towards a Schematic Model of Communications Media and Development in Latin America." In *Collapsing Space and Time: Geographic Aspects of Communications and Information*, edited by Stanley Brunn and Thomas Leinback, 278–301. New York: HarperCollins, 1991.

James, Henry P. "Harbors of South America." *Journal of Geography* 14 (1915): 43–48.

James, Preston E. "Geographic Factors in the Development of Trans-Andean Communications between Argentina and Chile." *Bulletin of the Geographical Society of Philadelphia* 21, no. 3 (1923): 1–9.

———. "Geographic Factors in the Development of Transportation in South America." *Economic Geography* 1 (1925): 221–30.

———. "The Upper Parana Lowland: A Problem in South American Railroad Development." *Journal of Geography* 22 (1923): 245–56.

Keeling, David J. "Latin America's Transportation Conundrum." *Journal of Latin American Geography* 7, no. 2 (2008): 133–54.

Rees, Peter W. "Transportation in Latin America." *Yearbook, Conference of Latin Americanist Geographers* 17–18 (1990): 191–98.

Schley, Julian L. "Ocean-to-Amazon Highway." *Journal of Geography* 43 (1944): 151–67.

Schoff, W. R. "Latin-American Railroads, Present and Projected." *Bulletin of the Geographical Society of Philadelphia* 11 (1913): 1–14, 209–223.

Warf, Barney. "Diverse Spatialities of the Latin American and Caribbean Internet." *Journal of Latin American Geography* 8, no. 2 (2009): 125–45.

Wilcox, H. Case. "Air Transportation in Latin America." *Geographical Review* 20, no. 4 (1930): 587–604.

Theses and Dissertations

Clutter, Lester. "Oil Fields of South America, Excluding Venezuela: A Geographic Analysis." Master's thesis, Clark University, Worcester, MA, 1951.

Dickinson, Joshua C. "The Cultivation and Utilization of the Eucalypt in the Peruvian Sierra and the Industrial Triangle of Brazil." PhD diss., University of Florida, Gainesville, 1967.

Field, Chris. "A Reconnaissance of Southern Andean Agricultural Terracing." PhD diss., University of California, Los Angeles, 1966.

Gillette, Lynn. "Business Opportunities in Argentina, Brazil, and Chile." Master's thesis, State University of New York, Buffalo, 1993.

Hegen, Edmund E. "Highways into the Upper Amazon Basin: A Study of Trans-Andean Roads in Southern Colombia, Ecuador, and Northern Peru." PhD diss., University of Florida, Gainesville, 1962.

Horst, Oscar H. "The Economic Orientation of Venezuela and Colombia as Conditioned by Geographic Factors." Master's thesis, Ohio State University, Columbus, 1951.

Howell, Doris M. "Hydroelectric Resources and Developments in Selected South American Countries." Master's thesis, Oklahoma State University, Stillwater, OK, 1949.

James, Preston E. "Geographic Factors in the Development of Transportation in South America." PhD diss., Clark University, Worcester, MA, 1923.

Kintz, Damian B. "Land Use and Land Cover Change between 1987 and 2001 in the Buffer Zone of a National Park in the Tropical Andes." Master's thesis, University of Texas, Austin, 2003.

McBride, G. M. "Land Tenure in Latin America." PhD diss., Yale University, New Haven, CT, 1921.

Pacheco, Pablo. "Populist and Capitalist Frontiers in the Amazon: Diverging Dynamics of Agrarian and Land-Use Change." PhD diss., Clark University, Worcester, MA, 2005.

Pinton, Celso. "Economic Integration in Latin America." Master's thesis, State University of New York, Buffalo, 1993.

Ruddle, Kenneth R. "The Yukpa Auto-subsistence System: A Study of Shifting Cultivation and Ancillary Activities in Colombia and Venezuela." PhD diss., University of California, Los Angeles, 1970.

Rutecki, Geoffrey. "Market Assessment for Spectrum Computer Technologies' Complaints Desk Software in Latin America." Master's thesis, State University of New York, Buffalo, 1995.

Salisbury, David S. "Overcoming Marginality on the Margins: Mapping, Logging, and Coca in the Amazon Borderland." PhD diss., University of Texas, Austin, 2006.

Wrigley, Gladys M. "Roads and Towns of the Central Andes." PhD diss., Yale University, New Haven, CT, 1917.

HISTORICAL GEOGRAPHY

General Works

Barham, Bradford L., and Oliver T. Coomes. "Reinterpreting the Amazon Rubber Boom: Investment and the Role of the State." *Latin American Research Review* 29, no. 2 (1994): 73–109.

———. "Wild-Rubber, Industrial Organization, and Microeconomics of Extraction during the Amazon Rubber Boom (1860–1920)." *Journal of Latin American Studies* 29, no. 1 (1994): 37–72.

Bergmann, John. "The Distribution of Cacao Cultivation in Pre-Columbian America." *Annals of the Association of American Geographers* 59, no. 1 (1969): 85–96.

Brannstrom, Christian. "The River of Silver and the Island of Brazil." *Terra Incognitae* 27 (1995): 1–14.

Bromley, Rosemary D. F. "Urban-Rural Interrelationships in Colonial Hispanic America: A Case Study of Three Andean Towns." *Swansea Geographer* 12 (1974): 15–22.

Burnett, D. Graham. *Masters of All They Surveyed: Exploration, Geography, and a British El Dorado*. Chicago: University of Chicago Press, 2000.

Butzner, Karl W. "The Americas before and after 1492: An Introduction to Current Geographical Research." *Annals of the Association of American Geographers* 82, no. 3 (1992): 345–68.
Cook, Noble D., and W. George Lovell, eds. *Secret Judgements of God: Old World Disease in Colonia Spanish America.* Norman: University of Oklahoma Press, 1992, 2001.
Coomes, Oliver T., and Bradford L. Barham. "The Amazon Rubber Boom: Labor Control, Resistance, and Failed Plantation Development Revisited." *Hispanic American Historical Review* 72, no. 2 (1994): 231–57.
———. *Prosperity's Promise: The Amazon Rubber Boom and Distorted Economic Development.* Boulder, CO: Westview, Dellplain Latin American Series, 1996.
Crist, Raymond E., and Carlos E. Chardon. "Intercultural Colonial Policies in the Americas: Iberians and Britons within the New World." *American Journal of Economics and Sociology* 6, no. 3 (1947): 371–85.
Davidson, William, and James J. Parsons, eds. *Historical Geography of Latin America: Papers in Honour of Robert C. West.* Baton Rouge: Louisiana State University, School of Geoscience, Geoscience and Man 21, 1980.
Davies, Arthur. "The First Discovery and Explorations of the Amazon." *Transactions, Institute of British Geographers* 22 (1956): 87–96.
———. "The 'Miraculous' Discovery of South American by Columbus." *Geographical Review* 44, no. 4 (1954): 573–82.
Denevan, William M. *A Bibliography of Latin American Historical Geography.* Washington, DC: Pan American Institute of Geography and History, Special Publication 6, 1971.
———. "A Bluff Model of Riverine Settlement in Prehistoric Amazonia." *Annals of the Association of American Geographers* 86, no. 4 (1996): 654–82.
———. "Comments on Prehistoric Agriculture in Amazonia." *Culture and Agriculture* 20, nos. 2–3 (1998): 54–59.
———. *The Native Population of the Americas in 1492.* Madison: University of Wisconsin Press, 1976.
———. "Prehistoric Roads and Causeways of Lowland Tropical America." In *Ancient Road Networks and Settlement Hierarchies in the New World,* edited by Charles D. Trombold, 230–42. Cambridge, England: Cambridge University Press, 1991.
———. "The Pristine Myth: The Landscape of the Americas in 1492." *Annals of the Association of American Geographers* 82, no. 3 (1992): 369–86.
———. "Rewriting the Late Pre-European History of Amazonia." *Journal of Latin American Geography* 11, no. 1 (2012): 9–24.
———. "Stone vs. Metal Axes: The Ambiguity of Shifting Cultivation in Prehistoric Amazonia." *Journal of the Steward Anthropological Society* 20 (1992): 153–65.
Denevan, William M., and Gregory W. Knapp, eds. *Pre-Hispanic Agricultural Fields in the Andes Region.* Oxford, England: British Archaeological Reports, International Series 359, 1987.
Dickenson, John P., and Timothy Unwin. *Viticulture in Colonial Latin America: Essays on Alcohol, the Vine, and Wine in Spanish America and Brazil.* Liverpool, England: Liverpool University, Institute of Latin American Studies, 1992.
DiFrieri, Horacio. *El virreinato del Rio de la Plata: Essayo de geografía historica.* Buenos Aires, Argentina: Editiones Universidad del Salvador, 1980.
Donkin, R. A. *Agricultural Terracing in the Aboriginal New World.* Tucson: University of Arizona Press, Viking Fund Publications in Anthropology 56, 1979.
———. "Pre-Columbian Field Implements and Their Distribution in the Highlands of Middle and South America." *Anthropos* 65 (1970): 505–29.
Doolitle, William E. "Gardens Are Us, We Are Nature: Transcending Antiquity and Modernity." *Geographical Review* 94, no. 3 (2004): 391–404.
———. "Terrace Origins: Hypotheses and Research Strategies." *Yearbook, Conference of Latin Americanist Geographers* 16 (1990): 94–97.
Duvall, Chris S. "A Maroon Legacy? Sketching African Contributions to Live Fencing Practices in Early Spanish America." *Singapore Journal of Tropical Geography* 30, no. 2 (2009): 232–47.

Edwards, Clinton R. *Aboriginal Watercraft of the Pacific Coast of South America.* Berkeley: University of California Press, Ibero-Americana 42, 1965.
————. "Geographical Coverage of the Sixteenth-Century 'Relaciones de Indias' from South America." *Geoscience and Man* 21 (1980): 75–82.
Friede, Juan. "Las Ideas Geograficas en la Conquista del Nuevo Reino de Granada." *Revista Geografica* 41 (1954): 45–66.
Gade, Daniel. "Inca and Colonial Settlement, Coca Cultivation, and Endemic Disease in the Tropical Forest." *Journal of Historical Geography* 5, no. 3 (1979): 263–80.
Galvin, Peter R. *Patterns of Pillage: A Geography of Caribbean-Based Piracy in Spanish America, 1536–1718.* New York: Lang, 1999.
Glassner, Martin I. "The Chibchas: A History and Re-evaluation." *Americas* 26 (1970): 302–27.
Hecht, Susanna B. "The Last Unfinished Page of Genesis: Euclides Da Cunha and the Amazon." *Historical Geography* 32 (2004): 43–70.
Holstein, Otto. "Chan-Chan: Capital of the Great Chimu." *Geographical Review* 17, no. 1 (1927): 36–41.
Houston, J. M. "The Foundation of Colonial Towns in Hispanic America." In *Urbanization and Its Problems*, edited by R. P. Beckinsdale and J. M. Houston, 352–98. Oxford, England: Blackwell, 1968.
Huetz de Lemps, Alain. "Principales plantes cultivees introduites en Amerique latine depuis 1492." *Les Cahiers d'Outre Mer* 53, nos. 209–210 (2000): 129–86.
Hurault, Jean. "Montagnes mythiques: Les Tumac-Humac." *Les Cahiers d'Outre Mer* 53 (2000): 367–92.
Jones, Clarence F. "Geographical Background of the Colonial Period in South America." In *Colonial Hispanic America*, edited by Curtis Wilgus, 35–66. Washington, DC: George Washington University Press, 1936.
Jordan, Terry G. "An Iberian Lowland/Highland Model for Latin American Cattle Ranching." *Journal of Historical Geography* 15, no. 2 (1989): 111–25.
Keeling, David J. "The Geopolitics of Transport in Pre-Hispanic America: A Comparison of Transport Systems in the Core-Periphery Structure of the Aztec and Inca Empires." *Proceedings, Oregon Academy of Science* 26 (1990): 49–55.
Kelly, Kenneth. "An Explanation of the Great North-South Extent of the Inca Empire in 1532 and of the Position of Its Eastern Boundary through Peru and Bolivia." *Journal of Tropical Geography* 28 (1968): 57–63.
————. "Land-Use Regions in the Central and Northern Portions of the Inca Empire." *Annals of the Association of American Geographers* 55 (1965): 327–38.
Knapp, Gregory W. "Tecnologia e intensificacion Agricola en los Andes Ecuatoriales Pre-Hispanicos." In *Area Septentrionel Andina Norte: Arqueologia y Ethnohistoria*, edited by Jose Echevarria and Maria V. Uribe, 317–33. Quito, Ecuador: Banco Central del Ecuador, 1995.
Kus, James S. "The Chicama-Moche Canal: Failure or Success? An Alternative Explanation for an Incomplete Canal." *American Antiquity* 49 (1984): 408–15.
Lovell, W. George, and David Cook, eds. *Secret Judgements of God: Old World Disease in Colonial Spanish America.* Norman: University of Oklahoma Press, 2001.
Mathis, F. "Historisches Erbe und wirtschaftliche Entwicklung: Eine Quadratur des Kreises." In *Lateinamerika-krise ohne Ende?* edited by Axel Borsdorf, 91–99. Innsbruck, Austria: Innsbrucker Geographische Studien, Band 21, 1994.
Meggers, Betty J. "The Early History of Man in Amazonia." In *Biogeography and Quaternary History in Tropical America*, edited by T. C. Whitman and G. T. Prance, 141–74. Oxford, England: Oxford Scientific Publications, 1987.
Meons, Philip A. "Note on the Guarana Invasions of the Inca Empire." *Geographical Review* 4 (1917): 482–84.
Merino, Olga, and Linda A. Newson. "Jesuit Missions in Spanish America: The Aftermath of the Expulsion." *Yearbook, Conference of Latin Americanist Geographers* 21 (1995): 133–48.

Mignolo, W. D. "The Movable Center: Geographical Discourses and Territoriality during the Expansion of the Spanish Empire." In *Coded Encounters: Writing, Gender, and Ethnicity in Colonial Latin America*, edited by F. J. Cevallos-Candou et al., 15–46. Amherst: University of Massachusetts Press, 1994.

Morris, Arthur S. "The Agricultural Base of the Pre-Inca Andean Civilizations." *Geographical Journal* 165, no. 3 (1999): 286–95.

Mukerji, A. B. "Sea Transport among the Indians of South America." *Indian Geographical Journal* 36, no. 1 (1961): 29–45.

Newson, Linda A. "Indian Population Patterns in Colonial Spanish America." *Latin American Research Review* 20, no. 3 (1985): 41–74.

———. "The Latin American Colonial Experience." In *Latin American Development: Geographical Perspectives*, edited by David A. Preston, 11–40. London: Longman, 1996.

———. "The Population of the Amazon Basin in 1492: A View from the Ecuadorian Headwaters." *Transactions, Institute of British Geographers* 21, no. 1 (1996): 5–26.

Nordenskiold, Erland. "The Guarani Invasion of the Inca Empire in the Sixteenth Century: An Historical Indian Migration." *Geographical Review* 4 (1917): 103–21.

Nowell, Charles L. "Some Comments on That Miraculous Discovery of South America by Columbus." *Geographical Review* 45, no. 2 (1955): 250–54.

Owens, David J. "Spanish-Portuguese Territorial Rivalry in Colonial Rio de la Plata." *Yearbook, Conference of Latin Americanist Geographers* 19 (1993): 15–24.

Parsons, James J., and William M. Denevan. "Pre-Columbian Ridged Fields." *Scientific American* 217, no. 1 (1967): 92–100.

Richardson, Miles. "The Speech of Habitus in the New World: Divergence and Convergence in Landscape Practices of Administrative Towns in Spanish America and the America South." In *Cultural and Physical Expositions: Geographic Studies in the Southern United States and Latin America*, edited by Michael K. Steinberg and Paul F. Hudson, 151–82. Baton Rouge: Louisiana State University, Geoscience and Man 36, 2002.

Robinson, David J. "Evolution of the Orinoco Trade in the Middle of the Nineteenth Century." *Revista Geografica* 72 (1970): 13–44.

———. "Liberty, Fragile Fraternity, and Inequality to Early-Republican Spanish America; Assessing the Impact of French Revolutionary Ideals." *Journal of Historical Geography* 16, no. 1 (1990): 51–75.

———, ed. *Migration in Colonial Spanish America*. New York: Cambridge University Press, 1990.

———, ed. *Social Fabric and Spatial Structure in Colonial Latin America*. Ann Arbor: University of Michigan, Department of Geography, Dellplain Latin American Studies 1, 1979.

Roosevelt, Cornelius. "Ancient Civilization of the Santa Valley and Chauvin." *Geographical Review* 25, no. 1 (1935): 21–42.

Sauer, Carl O. *South America: Notes on Lectures by Carl O. Sauer at the University of California–Berkeley, 1936.* Northridge: California State University, Occasional Paper 2, 1985.

Silverman, Helaine. "Beyond the Pampas: The Geoglyphs in the Valley of Nazca." *National Geographical Research* 6, no. 4 (1990): 435–56.

Sluyter, Andrew, and Kent Mathewson. "Intellectual Relations between Historical Geography and Latin Americanist Geography." *Journal of Latin American Geography* 6, no. 1 (2007): 25–41.

Smith, Clifford T. "Depopulation of the Central Andes in the Sixteenth Century." *Current Anthropology* 2 (1970): 453–64.

Smith, Clifford T., et al. "Ancient Ridged Fields in the Region of Lake Titicaca." *Geographical Journal* 134, no. 3 (1968): 353–66.

Smole, William J. "Musa Cultivation in Pre-Columbian South America." *Geoscience and Man* 21 (1980): 47–50.

Spate, O. H. K. *The Spanish Lake*. Minneapolis: University of Minnesota Press, 1979.

Sternberg, Rolf. "Occupance of the Humid Pampa, 1856–1914." *Revista Geografica* 76 (1972): 61–102.

Sternberg, Ruben, and N. Carvajal. "Road System of the Incas in the Southern Part of Their Tawantinsuyu Empire." *National Geographic Research* 4, no. 1 (1988): 74–87.

Turner, B. L., II, and William M. Denevan. "Prehistoric Manipulation of Wetlands in the Americas: A Raised Field Prospect." In *Prehistoric Intensive Agriculture in the Tropics*, edited by I. S. Farrington, 11–30. Oxford, England: British Archaeological Report 232, 1985.

Wester, L. "Invasions and Extinctions on Masa terra (Juan Fernandez Islands): A Review of Early Historical Evidence." *Journal of Historical Geography* 17, no. 1 (1991): 18–34.

Wilson, William J. "The Spanish Discovery of the South American Mainland." *Geographical Review* 31, no. 2 (1941): 283–99.

WinklerPrins, Antoinette M. "Jute Cultivation in the Lower Amazon, 1940–1990: An Ethnographic Account from Santarem, Para, Brazil." *Journal of Historical Geography* 32, no. 4 (2006): 818–38.

Zimmerer, Karl S. "Agricultural Biodiversity and Peasant Markets in the Central Andes during Inca Rule." *Journal of Historical Geography* 19, no. 1 (1993): 15–32.

———. "The Origins of Andean Irrigation." *Nature* 378 (1994): 481–83.

Theses and Dissertations

Knapp, Gregory W. "Soil, Slope, and Water in the Equatorial Andes: A Study of Prehistoric Agricultural Adaptation." PhD diss., University of Wisconsin, Madison, 1984.

PHYSICAL GEOGRAPHY

General Works

Acosta Solis, Misael A. "Por la Conservacion de la Amazonia." *Revista Geografica* 115 (1992): 77–84.

Baied, Carlos A., and Jane C. Wheeler. "Ecolution of High Andean Ecosystems, Environment, Climate, and Culture Change over the Last 12,000 Years in the Central Andes." *Mountain Research and Development* 13, no. 2 (1993): 145–56.

Baruth, B., et al. "Climate und Desertification Processes in Patagonia." In *Landschaftsentwicklung, Palaeokologie und Klimageschichta der Ariden Diagonale Sudamerikas im Jungquarter*, edited by K. Garleff, 307–20. Bamberg, Germany: Bamberger Geographische Schriften, Heft 15, 1998.

Caviedes, Cesar N. "Natural Environments and the Regional Paradigm." In *Future Directions in Latin Americanist Geography: Research Agendas for the Nineties and Beyond*, edited by Gary S. Elbow, 59–66. Auburn, AL: Conference of Latin Americanist Geographers, Special Publication 3, 1992.

———. "Natural Hazards in South America: In Search of a Method and a Theory." *GeoJournal* 6, no. 2 (1982): 101–9.

Coomes, Oliver T. "Blackwater Rivers, Adaptation, and Environmental Heterogeneity in Amazonia." *American Anthropologist* 94, no. 3 (1992): 698–700.

Dickinson, Robert E., ed. *The Geophysiology of Amazonia: Vegetation and Climate Interactions*. New York: Wiley, 1987.

Estremera, Maria, and Mercedes Lozano. "Los efectos de los desatres naturales en America Latina." *Anales de geografia de la Universidad Complutense* 20 (2000): 219–33.

Garleff, K., ed. *Landschaftsentwicklung, Palaookologie, und Klimageschichta der Ariden Diagonale Sudamerikas im Jungquarter*. Bamberg, Germany; Bamberger Geographische Schriften, Heft 15, 1998.

Goergen, J., et al. "Die Desertifikation in Patagonien und ihre Bekampfung-ein Bericht." In *Landschaftswicklungen, Palaookologie und Klimagesdeischte der Ariden Diagonale Sudamerika im Jungquarter*, edited by K. Garleff, 299–306. Bamberg, Germany: Bamberger Geographische Schriften, Heft 15, 1998.

Manners, R. B., et al. "Floodplain Development, El Niño, and Cultural Consequences in a Hyperarid Andean Environment." *Annals of the Association of American Geographers* 97, no. 2 (2007): 229–49.

Parsons, James J. "The Northern Andean Environment." *Mountain Research and Development* 2 (1982): 253–62.

Radtke, Ulrich. *Marine Terrassen und Korallenriffe-das Problem der quataren Meeresspiegalshwankungen erlautert an Fallstudien aus Chile, Argentinien und Barbados.* Dusseldorf, Germany: Dusseldorfer Geographische Schriften 27, 1989.

Salati, E. "The Climatology and Hydrology of Amazonia." In *Key Environments: Amazonia,* edited by G. T. Prace and Thomas E. Lovejoy, 18–48. Oxford, England: Pergamon, 1985.

Smith, Nigel J. H., et al. "Human-Induced Environmental Change in Amazonia." *Global Environmental Change* 1, no. 4 (1991): 313–20.

Van der Hammen, T., and A. M. Cleef. "Holocene Changes of Rainfall and River Discharge in Northern South America and the El Niño Phenomenon." *Erdkunde* 46, nos. 3–4 (1992): 252–56.

Veblen, Thomas T., et al., eds. *The Physical Geography of South America.* New York: Oxford University Press, 2007.

Whitbeck, R. H. "Adjustments to Environment in South America: An Interplay of Influences." *Annals of the Association of American Geographers* 16 (1926): 1–11.

Zimmerer, Karl S., and Robert P. Langstroth. "Physical Geography of Tropical Latin America: The Spatial and Temporal Heterogeneity of Environments." *Singapore Journal of Tropical Geography* 14, no. 2 (1993): 157–72.

Biogeography

Adams, Charles C. "The Zoogeography of Northwesternmost South America." *Geographical Review* 10 (1920): 101–7.

Archibold, John C. "The Lamoids of South America: A Search for Truth." *California Geographer* 15 (1975): 66–75.

Arzamendia, Vanesa, and Alejandro Giraudo. "Influence of Large South American Rivers of the Plata Basin on Distributional Patterns of Tropical Snakes: A Panbiogeographical Analysis." *Journal of Biogeography* 36, no. 6 (2009): 1739–49.

Bilbao, Bibiana, and Ernesto Medina. "Nitrogent-Use Efficiency for Growth in a Cultivated African Grass and a Native South American Pasture Grass." *Journal of Biogeography* 17, nos. 4–5 (1990): 421–26.

Bini, Luis M., et al. "Macroecological Explanations for Differences in Species Richness Gradients: A Canonical Analysis of South American Birds." *Journal of Biogeography* 31, no. 11 (2004): 1819–29.

Blackburn, Timothy, and Kevin Gaston. "What Determines the Probability of Discovering a Species? A Study of South American Oscine Passerine Birds." *Journal of Biogeography* 22, no. 1 (1995): 7–14.

Brinkman, W. L. F. "System Propulsion of an Amazonian Lowland Forest: An Outline." *GeoJournal* 19, no. 4 (1989): 369–80.

Budowski, G. "La Influencia Humanen la Vegetacion Natural de Montanes Tropicale Americanas." *Colloquium Geographicum* 9 (1968): 157–62.

Buermann, Wolfgang, et al. "Predicting Species Distributions across the Amazonian and Andean Regions Using Remote Sensing." *Journal of Biogeography* 35, no. 7 (2008): 1160–76.

Burns, Bruce R. "Fire-Induced Dynamics of *Araucaria araucana*–Nothofagus Anarctica Forest in the Southern Andes." *Journal of Biogeography* 20, no. 6 (1993): 669–86.

Bush, Mark B. "Amazonian Speciation: A Necessarily Complex Model." *Journal of Biogeography* 21, no. 1 (1994): 5–18.

Bush, Mark B., and Thomas E. Lovejoy. "Amazonian Conservation: Pushing the Limits of Biogeographical Knowledge." *Journal of Biogeography* 34, no. 8 (2007): 1291–93.

Butt, Nathalia, et al. "Floristic and Functional Affiliation of Wood Plants with Climate in Western Amazonia." *Journal of Biogeography* 35, no. 5 (2008): 939–50.

Camara Artigas, Rafael. "Concepts, approches bioclimatique et typologie des savanes: Applications aux savanes americaines." *Les Cahiers d'Outre Mer* 53, no. 246 (2009): 175–218.

Cepparo de Grosso, Maria E. "La Valoracion del Pastizal en el extreme Sur Americano." *Revista Geografica* 111 (1999): 171–206.

Cerqeira, Rui. "The Distribution of Didelphis in South America." *Journal of Biogeography* 12, no. 2 (1985): 135–46.

Colinaux, P. A. "Pleistocene Biogeography and Diversity in Tropical Forests of South America." In *Biological Relationships between Africa and South America*, edited by P. Goldblatt, 473–99. New Haven, CT: Yale University Press, 1993.

Coney, P. J. "Plate Tectonic Constraints on Biogeographic Connections between North and South America." *Annals of the Missouri Botanical Gardens* 69 (1982): 432–43.

Cooke, Georgina, et al. "Marine Incursions, Cryptic Species, and Ecological Diversification in Amazonia: The Biogeographic History of the Croaker Genus Plagioscion (Sciaenidae)." *Journal of Biogeography* 39, no. 4 (2012): 724–38.

Costa, M. P. F., et al. "Biophysical Properties and Mapping of Aquatic Vegetation during the Hydrological Cycle of the Amazon Floodplain Using JERS-1 and Radarsat." *International Journal of Remote Sensing* 23, no. 7 (2002): 1401–26.

Cowling, Sharon, et al. "Paleovegetation Simulations of Lowland Amazonia and Implications for Neotropical Allopadry and Speciation." *Quaternary Research* 55, no. 2 (2001): 140–49.

Cuatrecasas, J. "Paramos Vegetation and Its Life Forms." *Colloquium Geographicum* 9 (1968): 163–86.

Daniels, Lori D., and Thomas T. Veblen. "Altitudinal Treelines of the Southern Andes Near 40 Degrees S." *Forestry Chronicle* 79 (2003): 237–41.

———. "ENSO Effects on Temperature and Precipitation of the Patagonian-Andean Region: Implications for Biogeography." *Physical Geography* 20, no. 3 (2000): 223–43.

———. "Regional and Local Effects of Disturbance and Climate on Altitudinal Treelines in Northern Patagonia." *Journal of Vegetation Science* 14 (2003): 733–42.

———. "Spatio-Temporal Influences of Climate on Altitudinal Treeline in Northern Patagonia." *Ecology* 85, no. 5 (2004): 1284–96.

Denevan, William M. "Ecological Heterogeneity and Horizontal Zonation in the Amazon Floodplain." In *Frontier Expansion in Amazonia*, edited by M. Schmink and C. H. Wood, 311–36. Gainesville: University of Florida Press, 1984.

Diaz, Francisca, et al. "Rodent Middens Reveal Episodic, Long-Distance Plant Colonization across the Hyperarid Atacama Desert over the Last 34,000 Years." *Journal of Biogeography* 39, no. 3 (2012): 510–25.

Eva, Hugh, and Steffen Fritz. "Examining the Potential of Using Remotely Sensed Fire Data to Predict Areas of Rapid Forest Change in South America." *Applied Geography* 23, nos. 2–3 (2003): 189–204.

Fittkau, E. J., et al., eds. *Biogeography and Ecology in South America*. The Hague, Netherlands: W. Junk, 1968.

Franzinelli, Elena, et al. "Fragmentation of Ecosystem Owing to Neotectonics in the Amazon Basin." *Science Reports of the Tohoku University, 7th Series, Geography* 49, no. 2 (1999): 207–14.

Gade, Daniel. "Setting the Stage for Domestication: Brassica Weeds in Andean Peasant Ecology." *Proceedings of the Association of American Geographers* 4 (1972): 38–41.

Graciela Garcia, Veronica, et al. "Patterns of Population Differentiation in Annual Killifishes from the Parana-Uruguay-La Plata Basin: The Role of Vicariance and Dispersal." *Journal of Biogeography* 39, no. 9 (2012): 1707–19.

Hardin, P. J., and D. G. Long. "Discriminating between Tropical Vegetation Formations Using Reconstructed High-Resolution Seasat: A Scatterometer Data." *Photogrammetric Engineering and Remote Sensing* 60, no. 12 (1994): 1453–62.

Hayes, Floyd, and Jo-Anne Sewfol. "The Amazon River as a Dispersal Barrier to Passerine Birds: Effects of River Width, Habitat, and Taxonomy." *Journal of Biogeography* 31, no. 11 (2004): 1809–18.

Hildebrand-Vogel, Renate, et al. "Subantarctic-Andean Nothofagus Palilio Forests." *Vegetatio* 89, no. 1 (1990): 55–68.

Hills, Theo L. "The Ecology of Hazardousness: The Experience of South America." *GeoJournal* 6, no. 2 (1982): 151–56.

———. "Savanna Landscapes of the Amazon Basin." *McGill University Savanna Research Series* 14 (1969): 1–41.

Holz, Carlos A., and Thomas T. Veblen. "Tree Regeneration Responses to Chusquea Montana Bamboo Dieback in a Subalpine Nothofagus Forest in the Southern Andes." *Journal of Vegetation Science* 17, no. 1 (2006): 19–28.

Hubert, Nicolas, and Jean-Francois Renno. "Historical Biogeography of South American Freshwater Fishes." *Journal of Biogeography* 33, no. 8 (2006): 1414–36.

Joseph, Lee, et al. "Independent Evolution of Migration on the South American Landscape in a Long-Distance Temperate Tropical Migratory Bird, Swainson's Flycatcher (*Myiarchus swainsoni*)." *Journal of Biogeography* 30, no. 6 (2003): 925–38.

Kent, Robert B. "The Introduction and Diffusion of the African Honeybee in South America." *Yearbook, Association of Pacific Coast Geographers* 50 (1988): 21–44.

Kessler, Michael. "Altitudinal Zonation of Andean Cryptogam Communities." *Journal of Biogeography* 27, no. 2 (2000): 275–82.

———. "The Elevational Gradient of Andean Plant Endemism: Varying Influences of Taxon-Specific Traits and Topography at Different Taxonomic Levels." *Journal of Biogeography* 29, no. 9 (2002): 1159–66.

Kleiden, A., and H. Heimann. "Deep Rooted Vegetation, Amazonian Deforestation, and Climate: Results from a Modelling Study." *Global Ecology and Biogeography* 8, no. 5 (1995): 397–405.

Lowenberg-Neto, Peter, et al. "Areas of Endemism and Spatial Diversification of the Muscidae (Insecta: Diptera) in the Andean and Neotropical Regions." *Journal of Biogeography* 36, no. 6 (2009): 1750–59.

Lynch Alfaro, Jessica, et al. "Explosive Pleistocene Range Extension Leads to Widespread Amazonian Sympatry between Robust and Gracile Capuchin Monkeys." *Journal of Biogeography* 39, no. 2 (2012): 272–88.

Maslin, Mark, et al. "New Views on an Old Forest: Assessing the Longevity, Resilience, and Future of the Amazon Rainforest." *Transactions, Institute of British Geographers* 30, no. 4 (2005): 477–99.

Mayle, F. E. "A Review of Holocene Rainforest Ecotonal Dynamics at Opposite Ends of the Amazon: Bolivia versus Colombia." *Geographica Helvetica* 66, no. 3 (2011): 202–7.

Medina, Ernesto, and Juan Silva. "Savannas of Northern South America: A Steady State Regulated by Water-Fire Interactions on a Background of Low Nutrient Availability." *Journal of Biogeography* 17, nos. 4–5 (1990): 403–14.

Moran, Emilio, et al. "Effects of Soil Fertility and Land-Use on Forest Succession in Amazonia." *Forest Ecology and Management* 139, nos. 1–3 (2000): 93–108.

Morrone, Juan J., and Estela Lopretto. "Distributional Patterns of Freshwater Decapoda (Crustacea: Malacostraca) in Southern South America: A Panbiogeographic Approach." *Journal of Biogeography* 21, no. 1 (1994): 97–110.

Mota de Oliveira, Sylvia, et al. "Niche Assembly of Epiphytic Bryophyte Communities in the Guianas: A Regional Approach." *Journal of Biogeography* 36, no. 11 (2009): 2076–84.

Naku, L. N. "Avian Distribution Patterns in the Guiana Shield: Implications for the Delimitation of Amazonian Areas of Endemism." *Journal of Biogeography* 38, no. 4 (2011): 681–96.

Nemesio, Andre, and Fernando A. Silveira. "Deriving Ecological Relationships from Geographical Correlations between Host and Parasitic Species: An Example with Orchid Bees." *Journal of Biogeography* 33, no. 1 (2006): 91–97.

Noonan, Brice, and Kenneth Wray. "Neotropical Diversifications: The Effects of a Complex History on Diversity within the Poison Frog Genes Dendrobates." *Journal of Biogeography* 33, no. 6 (2006): 1007–20.

Nores, Manuel. "An Alternative Hypothesis for the Origin of Amazonian Bird Diversity." *Journal of Biogeography* 26, no. 3 (1999): 475–86.

Parsons, James J. "The Spread of African Pasture Grasses to the American Tropics." *Journal of Range Management* 25 (1972): 13–17.

Pastorino, M. J., and L. A. Gallo. "Quaternary Evolutionary History of *Austrocedrus chilensis*: A Cypress Native to the Andean Patagonian Forest." *Journal of Biogeography* 29, no. 9 (2002): 1167–78.

Pennington, R. T., et al. "Neotropical Seasonally Dry Forests and Quaternary Vegetation Changes." *Journal of Biogeography* 27, no. 2 (2000): 261–74.

Pineiro, Rosalia, et al. "Circumarctic Dispersal and Long-Distance Colonization of South America: The Moss Genus Cinclidium." *Journal of Biogeography* 39, no. 111 (2012): 2041–51.

Pirie, Michael D., et al. "Andean-Centred Genera in the Short-Branch Glade of Annonacene: Testing Biogeographical Hypotheses Using Phytogeny Reconstruction and Molecular Dating." *Journal of Biogeography* 33, no. 1 (2006): 31–46.

Place, Susan E., ed. *Tropical Rainforests: Latin American Nature and Society in Transition.* Wilmington, DE: Jaguar, 1993.

Prince, S. D., and Marc K. Steininger. "Biophysical Stratification of the Amazon Basin." *Global Change Biology* 5 (1999): 1–22.

Premoil, Andrea C. "Genetic Variations in a Geographically Restricted Area and Two Widespread Species of South American Nothofagus." *Journal of Biogeography* 24, no. 6 (1997): 883–92.

Quijano-Abril, Mario A., et al. "Areas of Endemism and Distribution Patterns for Neotropical Piperspeciles (Piperaceae)." *Journal of Biogeography* 33, no. 7 (2006): 1266–78.

Quiroga, M. P., and Andrea C. Premoil. "Genetic Patterns in *Podocarpus pariatorei* Reveal the Long-Term Persistence of Cold-Tolerant Elements in the Southern Yungas." *Journal of Biogeography* 34, no. 3 (2007): 447–55.

Reese, Carl A., and Kom Bui Lin. "A Modern Pollen Rain Study from the Central Andes Region of South America." *Journal of Biogeography* 32, no. 4 (2005): 709–18.

Riccardi, A. C. "Cretaceous Paleogeography of Southern South America." *Paleogeography, Paleoclimatology, and Paleoecology* 59, nos. 1–3 (1987): 169–95.

Richter, Michael. "Okologische Probleme Lateinamerika." In *Lateinamerika-Krise ohne Ende?* edited by Axel Borsdorf, 181–83. Innsbruck, Austria: Innsbrucker Geographische Studien, Band 21, 1994.

Roulet, M., et al. "Effects of Recent Human Colonization on the Presence of Mercury in Amazonian Ecosystems." *Water, Air, and Soil Pollution* 112 (1999): 297–313.

Ruggiero, Adriana. "Latitudinal Correlates of the Size of Mammalian Geographical Ranges in South America." *Journal of Biogeography* 21, no. 5 (1994): 545–60.

———. "Size and Shape of the Geographical Ranges of Andean Passerine Birds: Spatial Patterns in Environmental Resistance and Anistrophy." *Journal of Biogeography* 28, no. 10 (2001): 1281–94.

Ruggiero, Adriana, et al. "The Geographic Ranges of Mammalian Species in South America: Spatial Patterns in Environmental Resistance and Anistotrophy." *Journal of Biogeography* 25, no. 6 (1998): 1093–104.

Sarkinen, Tiina, et al. "Evolutionary Islands in the Andes: Persistence and Isolation Explain High Endemism in Andean Dry Tropical Forests." *Journal of Biogeography* 39, no. 5 (2012): 884–90.

Sarmiento, Fausto O., and L. M. Frolich. "Andean Cloud Forest Treelines: Naturainess, Agriculture, and the Human Dimension." *Mountain Research and Development* 22 (2002): 278–87.

Sarmiento, G. "The Dry Plant Formations of South America and Their Floristic Connections." *Journal of Biogeography* 2, no. 4 (1975): 233–52.

Schlupp, L., et al. "Biogeography of the Amazon Molly, Peocilia Formosa." *Journal of Biogeography* 29, no. 1 (2002): 1–6.

Seavey, Steven R., and Peter H. Raven. "Chromosomal Differentiation and the Sources of the South American Species of Epilobium (Onagraceae)." *Journal of Biogeography* 4, no. 1 (1977): 55–59.

Sennhauser, E. B. "The Concept of Stability in Connection with the Galley Forests of the Chaco Region." *Vegetatio* 194, no. 1 (1991): 1–13.

Serena, Melody. "Zoogeography of Parthenogenetic Whiptail Lizards in the Guianas: Evidence from Skin Grafts, Karyotypes, and Erythrocyte Areas." *Journal of Biogeography* 12, no. 1 (1985): 49–56.

Sicli, Harold. "The Effects of Deforestation in Amazonia." *Geographical Journal* 151, no. 2 (1985): 198–203.

Smith, Nigel J. H. *The Amazon River Forest: A Natural History of Plants, Animals, and People.* New York: Oxford University Press, 1999.

————. *The Enchanted Amazon Rain Forest: Stories from a Vanishing World.* Gainesville: University Press of Florida, 1996.

Speziale, Karina, et al. "Plant Species Richness-Environment Relationships across the Subantarctic-Patagonian Transition Zone." *Journal of Biogeography* 37, no. 3 (2010): 449–64.

Stadel, Christoph. "Altitudinal Belts in the Tropical Andes: Their Ecology and Human Utilization." *Yearbook, Conference of Latin Americanist Geographers,* 17–18 (1990): 45–60.

Steininger, Marc K. "Satellite Estimation of Tropical Secondary Forest Above-Ground Biomass: Data from Brazil and Bolivia." *International Journal of Remote Sensing* 21, nos. 6–7 (2000): 1139–58.

————. "Tropical Secondary Forest Regrowth in the Amazon: Age, Area, and Change Estimation with Thematic Mapper Data." *International Journal of Remote Sensing* 17, no. 1 (1996): 9–27.

Sternberg, Hilgard O. "Man and Environmental Change in South America." In *Biogeography and Ecology in South America,* edited by E. J. Fittkau et al., 413–45. The Hague, Netherlands: Junk, 1968.

Tackaberry, Rosanne, and Martin Kellman. "Patterns of Tree Species Richness along Peninsula Extensions of Tropical Forests." *Global Ecology and Biogeography Letters* 5, no. 2 (1996): 95–90.

De Toledo, Mauro, and Mark B. Bush. "A Mid-Holocene Environmental Change in Amazonian Savannas." *Journal of Biogeography* 34, no. 8 (2007): 1313–26.

Troll, Carl. "The Cordilleras of the Tropical Americas: Aspects of Climatic, Phytogeographical, and Agrarian Ecology." In *Geo-ecology of the Mountainous Regions of the Tropical Americas,* edited by Carl Troll, 9:15–56. Bonn, Germany: Colloquium Geographicum, Geographischen Institut der Universitat Bonn, 1968.

————, ed. *Geo-ecology of the Mountainous Regions of the Tropical Americas.* Vol. 9. Bonn, Germany: Colloquim Geographicum, Geographischen Institut der Universitat Bonn, 1968.

Van der Hammen, T. "The Pleistocene Changes of Vegetation and Climate in Tropical South America." *Journal of Biogeography* 1, no. 1 (1974): 3–26.

Vasconcelos, Heraldo, et al. "Long-Term Effects of Forest Fragmentation on Amazonian Ant Communities." *Journal of Biogeography* 33, no. 8 (2006): 1348–56.

————. "Patterns of Ant Diversity and Turnover across 2000 km of Amazonian Floodplain Forest." *Journal of Biogeography* 37, no. 3 (2010): 432–40.

Veblen, Thomas T. "Disturbance and Vegetation Dynamics in the Southern Andean Region of Chile and Argentina." In *Repeat Photography: Methods and Applications in the Natural Sciences,* edited by R. H. Webb et al., 167–85. Washington, DC: Island Press, 2008.

————. "Effects of Exotic Deer on Forest Regeneration and Composition in Northern Patagonia." *Journal of Applied Ecology* 26 (1986): 711–24.

————. "Nothofagus Regeneration in Treefall Gaps in Northern Patagonia." *Canadian Journal of Forest Research* 19, no. 3 (1989): 365–71.

————. "Temperate Broad-Leaved Evergreen Forests of South America." In *Temperate Broad-Leaved Evergreen Forests,* edited by John D. Overton, 5–31. Amsterdam, Netherlands: Elsevier, 1983.

Veblen, Thomas T., and P. B. Alabach. "A Comparative Review of Forest Dynamics and Disturbance in the Temperate Rainforests of North and South America." In *High-Latitude Rainforests and Associated Ecosystems of the West Coast of the Americas: Climate, Hydrology, Ecology, and Conservation,* edited by R. G. Lawford et al., 173–213. New York: Springer-Verlag, 1996.

Veblen, Thomas T., et al. "Disturbance and Forest Dynamics along a Transect from the Andean Rain Forest to Patagonian Schrublands." *Journal of Vegetation Science* 3 (1992): 507–20.

————. "Ecology of Southern Chilean and Argentine Nothofagus." In *Ecology and Biogeography of Nothofagus Forests*, edited by Thomas T. Veblen et al., 293–353. New Haven, CT: Yale University Press, 1996.

————. "Fire History in Northern Patagonia: The Role of Humans and Climate Variation." *Ecological Monographs* 69, no. 1 (1999): 47–67.

————. "The Historical Range of Variability of Fires in the Andean-Patagonian Nothofagus Forest Region." *International Journal of Wildlife Fire* 17, no. 6 (2008): 724–41.

Veblen, Thomas T., and D. C. Lorenz. "Post-fire Stand Development of Austro-cedrus-Nothofagus Forests in Northern Patagonia." *Vegetatio* 71 (1987): 113–26.

————. "Recent Vegetation Changes along the Forest/Steppe Ecotone of Northern Patagonia." *Annals of the Association of American Geographers* 78, no. 1 (1988): 93–111.

Veblen, Thomas T., and Very Markgraf. "Steppe Expansion in Patagonia." *Quaternary Research* 30, no. 3 (1988): 331–38.

Vetter, R. E., and P. C. Botosso. "El Nino May Affect Growth Behavior of Amazonian Trees." *GeoJournal* 19, no. 4 (1989): 419–22.

Villalba, Ricardo. "Regional Patterns of Tree Population Age Structure in Northern Patagonia: Climatic and Disturbance Influence." *Journal of Ecology* 85, no. 2 (1997): 113–25.

Villalba, Ricardo, and Thomas T. Veblen. "Influences of Large Scale Climatic Variability on Episodic Tree Mortality at the Forest-Steppe Ecotone in Northern Patagonia." *Ecology* 79 (1998): 2624–40.

Vose, P. B. "Amazon Bio-Geosciences: Towards a Synthesis through Amazonia, I." *GeoJournal* 19, no. 4 (1989): 361–68.

Vuilleumier, F. "Insular Biogeography in Continental Species: I. The Northern Andes of South America." *American Naturalist* 104 (1970): 373–88.

Webb, S. D., and L. G. Marshall. "Historical Biogeography of Recent South American Land Animals." In *Mammalia Biology in South America*, edited by M. A. Mares and H. H. Genoways, 39–52. Pittsburgh, PA: University of Pittsburgh, Laboratory of Ecology, Special Publication 6, 1982.

Winemiller, Kirk, et al. "Fish Assemblages of the Casiquiara River: A Corridor and Zoogeographical Filter for Dispersal between the Orinoco and Amazon Basins." *Journal of Biogeography* 35, no. 9 (2008): 1551–63.

Wittman, R., and W. J. Junk. "Sapling Communities in Amazonia White-Water Forests." *Journal of Biogeography* 30, no. 10 (2003): 1533–44.

Wittmann, Florian, et al. "Tree Species Composition and Diversity Gradients in Whitewater Forests across the Amazon Basin." *Journal of Biogeography* 33, no. 8 (2006): 1334–47.

Young, Kenneth R. "Deforestation in Landscapes with Humid Forests in the Central Andes." In *Nature's Geography: New Lessons for Conservation in Developing Countries*, edited by Karl S. Zimmerer and Kenneth R. Young, 75–99. Madison: University of Wisconsin Press, 1998.

————. "Roads and the Environmental Degradation of Tropical Montane Forests." *Conservation Biology* 8, no. 4 (1994): 972–76.

————. "Woody and Scamdent Plants on the Edge of an Andean Timberline." *Bulletin of the Torrey Botanical Club* 120 (1993): 1–18.

Zimmerer, Karl S., and Robert P. Langstroth. "Biological Diversity in Local Development: Popping Beans of the Central Andes." *Mountain Research and Development* 12, no. 1 (1992): 47–61.

Zimmerman, Barbara, and Daniel Simberloff. "An Historical Interpretation of Habitat Use by Frogs in a Central Amazonian Forest." *Journal of Biogeography* 23, no. 1 (1996): 27–46.

Climatology and Weather

Arain, M. A., et al. "Coparing Micro-meteorology of Rainforests in Biosphere 2 and Amazon Basin." *Agricultural and Forest Meteorology* 100 (2000): 273–89.

Bendix, Jorg. "Precipitation Dynamics in Ecuador and Northern Peru during the 1991/2 El Nino: A Remote Sensing Perspective." *International Journal of Remote Sensing* 21, no. 3 (2000): 533–48.

Boffi, Jorge A. "Effects of the Andes Mountains on the General Circulation over the Southern Part of South America." *Bulletin of the American Meteorological Society* 30 (1949): 242–47.

Bowman, Isaiah. "Man and Climate Change in South America." *Geographical Journal* 33 (1909): 267–78.

Caviedes, Cesar N. "The Effects of ENSO Events in Some Key Regions of the South American Continent." In *Recent Climatic Change; A Regional Approach*, edited by Stanley Gregory, 252–66. London: Belhaven, 1988.

———. "El Nino." *Geographical Review* 74, no. 3 (1984): 267–90.

———. "El Nino, 1972: Its Climatic, Ecological, Human, and Economic Implications." *Geographical Review* 65, no. 4 (1975): 493–509.

———. "On the Genetic Links of Precipitation in South America." In *Forschrifte Landschaftsolologisher und Klimatologischer Forschungen in den Tropen*, edited by D. Havlik and R. Mackel, 55–77. Freiberg, Germany: Freiberger Geographishce Hefte 18, 1982.

———. "Rainfall in South America: Seasonal Trends and Spatial Correlations." *Erdkunde* 35, no. 2 (1981): 107–18.

———. "Secas and El Nino: The Simultaneous Climatological Hazards in South America." *Proceedings of the Association of American Geographers* 5 (1973): 44–50.

———. "South American and World Climate History." In *Environmental History: Critical Issues in Comparative Perspective*, edited by Kendall E. Bailes, 135–62. Washington, DC: University Press of America, 1985.

Caviedes, Cesar N., and Peter R. Waylen. "Chapters for a Climatic History of South America." In *Festschrift zur Ehre von Professor Wolfgang Weische's 70*, 149–80. Beburtstag. Freiberg, Germany: Freiberger Geographische Heft 32, 1991.

Chebataroff, Jorge. "Posibilidades para el mejoramiento de clima local." *Revista Geografica* 44 (1956): 1–22.

Cutrim, E., et al. "Enhancement of Cumulus Clouds over Deforested Lands in Amazonia." *Bulletin of the American Meteorological Society* 76 (1995): 1801–5.

DeLiberty, Tracy I. "A Regional Scale Investigation of Climatological Tropical Convection and Precipitation in the Amazon Basin." *Professional Geographer* 52, no. 2 (2000): 258–71.

Foster, J. L., et al. "Seasonal Snow Extent and Snow Mass in South America Using SSMI Passive Microwave Data." *Polar Geography* 25, no. 1 (2001): 41–53.

Gomez, Darwin, et al. "Estimacion de la tasa diaria de precipitacion, par satellite, en America del Sur." *Revista Geografica* 146 (2009): 7–18.

Gooze, Charles, and Clarence F. Jones. "The Seasonal Distribution of Rainfall in South America." *Bulletin of the Geographical Society of Philadelphia* 26 (1928): 93–115.

Haas, William H. "Major Climatic Controls of South America." *Bulletin of the Geographical Society of Philadelphia* 28 (1930): 79–88.

Helmens, Karin F., et al. "Magnetic Polarity and Fission-Track Chronology of a Late Pliocene-Pleistocene Paleoclimatic Proxy Record in the Tropical Andes." *Quaternary Research* 48, no. 1 (1997): 15–28.

———. "Warming at 18000 Years BP in the Tropical Andes." *Quaternary Research* 47, no. 3 (1996): 289–99.

Hoffmann, Jose A. J. *Climatic Atlas of South America*. New York: WMO/UNESCO, 1975.

James, Preston E. "Air Masses and Fronts in South America." *Geographical Review* 29 (1939): 132–34.

Jefferson, Mark. "Actual Temperatures of South America." *Geographical Review* 16 (1926): 443–66.

———. "The Real Temperatures throughout North and South America." *Geographical Review* 6 (1918): 240–67.

Kleiden, A., and H. Heimann. "Assessing the Role of Deep Rooted Vegetation in the Climate System with Model Simulations: Mechanism, Comparison to Observations, and Implications for Amazonian Deforestation." *Climate Dynamics* 16, nos. 2–3 (2000): 183–99.

Kleiden, A., and S. Lorenz. "Deep Roots Sustain Amazonian Rainforest in Climate Model Simulation of the Last Ice Age." *Geophysical Research Letters* 28, no. 12 (2001): 2425–28.

Lahey, James F. "Rainfall Characteristics of the Dry Area of Northern South America and the Southern Caribbean." *Bulletin of the Illinois Geographical Society* 11, no. 2 (1969): 12–35.

Meggers, Betty J. "Archeological Evidence for the Impact of Mega-Nino Events on Amazonia during the Past Two Millenia." *Climate Change* 28, no. 4 (1994): 321–38.

Molion, L. C. "On the Dynamic Climatology of the Amazon Basin and Associated Rain Producing Mechanisms." In *The Geophysiology of Amazonia*, edited by Robert E. Dickinson, 391–408. New York: Wiley, 1987.

Moulton, Benjamin. "The Climates of the Amazon Lowland." *Journal of Geography* 44 (1945): 367–70.

Murphy, Robert C. "Oceanic-Climatic Phenomena along the West Coast of South America, 1925." *Geographical Review* 16, no. 1 (1926): 26–54.

Nishizawa, Toshie, and Minoru Tanaka. "The Annual Change in the Tropospheric Circulation and the Rainfall in South America." *Archiv fur Meteorologie, Geophisik, und Bioklimatologie* 33 (1983): 107–16.

———. "The Annual Change in the Tropospheric Circulation and the Rainfall in South America." *Latin American Studies, Japan* 6 (1983): 19–44.

———. "Climatic Environment in Tropical South America to the East of the Andes." *Latin American Studies, Japan* 6 (1983): 15–18.

Plancon, O. "Transitions entre climates tropicaux et temperes en Amerique du Sud: essai de regionalization climatique." *Les Cahiers d'Outre Mer* 56, no. 223 (2003): 259–80.

Schabitz, F., and H. Liebricht. "Landscape and Climate Development in the South-Eastern Part of the Arid Diagonal during the Last 13,000 Years." In *Landschatsentwicklung, Palaookologie, und Klimageschichata der Ariden Diagonale Sudamerika im Jungquarter*, edited by K. Garleff, 371–88. Bamberg, Germany: Bamberger Geographische Schriften, Heft 15, 1998.

Schneider, C., et al. "Weather Observations across the Southern Andes at 53 Degrees South." *Physical Geography* 24, no. 2 (2003): 97–119.

Schneider, C., and D. Gries. "Effects of El Nino-Southern Oscillation on Southern and South American Precipitation at 53 Degrees S Revealed from NCEP-NCAR Reanalyses and Weather Station Data." *International Journal of Climatology* 24 (2004): 1057–76.

Schwerdtfeger, Werner, ed. *Climates of Central and South America. World Survey of Climatology*. Vol. 12. Amsterdam, Netherlands: Elsevier, 1976.

Serra, Adalberto. "Clima da America do Sul." *Revista Geografica* 59 (1962): 93–126.

Seth, Anji, et al. "Making Sense of Twenty-First Century Climate Change in the Altiplano: Observed Trends and CMiP3 Projections." *Annals of the Association of American Geographers* 100, no. 4 (2010): 835–47.

Shuttleworth, W. J., et al. "Daily Variation in Temperature and Humidity within and above Amazonian Forest." *Weather* 40, no. 4 (1985): 102–8.

———. "Post-deforestation Amazonian Climate: Anglo-Brazilian Research to Improve Prediction." *Journal of Hydrology* 129 (1991): 71–85.

Sturman, A. P. "An Unusual Effect of the Andes Mountains on Atmospheric Circulation of the Southern South American Sector." In *Proceedings, Ninth New Zealand Geography Conference*, edited by T. J. Hearn and R. P. Hargreaves, 113–15. Dunedin: New Zealand Geographical Society, 1977.

Suzuki, Hideo. "Recent and Warm Climates of the West Coast of South America [in Japanese]." *Bulletin, Tokyo University Department of Geography* 5 (1973): 3–32.

Tsuchiya, Akio. "Comparison of Micrometeorological Components between Dry and Rainy Seasons in Amazonian Dense Tropical Forests." *Papers of the Applied Geography Conferences* 26 (2003): 286–96.

Veit, Heinz. "Holocene Climatic Changes and Atmospheric Paleocirculation in the Northwestern Part of the Arid Diagonal of South America." In *Landschaftsentwicklung, Palaookologie, und Klimageschichta der Ariden Diagonale Sudamerikas*, edited by K. Garleff, 355–70. Bamberg, Germany: Bamberger Geographische Schriften, Helt 15, 1998.

Villalba, Ricardo. "Climatic Fluctuations in Northern Patagonia during the Last 1000 Years as Inferred from Tree-Ring Records." *Quaternary Research* 34, no. 3 (1990): 346–60.

Ward, Robert. "The Climate of South America." *Bulletin of the American Geographical Society* 35, no. 4 (1903): 353–60.

Weischet, W. "Climatic Constraints for the Development of the Far South of Latin America." *GeoJournal* 11, no. 1 (1985): 79–87.
Willmott, Cort, et al. "Estimating Continental and Terrestrial Precipitation Averages from Rain Gauge Networks." *International Journal of Climatology* 14, no. 4 (1994): 403–14.
Yazawa, Taiji. "A Study of the Andes-Especially on Their Climate [in Japanese]." *Jimbun Chiri/Human Geography* 11, no. 5 (1959): 77–87.
Young, Kenneth R. "The Tropical Andes as a Morphoclimatic Zone." *Progress in Physical Geography* 13, no. 1 (1989): 13–22.

Geomorphology, Landforms, and Volcanism

Bowman, Isaiah. "Asymmetrical Crest Lines and Abnormal Valley Profiles in the Central Andes." In *Sonder-Abdruck aus der Zeitschrift fur Gletscherkunde fur Eisezeitforschung und Geschichte des Klimas*, 119–29. Berlin, Germany: Band 7, 1913.
———. "The Physiography of the Central Andes: I, the Maritime Andes." *American Journal of Science* 28 (1909): 197–217.
———. "The Physiography of the Central Andes: II, the Eastern Andes." *American Journal of Science* 28 (1909): 373–402.
Craig, Alan K. "Status of Geomorphic Research in South America." *Proceedings of the Conference of Latin Americanist Geographers* 1 (1970): 240–49.
Glasser, N. F., et al. "Evidence from the Rio Bayo Valley on the Extent of the North Patagonia Icefield during the Late Pleistocene-Holocene Transition." *Quaternary Research* 65 (2006): 70–77.
———. "Geomorphological Evidence for Variations of the North Patagonian Icefield during the Holocene." *Geomorphology* 71, nos. 3–4 (2005): 203–27.
Glasser, N. F., and M. J. Hambrey. "Sedimentary Facies and Landform Genesis at a Temperate Outlet Glacier: Soler Glacier, North Patagonian Icefield." *Sedimentology* 49, no. 1 (2002): 43–64.
Grosso, S. A., and A. E. Carte. "Cryoplanation Surfaces in the Central Andes at Latitude 35 Degrees South." *Permafrost and Periglacial Processes* 2 (1991): 49–58.
Harden, Carol. "Human Impacts on Headwater Fluvial Systems in the Northern and Southern Andes." *Geomorphology* 79, nos. 3–4 (2006): 249–63.
Marbut, C. "The Topography of the Amazon Valley." *Geographical Review* 15, no. 4 (1925): 617–42.
Martinelli, I. A., et al. "Suspended Sediment Load in the Amazon Basin: An Overview." *GeoJournal* 19, no. 4 (1989): 381–90.
Neller, R. J., et al. "In the Formation of Blocked Valley Lakes by Channel Avulsion in Upper Amazon Forelands Basins." *Zeitschrift fur Geomorphologie* 36, no. 4 (1992): 401–11.
Perez, Francisco L. "Talus Texture and Particle Morphology in a North Andean Paramo." *Zeitschrift fur Geomorphologie* 30, no. 1 (1986): 15–34.
Populizio, Eliseo. "El Parana, un rio y histira geomorfologica." *Revista Geografica* 140 (2006): 79–90.
Reichert, T., et al. "Morphodynamische Aktivitatsphasen in Nordpatagonien und am Sundrand der Pamaerste Ergebnisse." In *Landschaftsentwicklung, Palaookologie, und Klimageschichta der Ariden Diagonale Sudamerika im Jungquarter*, edited by K. Garleff, 109–32. Bamberg, Germany: Bamberger Geographische Schriften, Heft 15, 1998.
Satoh, Hisashi. "Geomorphology of the Central Andes [in Japanese]." *Bulletin, Tokyo University Department of Geography* 12 (1980): 1–34.
———. "Geomorphology of the Central Andes [in Japanese]." *Bulletin, Tokyo University Department of Geography* 14 (1982): 75–121.
Shipton, Eric. "Volcanic Activity on the Patagonian Ice Cap." *Geographical Journal* 126, no. 4 (1960): 389–96.
Sternberg, Hilgard O. "Radiocarbon Dating as Applied to a Problem of Amazonian Morphology." In *Comptes Rendus du XVIII Congres International de Geographie*, 2:399–424. Rio de Janeiro, Brazil: International Geographical Union, 1960.

Street-Perrott, F. A., and P. A. Baker. "Biogenic Silica: A Neglected Component of the Coupled Global Continental Biogeochemical Cycles of Carbon and Silicon." *Earth Surface Processes and Landforms* 33, no. 9 (2008): 1436–57.

Vanacker, Veerle, et al. "River Channel Response to Short-Term Human-Induced Change in Landscape Connectivity in Andean Ecosystems." *Geomorphology* 72, nos. 1–4 (2005): 340–53.

Vis, M. *Processes and Patterns of Erosion in Natural and Disturbed Andean forest Ecosystems*. Amsterdam, Netherlands: Honinklijk Nederlands Aardrijkskindig Ge, 1991.

Yim, W. S. "Some Problems in Evolutionary Geomorphology from a Review of Evidence for Multiple Glaciations in South America and Methods Used for Reconstructing Late Cainozoic Climate Events." *Asian Geographer* 1, no. 2 (1982): 57–60.

Hydrology and Glaciology

Aniya, Masamu. "Holocene Glacial Chronology in Patagonia: Tyndall and Upsala Glaciers." *Arctic and Alpine Research* 27, no. 4 (1995): 311–22.

Aniya, Masamu, et al. "Morphology, Surface Characteristics, and Flow Velocity of Soler Glacier, Patagonia." *Arctic and Alpine Research* 20, no. 4 (1988): 414–21.

———. "Variations of Some Patagonia Glaciers, South America, Using RADARSAT and Landsat Images." *Science Reports, Tohoku University, 7th Series, Geography* 21 (2000): 23–38.

Araya, Jose F. "Relacones entre las Piedmonts Glacigenicas y las Formes Marinas Litorals en el Sur de Sudamerica." *Revista Geografica* 81 (1974): 115–38.

Bell, C. M. "Punctuated Drainage of an Ice-Dammed Quaternary Lake in Southern South America." *Geografiska Annaler* 90A, no. 1 (2008): 1–17.

Bentley, Michael, et al. "The Landforms and Pattern of Deglaciation in the Strait of Magellan and Bahia Inutil, Southernmost South America." *Geografiska Annaler* 87A, no. 2 (2005): 313–33.

Brinkman, W. L. F., and A. Dos Santos. "Natural Waters in Amazonia, III." *Acta Amazonia* 2 (1970): 443–48.

Clapperton, Chalmers M. "A Younger Dryas Icecap in the Equatorial Andes." *Quaternary Research* 47, no. 1 (1997): 13–28.

Elsenbeer, Helmut, and Andreas Lech. "Hydrometric and Hydrochemical Evidence for Fast Flow Paths at La Cuenca, Western Amazonia." *Journal of Hydrology* 180 (1996): 237–50.

Galloway, R. W. "Fossil Ice Wedges in Patagonia and Their Paleoclimatic Significance." *Zeitschrift fur Geomorphologie* 29, no. 4 (1985): 389–96.

Glasser, N. F., et al. "An Advance of Soler Glacier, North Patagonia Icefield at c. AD 1222–1342." *The Holocene* 12, no. 1 (2002): 113–20.

———. "Late Pleistocene and Holocene Paleoclimate and Glacier Fluctuations in Patagonia." *Global and Planetary Change* 43, nos. 1–2 (2004): 79–101.

Glasser, N. F., and K. Jansson. "Fast-Flowing Outlet Glaciers of the Last Maximum Patagonia Icefield." *Quaternary Research* 63, no. 2 (2005): 206–11.

Grosso, S. A., and A. E. Carte. "Pleistocene Ice Wedge Casts at 34 Degrees South, Eastern Andes Piedmont, Southwest of South America." *Geografiska Annaler* 71A, nos. 3–4 (1989): 125–36.

Harrison, Stephen, et al. "A Glacial Lake Outburst Flood Associated with a Recent Mountain Glacier Retreat, Patagonian Andes." *The Holocene* 16, no. 4 (2006): 611–20.

Hatcher, J. B. "The Lake Systems of Southern Patagonia." *Bulletin of the Geographical Society of Philadelphia* 2, no. 6 (1900): 139–45.

Hubbard, Alan L., et al. "A Modelling Reconstruction of the Last Glacial Maximum Ice Sheet and Its Deglaciation in the Vicinity of the Northern Patagonian Icefield, South America." *Geografiska Annaler* 87A, no. 2 (2005): 375–91.

Hulton, Nick, et al. "Glacier Modeling and the Climate of Patagonia during the Last Glacial Maximum." *Quaternary Research* 42, no. 1 (1994): 1–19.

Klein, A. G., et al. "Modern and Last Glacial Maximum Snowlines in the Peruvian-Bolivian Andes." *Quaternary Science Reviews* 18 (1999): 63–84.

Lawrence, D. B., and E. G. Lawrence. *Recent Glacier Variations in Southern South America.* New York: American Geographical Society, Southern Chile Expedition Technical Report 1, 1959.

Lesach, Lance. "Seepage Exchange in an Amazon Floodplain Lake." *Limnology and Oceanography* 40, no. 3 (1995): 598–609.

Marden, Christopher J. "Factors Affecting the Volume of Quaternary Glacial Deposits in Southern Patagonia." *Geografiska Annaler* 76A, no. 4 (1994): 261–69.

———. "Late Glacial and Holocene Variations of the Grey Glacier, an Outlet of the Southern Patagonian Icefield." *Scottish Geographical Magazine* 109, no. 1 (1993): 27–31.

Marin, C. T., et al. "Gross Rainfall and Its Positioning into Throughfall, Stemflow, and Evaporation of Intercepted Water in Four Forest Systems in Western Amazonia." *Journal of Hydrology* 237, nos. 1–2 (2000): 40–57.

McCulloch, R. D., et al. "Chronology of the Glaciation in Central Strait of Magellan and Bahia Inutil, Southernmost South America." *Geografiska Annaler* 87A, no. 2 (2005): 289–312.

———. "Evidence for Late-Glacial Ice Dammed Lakes in the Central Strait of Magellan and Bahia Inutil, Southernmost South America." *Geografiska Annaler* 87A, no. 2 (2005): 335–62.

Mercer, John H. "Glacier Variations in the Andes." *Glaciological Notes* 12 (1962): 9–31.

———. "Glacier Variations in Southern Patagonia." *Geographical Review* 55 (1965): 390–413.

———. "Variations in Some Patagonian Glaciers since the Late Glacial, I." *American Journal of Science* 266 (1968): 91–109.

———. "Variations in Some Patagonian Glaciers since the Late Glacial, II." *American Journal of Science* 269 (1970): 1–25.

Mormer, N. A. "Tidal Investigations of the West Coast of South America." *Geographical Review* 33 (1943): 299–303.

Orfeo, O., and J. Stevaux. "Hydraulic and Morphological Characteristics of Middle and Upper Reaches of the Parana River (Argentina and Brazil)." *Geomorphology* 44, nos. 3–4 (2002): 309–22.

Page, John. "The Gran Chaco and Its Rivers." *Proceedings of the Royal Geographical Society* 3 (1889): 129–52.

Rabassa, Jorge, and Chalmers M. Clapperton. "Quaternary Glaciations of the Southern Andes." *Quaternary Science Review* 9, nos. 2–3 (1990): 153–74.

Rabatel, A., and V. Jomelli. "Palaeoclimatic Interpretation of Glacial Variations in the Intertropical Andes over the Last Millennium [in French]." *Bulletin de l'Association de Geographes Francais* 88, no. 1 (2011): 7–16.

Rivera, Andres, et al. "Recent Fluctuations of Glacier Pio XI, Patagonia: Discussion of a Glacial Surge Hypothesis." *Mountain Research and Development* 17, no. 4 (1997): 309–22.

Rivera, Andres, and Gino Casassa. "Ice Elevation, Areal, and Frontal Changes of Glaciers from National Park Torres del Paine, Southern Patagonian Icefield." *Arctic, Antarctic, and Alpine Research* 36, no. 4 (2004): 379–89.

Schoewe, Walter H. "Glaciers and Glaciation of South America." *Journal of Geography* 14 (1915): 49–54.

Selzer, Geoffrey O. "Late Quaternary Glaciation as a Proxy for Climate Change in the Central Andes." *Mountain Research and Development* 13, no. 2 (1993): 129–38.

———. "Recent Glacial History and Paleoclimate of the Peruvian-Bolivian Andes." *Quaternary Science Reviews* 9, nos. 2–3 (1990): 137–52.

Sugden, D. E., et al. "Late-Glacial Glacier Events in Southernmost South America: A Blend of 'Northern' and 'Southern' Hemispheric Climatic Signals?" *Geografiska Annaler* 87A, no. 2 (2005): 273–88.

Tilman, H. W. "The Patagonia Ice Cap." *Geographical Journal* 123, no. 2 (1957): 148–56.

Tower, Walter S. "The World's Great Rivers: The Parana." *Journal of Geography* 11 (1912): 119–25.

Turner, K. J., et al. "Deglaciation of the Eastern Flank of the North Patagonian Icefield and Associated Continental-Scale Lake Diversions." *Geografiska Annaler* 87A, no. 2 (2005): 363–74.

Warren, C. R. "Freshwater-Calving and Anomalous Glacier Oscillations: Recent Behaviour of Moreno and Ameghino Glaciers, Patagonia." *Holocene* 4, no. 4 (1994): 422–29.

Warren, C. R., and D. E. Sugden. "The Patagonian Icefields: A Glaciological Review." *Arctic and Alpine Research* 25, no. 4 (1994): 316–31.

Soils

Correa, J. C., and K. Reichandt. "The Spatial Variability of Amazonian Soils under Natural Forest and Pasture." *GeoJournal* 19, no. 4 (1989): 423–28.

Duran, Artigas, et al. "Distribution, Properties, Land Use, and Management of Mollisols in South America." *Chinese Geographical Sciences* 21, no. 1 (2011): 511–30.

Harden, Carol. "Andean Soil Erosion." *National Geographic Research and Exploration* 7, no. 2 (1991): 216–31.

Imeson, A. C., and M. Vis. "A Survey of Soil Erosion Processes in Tropical Forest Ecosystems on Volcanic Ash Soils in the Central Andean Cordillera." *Geografiska Annaler* 64A, nos. 3–4 (1982): 181–98.

Magnusson, W. E., et al. "A Comparison of 13C Ratios of Surface Soils in Savannas and Forests in Amazonia." *Journal of Biogeography* 27, no. 7 (2002): 857–64.

Marbut, C., and C. B. Manifold. "The Soils of the Amazon Basin in Relation to Agricultural Possibilities." *Geographical Review* 16, no. 3 (1926): 414–42.

Nortcliff, Stephen. "A Review of Soil and Soil-Related Constraints to Development in Amazonia." *Applied Geography* 9, no. 3 (1989): 147–60.

Perez, Francisco L. "Alpine Turf Destruction by Cattle in the High Equatorial Andes." *Mountain Research and Development* 13, no. 1 (1993): 107–10.

———. "Microbiotic Crusts in the High Equatorial Andes and Their Influence on Paramo Soils." *Catena* 31, no. 3 (1997): 173–98.

———. "Talus Movement in the High Equatorial Andes: A Synthesis of Ten Years of Data." *Permafrost and Periglacial Processes* 4, no. 3 (1993): 199–216.

Tambotto, D. "Palaeo-Permafrost in Patagonia." In *Landschaftsentwicklung, Palaookologie, und Climageschichta der Ariden Diagonale Sudamerikas im Jungquarter*, edited by K. Garleff, 133–48. Bamberg, Germany: Bamberger Geographische Schriften, Heft 15, 1998.

Tsuchiya, Akio, et al. "Changes in Soil CO^2 Concentration Accompanying Infiltration and Evaporation at a Primary Forest and Grassland in Central Amazonia." *Geographical Review of Japan* 78, no. 12 (2005): 794–811.

Victoria, R. I., et al. "Spatial and Temporal Variations in Soil Chemistry on the Amazon Floodplain." *GeoJournal* 19, no. 1 (1989): 45–52.

WinklerPrins, Antoinette M., and Stephen P. Aldrich. "Locating Amazonian Dark Earths: Creating an Interaction GIS of Known Locations." *Journal of Latin American Geography* 9, no. 3 (2010): 33–50.

WinklerPrins, Antoinette M., and Edmundo Barrios. "Ethnopedology along the Amazon and Orinoca Rivers: A Convergence of Knowledge and Practice." *Revista Geografica* 142 (2007): 111–30.

Woods, W. I. "Comments of the Black Earths of Amazonia." *Papers of the Applied Geography Conferences* 18 (1995): 159–66.

Woods, W. I., and Joseph McCann. "The Anthropogenic Origin and Persistence of Amazonia Dark Earths." *Yearbook, Conference of Latin Americanist Geographers* 25 (1999): 7–14.

Theses and Dissertations

Abizaid, Christian. "Floodplain Dynamics and Rural Livelihoods in the Upper Amazon." PhD diss., McGill University, Montreal, Quebec, 2007.

Barrett, Kirsten. "Carbon Accumulation and Storage in Amazonian Ecosystems." PhD diss., Clark University, Worcester, MA, 2008.

Durkee, Joshua. "Assessing the Role of Mesoscale Convection Complexes in Subtropical South American Precipitation Variability." PhD diss., University of Georgia, Athens, 2007.

Edger, Timothy. "Analysis of Radar Glacier Zones in the Southern Patagonia Icefield." Master's thesis, University of Utah, Salt Lake City, 2008.

Grady, Mary. "Human Impact on Ecological Degradation in the Andes Mountains." Master's thesis, University of Hawaii, Manoa, 1990.

Lahey, James F. "On the Origin of the Dry Climate in Northern South America and the Southern Caribbean." PhD diss., University of Wisconsin, Madison, 1958.

Little, Dana A. "Classification of Climates of South America." Master's thesis, Clark University, Worcester, MA, 1951.

Menzies, John. "A Change Detection Comparison of Deforestation in Rondonia, Brazil and Southern Bolivia Using Satellite Image Classification (1975–1998)." Master's thesis, University of North Dakota, Grand Forks, 2001.

Morton, Douglas. "Changes in Amazon Forest Structure from Land-Use Fires: Integrating Satellite Remote Sensing and Ecosytem Modeling." PhD diss., University of Maryland, College Park, 2008.

Ping, Jiang. "Biomass Estimation and Classification of Secondary Success on Using Radar and Optical Remote Sensing Data Based on Textural and Spectral Analysis in Amazonia." PhD diss., Indiana State University, Terre Haute, 2007.

Reesem, Carl A. "Pollen Dispersal and Deposition in the High-Central Andes, South America." PhD diss., Louisiana State University, Baton Rouge, 2003.

Schroeder, Wilfrid. "Towards an Integrated System for Vegetation Fire Monitoring in the Amazon Basin." PhD diss., University of Maryland, College Park, 2008.

Urrutia, Rocio. "Assessment of Twenty-First Century Climate Change in Tropical South America and the Tropical Andes." Master's thesis, University of Massachusetts, Amherst, 2008.

Villalba, Ricardo. "Climatic Influences on Forest Dynamics along the Forest-Steppe Ecotone in Northern Patagonia." PhD diss., University of Colorado, Boulder, 1995.

Vina, Andres. "Analysis of the Spatial and Temporal Patterns of Forest Fragmentation in the Colombia-Ecuador Border." Master's thesis, University of Nebraska, Lincoln, 2000.

Wright, Emily. "Role of Cold Fronts in South American Monsoon Onset." Master's thesis, East Carolina University, Greenville, 2010.

POLITICAL GEOGRAPHY

General Works

Anderson, Thomas D. "The Gulf of Venezuela Sea Boundary: A Problem between Friends." *Yearbook, Conference of Latin Americanist Geographers* 15 (1989): 97–106.

Avery, W. P. "Origins and Consequences of the Border Dispute between Ecuador and Peru." *Inter-American Economic Affairs* 38, no. 1 (1984): 65–77.

Barton, Jonathan R. *A Political Geography of Latin America.* London: Routledge, 1997.

Bebbington, Anthony J. "New States, New NGOs? Crises and Transitions among Rural Development NGOs in the Andean Region." *World Development* 25, no. 11 (1997): 1755–65.

———. "Rural Development: Policies, Programmes, and Actors." In *Latin American Development: Geographical Perspectives*, edited by David A. Preston, 116–45. London: Longman, 1996.

———. "Social Capital and Rural Intensification: Local Organizations and Islands of Sustainability in the Rural Andes." *Geographical Journal* 163, no. 2 (1997): 189–97.

Bebbington, Anthony J., and G. Thiele. *NGOs and the State in Latin America: Rethinking Roles in Sustainable Agricultural Development.* London: Routledge, 1993.

Borsdorf, Axel. "Conception of Regional Planning in Latin America." *Applied Geography and Development* 15 (1980): 28–40.

———. "Raumliche Dimensionen der krise Lateinamerika." In *Latinamerika-krise ohne Ende?* edited by Axel Barsdorf, 27–42. Innsbruck, Austria: Innsbrucker Geographische Studien, Band 21, 1994.

Bowman, Isaiah. "The Ecuador-Peru Boundary Dispute." *Foreign Affairs* 20 (1942): 757–61.

———. "The Military Geography of Atacama." *Educational Bi-Monthly* 6 (1911): 1–21.

Braveboy-Wagner, Jacqueline A. *The Venezuelan-Guyana Border Dispute: Britain's Colonial Legacy in Latin America*. Boulder, CO: Westview, 1979.

Bromley, Raymond J. "Informality, de Soto Style: from Concept to Policy." In *Contrapuuto: The Informal Sector Debate in Latin America*, edited by C. A. Rakowski, 131–52. Albany: State University of New York Press, 1994.

Brown, E. "Articulating Opposition in Latin America: The Consolidation of Neo-liberalism and the Search for Radical Alternatives." *Political Geography* 15, no. 2 (1996): 169–92.

Caviedes, Cesar N. "The Emergence and Development of Geopolitical Doctrines in the Southern Cone Countries." In *Geopolitics of the Southern Cone and Antarctica*, edited by Philip Kelly and Jack Child, 13–29. Boulder, CO: Lynne Rienner, 1988.

———. *The Southern Cone: Realities of the Authoritarian States*. Totowa, NJ: Rowman & Littlefield, 1984.

Chermakian, Jean. "Le Canada, Le Quebec, et L'Amerique Latine: Une relation geopolitique a redefinie." *Zeitschrift fur Kanada-Studien* 26, no. 2 (2006): 9–16.

Child, Jack. "The American Southern Cone: Geopolitics and Conflict." *Proceedings, Conference of Latin Americanist Geographers* 9 (1983): 200–213.

———. *Antarctica and South American Geopolitics*. New York: Praeger, 1988.

———. "Geopolitical Thinking of Latin America." *Latin American Research Review* 14 (1979): 89–111.

———. *Geopolitics and Conflict in South America: Quarrels among Neighbors*. Westport, CT: Praeger, 1985.

———. "Latin Lebensraum: The Geopolitics of Ibero-American Antarctica." *Applied Geography* 10, no. 4 (1990): 287–306.

———. "South American Geopolitics and Antarctica: Confrontation or Cooperation?" In *Geopolitics of the Southern Cone and Antarctica*, edited by Philip Kelly and Jack Child, 187–202. Boulder, CO: Lynne Rienner, 1988.

———. "The Status of South American Geopolitical Thinking." In *South America into the 1990s: Evolving International Relations in a New Era*, edited by G. P. Atkins, 53–85. London: Westview, 1990.

Cochius, F. "Maritime Relations between the Netherlands and the South American Continent." *Revista Geografia* 52 (1960): 87–93.

Crist, Raymond E. "Politics and Geography: Some Aspects of Centrifugal and Centripetal Forces Operating in Andean America." *American Journal of Economics and Sociology* 25 (1966): 349–58.

Dieguez, Pedro A. "Geopolitica y Narco Trafico en la America Andina." *Anales de Geografia de la Universidad Complutense* 12 (1992): 119–33.

Dodds, Klaus. "Geopolitics, Cartography, and the State in South America." *Political Geography* 12, no. 4 (1993): 361–81.

Dowler, Lorraine. "Amazonian Landscapes: Gender, War, and Historical Repetition." In *The Geography of War and Peace*, edited by Colin Flint, 133–48. New York: Oxford University Press, 2005.

Eden, Michael J. "Environment, Politics, and Amazonian Deforestation." *Land Use Policy* 11 (1994): 55–66.

Faria-Nunes, Paulo H. "Integracao logistica na America do Sul: Uma Analise do Papel da Iniciativa para a Integracao Regional Sud Americana (IIRSA)." *Revista Geografica* 147 (2010): 43–62.

Fifer, J. Valerie. "Bolivia's Boundary with Brazil: A Century of Evolution." *Geographical Journal* 132, no. 3 (1966): 360–71.

Forest, Ronald A. "Amazonia and the Politics of Geopolitics." *Geographical Review* 82, no. 2 (1992): 128–42.

Garcia Negrete, J. "Vision geopolitica de la faja fronteriza en la region Ecuador-Colombia." *Revista Geografica del Insituto Geografico Miliatar* 32 (1993): 59–70.

George, Pierre. "Originality of the Capitals of Temperate Countries in Latin America." *Revista Geografica* 67 (1967): 31–42.

Giehake, K. "Regional Development Planning in Marginalized Andean Latin America: Examples from Bolivia, Colombia, and Peru." *Applied Geography and Development* 42 (1993): 7–25.

Glassner, Martin I. "The Rio Lauca: Dispute over an International River." *Geographical Review* 60, no. 2 (1970): 192–207.

———. "The Transit Problems of Land Locked States: The Cases of Bolivia and Paraguay." In *Ocean Yearbook 4*, edited by E. Mann and N. Ginsberg, 154–72. Chicago: University of Chicago Press, 1983.

Gwynne, Robert N. "Political Economy, Resource Use, and Latin American Environments." *Singapore Journal of Tropical Geography* 25, no. 3 (2004): 247–60.

Hardy, Osgood. "South American Alliances: Some Potential and Geographical Considerations." *Geographical Review* 8 (1919): 259–65.

Hawkins, Adrian. "Reluctant Collaborators: Argentina and Chile in Antarctica during the International Geophysical Year, 1957–1958." *Journal of Historical Geography* 34, no. 4 (2008): 596–617.

Hepple, Leslie W. "Metaphor, Geopolitical Discourse, and the Military in South America." In *Writing Worlds: Discourse, Text, and Metaphor in the Representation of Landscape*, edited by Trevor Barnes and James S. Duncan, 136–54. London: Routledge, 1992.

———. "South American Heartland: The Charcas, Latin American Geopolitics, and Global Strategies." *Geographical Journal* 170, no. 4 (2004): 359–267.

James, Preston E. "Geopolitical Structures in Latin America." *Papers of the Michigan Academy of Science, Arts, and Letters* 27 (1941): 369–76.

———. "Latin America: State Patterns and Boundary Problems." In *The Changing World: Studies in Political Geography*, edited by W. Gordon East and Arthur E. Moodie, 881–97. Yonkers, NY: World Book Company, 1956.

Kelly, Philip. *Checkerboards and Shatter Belts: The Geopolitics of South America*. Austin: University of Texas Press, 1997.

Kelly, Philip, and Jack Child, eds. *Geopolitics in the Southern Cone and the Antarctic*. Boulder, CO: Lynne Rienner, 1988.

Kruijt, Dirk, and Wim Hoogbergen. "Peaceful Relations in a Stateless Region: The Post War River Borders in the Guianas." *Tijdschrift voor Economische en Sociale Geographie* 96, no. 2 (2005): 199–208.

Long, Brian. "An Inventory of US Non-government, Non-profit Development Organizations Operating in Latin America." *Middle States Geographer* 20 (1987): 64–75.

Machado, Lia O. "The Intermittant Control of the Amazonian Territory, 1616–1960." *International Journal of Urban and Regional Research* 13, no. 4 (1989): 652–65.

Manero, Edgardo A. "Strategic Representations, Territory, and Border Areas: Latin America and Global Disorder." *Geopolitics* 12, no. 1 (2007): 19–56.

Mitchell, Jerry T., and William C. Terry. "Contesting Pisco: Chile, Peru, and the Politics of Trade." *Geographical Review* 101, no. 4 (2011): 518–35.

Morris, Arthur S. "Geopolitics in South America." In *Latin American Development: Geographical Perspectives*, edited by David A. Preston, 272–95. Harlow, England: Longman, 1996.

Naughton-Trevez, Lisa, et al. "Expanding Protected Areas and Incorporating Human Resource Use: A Study of 15 Forest Parks in Ecuador and Peru." *Sustainability, Science, Practice, and Policy* 2, no. 2 (2006): 32–44.

Newson, Linda A. "The Latin American Colonial Experience." In *Latin American Development: Geographical Perspectives*, edited by David A. Preston, 11–41. Harlow, England: Longman, 1996.

Perreault, Thomas A. "Latin American Social Movements: A Review and Critique of the Geographical Literature." *Geography Compass* 2, no. 5 (2008): 1363–85.

Perreault, Thomas A., and Gabriela Valdivia. "Hydrocarbons, Popular Protest, and National Imaginaries: Ecuador and Bolivia in Comparative Context." *Geoforum* 41, no. 5 (2010): 689–99.

Pieck, Sonja. "Beyond Postdevelopment Civic Responses to Regional Integration in the Amazon." *Journal of Cultural Geography* 28, no. 1 (2011): 179–202.

Pieck, Sonja, and Sandra Moog. "Competing Entanglements in the Struggle to Save the Amazon: The Shifting Terrain of Transnational Civil Society." *Political Geography* 28, no. 7 (2009): 416–25.

Pittman, Howard T. "Geopolitics in the ABC Countries: A Comparison." *Proceedings, Conference of Latin Americanist Geographers* 9 (1983): 214–15.

Platt, Raye R. "Present Status of International Boundaries in South America." *Geographical Review* 14, no. 4 (1924): 622–38.

Platt, Robert S. "Conflicting Territorial Claims in the Upper Amazon." In *Geographic Aspects of International Relations*, edited by C. C. Colby, 243–76. Chicago: University of Chicago Press, 1938.

Poulsen, Thomas M. "Latin America." In *Nations and States: A Geographical Background to World Affairs*, 322–63. Englewood Cliffs, NJ: Prentice Hall, 1995.

Pradeau, C. "Bresil et Cone Sud: des frontiers entre Regnum et Dominicum." *Les Cahiers d'Outre Mer* 56, no. 222 (2003): 125–48.

Price, Marie D. "Ecopolitics and Environmental Nongovernmental Organizations in Latin America." *Geographical Review* 84, no. 1 (1994): 42–58.

Price, Marie D., and Catherine Cooper. "Competing Visions, Shifting Boundaries: The Construction of Latin America as a World Region." *Journal of Geography* 106, no. 3 (2007): 113–22.

Quagliotti de Bellis, Bernardo. "The La Plata Basin in the Geopolitics of the Southern Cone." In *Geopolitics of the Southern Cone and Antarctica*, edited by Philip Kelly and Jack Child, 125–43. Boulder, CO: Lynne Rienner, 1988.

Radcliffe, Sarah A. "Frontiers and Popular Nationhood: Geographies of Identity in the 1995 Ecuador-Peru Border Dispute." *Political Geography* 17, no. 3 (1998): 273–94.

———. *Remaking the Nation: Place, Identity, and Politics in Latin America*. London: Routledge, 1996.

Salisbury, David S., et al. "Transboundary Political Ecology in Amazonia: History, Culture, and Conflicts of the Borderland Ashaninka." *Journal of Cultural Geography* 28, no. 1 (2011): 147–77.

Slater, David. "Democracy, Decentralization, and State Power: On the Politics of the Regional in Chile and Bolivia." *Yearbook, Conference of Latin Americanist Geographers* 21 (1995): 49–66.

Sternberg, Rolf. "Changing Patterns of Resource Management in Amazonia." *Revista Geografica* 121 (1995): 127–56.

Sundberg, Juanita. "Masculinist Epistemologies and the Politics of Fieldwork in Latin Americanist Geography." *Professional Geographer* 55, no. 2 (2003): 180–90.

Szary, Anne-Laure A. "Are Borders More Easily Crossed Today? The Paradox of Contemporary Trans-border Mobility in the Andes." *Geopolitics* 12, no. 1 (2007): 1–18.

Velilla de Arrillaga, Julia. "An Energy and Iron Community in the La Plata Basin." In *Geopolitics of the Southern Cone and Antarctica*, edited by Philip Kelly and Jack Child, 144–53. Boulder, CO: Lynne Rienner, 1988.

Whitesell, Edward A. "Local Struggles over Rain-Forest Conservation in Alaska and Amazonia." *Geographical Review* 86, no. 3 (1996): 414–36.

Williams, John H. "The Undrawn Line: Three Centuries of Strife on the Paraguayan-Mato Grosso Frontier." *Luso-Brazilian Review* 17, no. 1 (1980): 17–40.

Zibecki, Raul. "The Impact of Zapatismo in Latin America." *Antipode* 36, no. 3 (2004): 392–400.

Theses and Dissertations

Pieck, Sonja. "Crossed Paths to Eden: Transnational Entanglements and the Amazon Alliance." PhD diss., Clark University, Worcester, MA, 2006.

True, David W. "State Power Indices: A Case Study of South America." Master's thesis, Ohio University, Athens, 1995.

URBAN GEOGRAPHY

General Works

Aguilar, Jose C. "Cities on Edge: Smuggling and Neoliberal Policies at the Iguazu Triangle." *Singapore Journal of Tropical Geography* 33, no. 2 (2012): 171–83.

Bahr, Jurgen, and Gunter Mertins. "The Latin American City." *Colloquium Geographicum* 2 (1992): 65–75.

———. "A Model of the Social and Spatial Differentiation of Latin American Metropolitan Cities." *Applied Geography and Development* 19 (1982): 22–45.

———. "Urbanization in Latin America." *Applied Geography and Development* 41 (1993): 89–109.

Belden, Allen. "The Distribution of Major Cities in South America." *Bulletin of the Geographical Society of Philadelphia* 34 (1930): 30–44.

Browning, Clyde E. "Urban Primacy in Latin America." *Yearbook, Conference of Latin Americanist Geographers* 15 (1989): 71–78.

Eiselen, Elizabeth. "Impressions of Belem, Manaus, and Iquitos: Amazonia's Largest Cities." *Journal of Geography* 56, no. 2 (1957): 51–56.

Gade, Daniel. "Translocation of Latin American Capitals and the Concept of Functional Centralization." *Proceedings, New England-St. Lawrence Valley Division, Association of American Geographers* 1 (1971): 29–32.

Galvao, Marilia, et al. "Research Areas for the Definition of Metropolical Areas." *Revista Geografica* 70 (1969): 57–90.

Gilbert, Alan. *The Latin American City*. London: Latin American Bureau and Monthly Revision Press, 1994.

———. "The Latin American Mega-city: An Introduction." In *The Mega-city in Latin America*, edited by Alan Gilbert, 1–24. Tokyo, Japan: UN University Press, 1996.

———, ed. *The Mega-city in Latin America*. Tokyo, Japan: UN University Press, 1996.

Griffin, Ernst C. "Urban Geography." In *Future Directions in Latin American Geography: Research Agendas for the Nineties and Beyond*, edited by Gary S. Elbow, 25–30. Auburn, AL: Conference of Latin Americanist Geographers, Special Publication 3, 1992.

Griffin, Ernst C., and Larry Ford. "Cities of Latin America." In *Cities of the World: World Regional Urban Development*, edited by T. D. Brown and J. F. Williams, 198–240. New York: Harper & Row, 1983.

Hardy, S., and A. Sierra. "Understanding Challenges to Assess the Vulnerabilities of High-Altitude Urban Areas: The Cases of La Paz and Quito [in French]." *Bulletin de l'Association de Geographes Francais* 88, no. 1 (2011): 46–66.

Hays-Mitchell, Maureen, and Brian J. Godfrey. "Cities of South America." In *Cities of the World: World Regional Urban Development*, 3rd ed., edited by Stanley Brunn et al., 123–58. Lanham, MD: Rowman & Littlefield, 2003.

Klak, Thomas, and Victoria A. Lawson. "An Introduction to Current Research on Latin American Cities." *Economic Geography* 66, no. 4 (1990): 305–9.

Kurihara, Hisako. "The Studies of Urbanization in Latin America and the Marginality Concept [in Japanese]." *Annals, Japan Association of Economic Geographers* 24, no. 1 (1978): 1–18.

Pederson, P. O. *Urban Regional Development in South America: A Process of Diffusion and Integration*. The Hague, Netherlands: Mouton, 1975.

Sargent, Charles S. "The Latin American City." In *Latin America*, edited by Brian Blouet and Olwyn Blouet, 201–49. New York: Wiley, 1982.

Ward, Peter. "Contemporary Issues in the Government and Administration of Latin American Mega-cities." In *The Mega-city in Latin America*, edited by Alan Gilbert, 53–72. Tokyo, Japan: UN University Press, 1996.

Urban Social Geography

Bahr, Jurgen. "Intraurban Migration of Lower Income Groups and Peripheral Growth of Latin American Metropolitan Areas: The Impact of Political and Socio-economic Factors." *Applied Geography and Development* 36 (1990): 17–30.

Bahr, Jurgen, and Gunter Mertins. "Un modelo de la differenciacion socio-espacial de las metropolis de America Latina." *Revista Geografica* 98 (1983): 23–29.

Capron, G. "Territorialites urbaines et territorialisation en Amerique Latine: Les residences securisees ou fermees et la fragmentation sociospatiale." *Cahiers de Geographie du Quebec* 50, no. 141 (2006): 499–506.

Gilbert, Alan, and Peter Ward. "Latin American Migrants: A Tale of Three Cities." In *People and Environments: Issues and Enquiries*, edited by F. Slater, 24–42. London: Collins Educational, 1986.

Smith, Betty E. "Urban Population Density in Three Intermediate Size Cities of South America." *Papers of the Applied Geography Conferences* 32 (2009): 437–46.

Smith, Richard V., and Robert N. Thomas. "Population Crisis and the March to the Cities." In *Focus on South America*, edited by Alice Taylor, 3–14. New York: Praeger, 1973.

Trivelli, Pablo. "Access to Land by the Urban Poor: An Overview of the Latin American Experience." *Land Use Policy* 3, no. 2 (1986): 101–21.

Yamazaki, T. "Distribution of Population and Development of Cities in South America [in Japanese]." *Jimbun Chiri/Human Geography* 11, no. 5 (1959): 41–76.

Morphology and Neighborhoods

Borsdorf, Axel. "The Latin American City and the Symbolic Impact of Built Environment." *GeoJournal* 15, no. 1 (1987): 57–62.

Bromley, Rosemary D. F., and Gareth A. Jones. "The Conservation Cycle in Cities of the Developing World: Implications for Authenticity and Policy." *Urban Geography* 17, no. 7 (1996): 650–69.

———. "Identifying the Inner City in Latin America." *Geographical Journal* 162, no. 2 (1996): 179–90.

Capron, G. "Accessibility to Modern Public Spaces in Latin-American Cities: A Multidimentional Idea." *GeoJournal* 58, nos. 2–3 (2003): 217–23.

Geisse, Guillermo. "Conflicting Land Strategies in Large Latin American Cities." *Land Use Policy* 1, no. 4 (1984): 309–29.

Gilbert, Alan. "Land, Housing, and Infrastructure in Latin American Major Cities." In *The Mega-city in Latin America*, edited by Alan Gilbert, 73–109. Tokyo, Japan: UN University Press, 1996.

———. "Pirates and Invaders: Land Acquisition in Urban Colombia and Venezuela." *World Development* 9 (1981): 657–78.

———. "Renting and the Transition to Owner Occupation in Latin American Cities." *Habitat International* 15, nos. 1–2 (1991): 87–100.

———. "Urban Growth, Employment, and Housing." In *Latin American Development: Geographical Perspectives*, edited by David A. Preston, 169–98. London: Longman, 1988.

———. "Urban Growth, Employment, and Housing." In *Latin American Development: Geographical Perspectives*, edited by David A. Preston, 246–71. New York: Longman, 1996.

Gilbert, Alan, et al. *Urbanization in Contemporary Land America: Critical Approaches to the Analysis of Urban Issues*. Chichester, England: Wiley, 1982.

Gilbert, Alan, and Peter Ward. "Housing in Latin American Cities." In *Geography and the Urban Environment: Progress in Research and Applications*, edited by D. T. Herbert and Ronald J. Johnston, 1:285–318. Chichester, England: Wiley, 1978.

———. *Housing, the State, and the Poor: Policy and Practice in Three Latin American Cities.* Cambridge, England: Cambridge University Press, 1985.

Ingram, G. D., and A. Carroll. "The Spatial Structure of Latin American Cities." *Journal of Urban Economics* 9, no. 2 (1981): 257–73.

Kunihiro, Narumi. "Spatial Structure of South American Cities: A Preliminary Consideration on Relation between Street and Dwelling." *Latin American Studies, Japan* 7 (1983): 103–20.
Mertins, Gunter, and B. Thomas. "Suburbanisierung prozesse in Latinamerika." *Zeitschrift fur Wirtschaftsgeographie* 39, no. 1 (1995): 1–13.
Sequin, Anne-Marie. "Les quartiers residentials fermes: une forme segregative qui menace la cohesion sociale a l'echelle locale dans les villes latino-americaines?" *Cahiers de Geographie du Quebec* 47, no. 131 (2003): 179–200.
Ward, Peter. "The Latin American Inner City: Differences of Degree or of Kind?" *Environment and Planning A* 25, no. 8 (1993): 1131–60.

Urban Economic Geography

Bromley, Rosemary D. F. "Informal Commerce: Expansion and Exclusion in the Historic Centre of the Latin American City." *International Journal of Urban and Regional Research* 22, no. 2 (1998): 245–63.
———. "Market-Place Trading and the Transformation of Retail Space in the Expanding Latin American City." *Urban Studies* 35, no. 8 (1998): 1311–33.
Gans, Paul. *Die Innenstadte von Buenos Aires und Montevideo. Dynamik der Nutzungsstruktur, Wohnedingungen und informeller Sektor.* Kiel, Germany: Kieler Geographische Schriften, Band 77, 1990.
Gwynne, Robert N. "Industrialization and Urbanization." In *Latin American Development: Geographical Perspectives*, edited by David A. Preston, 216–45. London: Longman, 1996.
Lawson, Victoria A., and Thomas Klak. "Conceptual Linkages in the Study of Production and Reproduction in Latin American Cities." *Economic Geography* 66, no. 4 (1990): 310–27.
Wright, Charles. "The Future of Latin American Urban Transport." *Cities* 1, no. 5 (1984): 464–68.

Urban Environments

Iglesias, Alicia N. "Conflictos y escenarios del cambio climatico mundial en Latinamerica: Reflexiones para el Analisis de los Espacios Urbanas y Metropolitanos." *Anales de Geografia de la Universidad Complutense* 15 (1995): 401–8.

Theses and Dissertations

Gaffney, Christopher T. "Dynamic Sites and Cultural Symbols: The Stadiums of Rio de Janeiro and Buenos Aires." PhD diss., University of Texas, Austin, 2006.
Smith, Betty E. "Urban Morphology in South America." PhD diss., State University of New York, Buffalo, 1994.
Smole, William J. "Functions and Functional Organization of the Ports of Rosario, Argentina and Antofagasta, Chile." Master's thesis, University of Chicago, Chicago, 1955.

Chapter Two

Argentina

GENERAL WORKS

Atlases and Graphic Presentations

Chiozza, Elena, and Ricardo Figueria. *Atlas total de la Republica Argentina*. Buenos Aires, Argentina: Centro Editor de America Latina, 1982.

Czibener, Daniela, and Paulina E. Nabel. "Teledeteccion y Redes Neuronales Oplicadas al Mapeo de Cobertures del Suelo de la Cuenca del Matanza-riachuelo, Buenos Aires, Argentina." *Revista Geografica* 146 (2009): 125–52.

Kurtz, Ditmar, et al. "Ground and Satellite Based Assessment of Rangeland Management in Sub-tropical Argentina." *Applied Geography* 30, no. 2 (2010): 210–20.

Peng, Xu Long, et al. "Land Cover Mapping from RADARSAT Stereo Images in a Mountainous Area of Southern Argentina." *Canadian Journal of Remote Sensing* 29, no. 1 (2003): 75–87.

Perez, Analia M., et al. "Carta dinamica del medio ambiente del partido de Lujan, provincial de Buenos Aires, Republica Argentina." *Investigaciones Geograficas*, Especial (1996): 65–75.

Peyfet, Alejo. *Cartes Sobre Misiones*. Buenos, Aires, Argentina: Impresta de la Tribuna Nacional, 1881.

Books, Monographs, and Texts

Carlevari, Isidro. *La Argentina: Geografia humana y economica*. Buenos Aires, Argentina: Ergon, 1979.

Crooker, Richard A. *Argentina*. Broomall, PA: Chelsea, 2004.

Daus, Federico A. *Geografia y unidad Argentina*. Buenos Aires, Argentina: Editorial Nova, 1957.

———. *Resena geografica de las islas Malvinas*. Buenos Aires, Argentina: Universidad de Buenos Aires, Instibudo de Geografia, Publicacions Series A 19, 1955.

Daus, Federico A., and Raul C. Rey Balmaceda. *Islas Malvinas: Resena geografica*. Buenos Aires, Argentina: OIKOS, 1982.

Daus, Federico A., et al. *Geografia de la Argentina*. Buenos Aires, Argentina: A. Estrada, 1954, 1978, 1984, 1988.

Daus, Federico A., and Juan A. Roccatagliata. *Fundamentos para una division regional de la Argentina*. Buenos Aires, Argentina: Sociedad Argentina de Estudios Geograficos Gaea, 1982.

Duran, Diana, et al. *Geografia de la Argentina*. Buenos Aires, Argentina: Troqvel, 1995.

Hatcher, J. B. *Reports of the Princeton University Expeditions to Patagonia, 1896–99: Narrative and Geography*. Vol. 1. Princeton, NJ: Princeton University Press, 1903.

Keeling, David J. *Contemporary Argentina: A Geographical Perspective*. Boulder, CO: Westview, 1995.

Lorenzini, Horacio, et al. *Geografia de la Argentina*. Buenos Aires, Argentina: A-Z Editora, 1993.

Margalot, Jose A. *Geografia de Misiones*. Buenos Aires, Argentina: Cultural Argentina, 1971.

McGann, Thomas F. *Argentina: The Divided Land*. New York: Van Nostrand, 1966.

Morris, Arthur S. *Regional Disparities and Policy in Modern Argentina*. Glasgow, Scotland: Institute of Latin American Studies, 1975.

Roccatagliata, Juan A., ed. *La Argentina: Geografia general y los marcos regionales*. Buenos Aires, Argentina: Planeta, 1992.

———. *La Patagonia: Bases estrategicas para el desarrollo sustentable*. Buenos Aires, Argentina: Editorial Docencia, 2010.

———. *Region Centro: Desarrollo endogeno, cohesion territorial e insercion transnacional*. Buenos Aires, Argentina: Editorial Docencia, 2010.

———. *Region noroeste: De los mandates de la historia a los desafinos del futuro*. Buenos Aires, Argentina: Docencia, 2010.

Articles and Book Chapters

Bergmann, John. "Patagonia: A Diminishing Frontier." *Yearbook, Conference of Latin Americanist Geographers* 10 (1984): 62–67.

Capitanelli, Ricardo G. "Patagonia." *Revista Geographica* 95 (1982): 30–45.

Carter, Adams. "The American Alpine Club Expedition to the Ojos del Salado." *Geographical Review* 47, no. 2 (1957): 240–50.

Church, J. E. "In Argentine Tierra del Fuego: Notes on a Tour." *Geographical Review* 38, no. 3 (1948): 392–413.

Clifford, Miles. "The Falkland Islands and Their Dependencies." *Geographical Journal* 121, no. 4 (1955): 405–16.

Davis, Jehiel S. "Catamarca, the Arizona of Argentina." *Journal of Geography* 19 (1920): 327–35.

Deffontaines, Pierre. "Les Oasis du Piedmont Argentin des Andes." *Les Cahiers d'Outre Mer* 5 (1952): 42–69.

Escobar, Marcelo. "Promotion and Diffusion of Geographical Knowledges: Argentine Editorial Policies and the Nation's Geographical Body Representation (1863–1916)." In *Text and Image: Social Construction of Regional Knowledges*, edited by Anne Buttimer et al., 91–99. Leipzig, Germany: Institute für Landerkunde, Beitrage zur Regionalen Geographie, Heft 49, 1999.

Freeman, Otis W. "The Geographical Provinces of Argentina." *Journal of Geography* 24 (1925): 300–313.

Houk, Richard. "Some Geographic Reflections on a Recent Reconnaissance in Argentina." *Bulletin, Illinois Geographical Society* 28, no. 2 (1986): 41–54.

Letticia Zamorano, Gloria. "Una clasificacion regional systemica de la Argentina." *Revista Geografica* 143 (2008): 125–62.

Martinez, R. P. "Les Petites Oasis Argentines du Piedmont Andin: Medonitos." *Les Cahiers d'Outre Mer* 10 (1957): 107–16.

Ostuni, Josefina. "La Organizacion del espacio en la Faja de las Grandes Alturas del Oeste Argentina." *Revista Geografica* 95 (1982): 55–93.

Roccatagliata, Juan A. "Regionalizacion." In *La Argentina: Geografia general y los marcos regionales*, edited by Juan A. Roccatagliata, 427–50. Buenos Aires, Argentina: Planeta, 1992.

Rudolph, William E. "Southern Patagonia." *Geographical Review* 24 (1934): 251–71.
Santillan de Andres, Selva E. "La Region del Valle Calchaqui." *Revista Geografica* 95 (1982): 94–109.
Souto, Patricia G. "Publication as a Scientific Legitimation Strategy: Publications Argentinos de Geografia Cientifica, 1947." In *Text and Image: Social Construction of Regional Knowledges*, edited by Anne Buttimer et al., 161–66. Leipzig, Germany: Institut fur Landerkunde, Beitrage zur Regionalen Geographie, Heft 49, 1999.
Tower, Walter S. "A Journey through Argentina." *Bulletin of the Geographical Society of Philadelphia* 12, no. 3 (1914): 89–112.
———. "The Pampa of Argentina." *Geographical Review* 5 (1918): 293–315.
Wilcox, Marrion. "Argentina." In *Focus on South America*, edited by Alice Taylor, 228–41. New York: Praeger, 1973.
———. "Argentine Patagonia: A Land of the Future." *Bulletin of the American Geographical Society* 42, no. 12 (1910): 903–9.
Winsberg, Morton D. "Statistical Regionalization in the Argentine Pampa." *Revista Geografica* 72 (1970): 45–61.

Theses and Dissertations

Gooze, Charles. "Northern Mendoza: An Argentine Oasis." PhD diss., Clark University, Worcester, MA, 1929.
Jones, Wellington D. "Studies in the Geography of Northern Patagonia." PhD diss., University of Chicago, Chicago, 1914.

CULTURAL AND SOCIAL GEOGRAPHY

General Works

Albaladejo, Christophe. "Les transformations de l'espace rural pampeen face a la mondialisation." *Annales de Geographie* 121, no. 4 (2012): 387–409.
Bolsi, Alfredo S. "El hombre y el redio en la pura argentina." *Revista Geografica* 95 (1982): 46–54.
Ercolani, Patricia, and Miguel Segui. "Elocio en el Contexto Posmodernista: De un Derecho a la Satisfacion de una Necesidad. Estudio de caso: Bahia Blanca (Argentina)." *Anales de Geografia de la Universidad Complutense* 28, no. 1 (2008): 29–51.
Miller, Elbert E. "The Frontier and the Development of Argentine Culture." *Revista Geografica* 90 (1979): 183–98.
Naylor, Simon. "Discovering Nature, Rediscovering the Self: Natural Historians and the Landscapes of Argentina." *Environment and Planning D* 19, no. 2 (2001): 227–48.
Prat, M. C., and J. M. Sayago. "Risques naturels, actions anthroiques, et enjeux a San Miguel de Tucuman (Nord. Ouest Argentina)." *Les Cahiers d'Outre Mer* 56, no. 223 (2003): 301–26.

The Built Environment

Klak, Thomas. "Contextualizing State Housing Programs in Latin America: Evidence from Leading Housing Agencies in Brazil, Ecuador, and Jamaica." *Environment and Planning A* 25, no. 5 (1993): 653–76.

Medical Geography

Carter, Eric D. "Malaria, Landscape, and Society in Northwest Argentina in the Early Twentieth Century." *Journal of Latin American Geography* 7, no. 1 (2008): 7–38.

Curton de Casas, Susana I., and R. Boffi. "Malaria Reinfestation on the Argentine North Border." *GeoJournal* 26, no. 1 (1992): 65–68.

Falasca, Silvia, and Maria A. Bernabe. "El Arbol del Neem (Azadirachta indica) para controlar enfermedoles endencicas en Argentina." *Revista Geografica*, 146 (2009): 111–24.

Pickenhayn, J., and S. Curto. "La Geografia de la Salud en la Argentina." *Revista Geografica* 138 (2005): 89–108.

Ethnic, Social, and Population Geography

Bolsi, Alfredo S. "Poblacion y territorio del noroeste argentine durante el siglo XX." *Revista Geografica* 135 (2004): 137–61.

Bosco, Fernando J. "Global Aid Networks and Hungry Children in Argentina: Thinking about Geographies of Responsibility and Care." *Children's Geographies* 5 (2007): 55–76.

De Marco, Graciela M., and Dario C. Sanchez. "Inmigrantes Limitrofes en el Gran Buenos Aires: Un Analisis Socio-economico Espacial." *Revista Geografica*, 117 (1993): 19–48.

Diez Tetananti, Juan M. "Accion del estado: Vicnulaciones con procesos de despoblamiento y cambios socio-economicas locales en el sudeste Bonaerense." *Revista Geografica de America Central*, 43 (2009): 79–108.

Formiga, Nidia. "Distribution of Argentina's Population since 1947." *Population Geography, India* 24, nos. 1–2 (2002): 1–20.

Foschiatti, Ana Maria. "Aproximaciones y tendencias zcerca de la vulnerabilidad sociodemografia Misionera (Republica Argentina)." *Revista Geografica* 140 (2006): 29–60.

———. "Evaluacion de la dinamica demografica y su vulnerabilidad en la provincia del Chaco (Argentina)." *Anales de Geografia de la Universidad Complutense* 29, no. 2 (2009): 83–126.

———. "Variaciones demograficas del nordeste argentine a partir de 1960." *Revista Geografica* 136 (2004): 85–107.

———. "Vulnerability, Poverty, and Exclusion: Problems with a Great Impact on the Inhabitants of the Northeast of Argentina [in Spanish]." *Anales de Geografia de la Universidad Complutense* 27, no. 2 (2007): 9–40.

Foschiatti, Ana Maria, and Mirta L. Ramirez. "Analisis de variables demograficas en la provincia del Chaco (Republica Argentina) en el Trienio 1996, 1997, y 1998." *Revista Geografica* 132 (2002): 45–60.

Furloni de Civit, M. E., and M. I. Guitierrez de Manchon. "Poets and Narratives from Mendoza and Their Local Perceptions." *GeoJournal* 38, no. 1 (1996): 109–18.

Gaignard, Romain. "La Mise en Valeur Pionniere de la Terre de Fer (Argentine)." *Les Cahiers d'Outre Mer* 15 (1962): 105–37.

———. "La Montec Demografique Argentine: Le Recensement du 30 Septembre 1960." *Les Cahiers d'Outre Mer* 14 (1961): 85–97.

Gonzalez, Myriam S. "Una Aproximacion al Paisaje de Neorrurales y Otros Migrantes en Una Comarca Cordillerana: El Case do El Bolsan en la Patagonia Andina." *Revista Geografica* 133 (2003): 5–25.

Jefferson, Mark. *Peopling the Argentine Pampa*. New York: American Geographical Society, 1926, 1930.

Longhi, Fernando. "Misery Rate of Homes: A Methodological and Conceptual Approximation to Measuring Spatial Distribution and Varieties of Extreme Poverty in Argentina during the Decade of the Nineties [in Spanish]." *Estudios Geograficos* 72, no. 271 (2011): 505–33.

Orzoanco, Maria G. "Problemes ambientales detectados por la poblacion de Ushuaia (Tierra del Fuego, Argentina)." *Investigaciones Geograficas* 40 (1999): 85–98.

Pearson, Ross N. "Mapping Population Change in Argentina." *Revista Geografica* 59 (1963): 63–77.

———. "Recent Changes in the Distribution of Population in Argentina." *Papers of the Michigan Academy of Science, Arts, and Letters* 49 (1963): 367–81.

PRIGEO. "Argentina: The Geodemographic Contrasts of a Latin American Country." *Revue Belge de Geographie* 117, nos. 1–2 (1993): 75–89.

Rey Balmaceda, Raul C. "Poblacion y poblamiento." In *La Argentina: Geografia general y los marcos regionales*, edited by Juan A. Roccatagliata, 189–202. Buenos Aires, Argentina: Planeta, 1992.

Tata, Robert J. "Population Geography of Argentina." *Revista Geografica* 85 (1977): 79–96.

Turner, Thomas. "Peopling Mountain Environments: Changing Andean Livelihoods in Northwest Argentina." *Geographical Journal* 169, no. 3 (2003): 205–14.

Velazquez, Guillermo A. "Environmental Risks, Demographic Dynamics and Life Quality: Argentina at the Beginning of the Twenty-First Century." *Acta Universitatis Carolinae, Geografica* 45, no. 2 (2010): 115–34.

Whitson, Risa. "The Reality of Today Has Required Us to Change: Negotiating Gender through Informal Work in Contemporary Argentina." *Annals of the Association of American Geographers* 100, no. 1 (2010): 159–81.

Wilkie, Richard W. "Selectivity in Peasant Spatial Behavior: Regional Interactions in Entre Rios, Argentina." *Proceedings, New England-St. Lawrence Valley Division, Association of American Geographers* 2 (1972): 10–20.

———. "Toward a Behavioral Model of Peasant Migration: An Argentina Case of Spatial Behavior by Social Class Level." *Proceedings, Conference of Latin Americanist Geographers* 2 (1971): 83–114.

Williams, Glyn B. "Welsh Contributions to Explorations in Patagonia." *Geographical Journal* 135, no. 2 (1969): 213–27.

———. "The Welsh in Patagonia: A Geographic Perspective." *Revista Geografica* 69 (1968): 121–44.

———. *The Welsh in Patagonia: The State and the Ethnic Community*. Cardiff, Wales: University of Wales Press, 1991.

Winsberg, Morton D. "Colonia Baron Hirsch: Una colonia israelita agricola en Argentina." *Revista Geografica* 65 (1966): 45–56.

Community and Settlement Studies

Bortagaray, Lucia L. "Las etapas de ocupacion del territorio argentino, una rapida expansion con valoracion parcial del territorio." In *La Argentina: Geografia general y los marcos regionales*, edited by Juan A. Roccatagliata, 145–68. Buenos Aires, Argentina: Planeta, 1992.

Capitanelli, Ricardo G. "Patagonia, un medio duro, domino de ovejas, con focus pioneros de ocupacion e industias promovides." In *La Argentina: Geografia general y los marcos regionales*, edited by Juan A. Roccatagliata, 697–746. Buenos Aires, Argentina: Planeta, 1992.

Eidt, Robert C. "Japanese Agricultural Colonization: A New Attempt at Land Opening in Argentina." *Economic Geography* 44, no. 1 (1968): 1–20.

———. *Pioneer Settlement in Northeast Argentina*. Madison: University of Wisconsin Press, 1977.

Foschiatti, Ana Maria. "El medio natural y el proceso de ocupacion del espacio en el nordeste argentine." *Revista Geografica* 135 (2004): 9–28.

Gaignard, Romain. "La Faillite de l'Experience de colonization agricole des 'Pieds-Noirs' en Argentine." *Les Cahiers d'Outre Mer* 21 (1968): 308–17.

Haynes, Kingsley, and Wayne T. Enders. "Settlement Pattern Evolution in Central Argentina: Spatial Dynamics in a Developing Nation." *Proceedings of the Association of American Geographers* 5 (1974): 97–106.

James, Preston E. "The Process of Pastoral and Agricultural Settlement on the Argentine Humid Pampa." *Journal of Geography* 40 (1941): 121–37.

Morris, Arthur S. "The Failure of Small Farmer Settlement in Buenos Aires Province, Argentina." *Revista Geographica* 85 (1977): 63–78.

Platt, Robert S. "Peripheral Items in the Argentine Pattern of Terrene Occupancy." *Transactions of the Illinois State Academy of Science* 24 (1931): 410–23.

———. "Pirovano: Items in the Argentine Pattern of Terrace Occupance." *Annals of the Association of American Geographers* 21 (1931): 215–37.

Tanner, T. "Peopling Mountain Environments: Changing Andean Livelihoods in North West Argentina." *Geographical Journal* 169, no. 3 (2003): 205–14.

Winsberg, Morton D. "Colonia Baron Hirsch: An Israelite Agricultural Colony in Argentina." *Revista Geografica* 65 (1966): 45–56.

————. *Colonia Baron Hirsch: A Jewish Agricultural Colony in Argentina.* Gainesville: University of Florida, Monograph in the Social Sciences 19, 1963.

————. "Jewish Agricultural Colonization in Entre Rios, Argentina." *American Journal of Economics and Sociology* 27, no. 3 (1968): 285–96.

————. "Jewish Agricultural Colonization in Entre Rios, Argentina, II." *American Journal of Economics and Sociology* 27, no. 4 (1968): 423–32.

————. "Jewish Agricultural Colonization in Entre Rios, Argentina, III." *American Journal of Economics and Sociology* 28, no. 2 (1969): 179–92.

Tourism and Recreation

Almiron, Analia, et al. "Promocion turistica y cartografia. La Argentina turistica en los mapas de la Secretaria de Turismo de la Nacion (1996–2004)." *Investigaciones Geograficas* 62 (2007): 138–54.

Denis, Paul-Yves. "Desarrollo Turistico e Integracion en Neuquen (Argentina): Un Estudio de Geografia Aplicado." *Revista Geografica* 80 (1974): 111–28.

Fernandez, Guillermina, and Aldo G. Ramos. "Preservacion del Patrimanio Industrial y Dinamizacion del Turismo a Partir de Rutas: Algunas pro Puestas para Argentina." *Revista Geografica* 138 (2005): 5–22.

Theses and Dissertations

Campney, Thomas. "National Community Formation as Examined through Newspaper Content Analysis: The Example of Argentina." PhD diss., University of Western Ontario, London, Ontario, 1977.

Rhys, David H. "A Geographic Study of the Welsh Colonization in Chubut, Patagonia." PhD diss., University of California, Riverside, 1976.

Seipt, Clark K. "Understanding Argentine Farmers' Perceptions of the Utility of Seasonal Climatic Forecasts." Master's thesis, Pennsylvania State University, University Park, 2007.

Wilkie, Richard W. "On the Theory of Process in Human Geography: A Case Study of Migration in Rural Argentina." PhD diss., University of Washington, Seattle, 1968.

ECONOMIC GEOGRAPHY

General Works

Ackerman, William V. "Development Strategy for Cuyo, Argentina." *Annals of the Association of American Geographers* 65, no. 1 (1975): 36–47.

Alderman, Neil. "Innovation in Complex Capital Projects Clustering and Dispersion in Two Cases from Argentina and the UK." *Journal of Economic Geography* 4, no. 1 (2004): 65–82.

Bortagaray, Lucia L. "Recursos naturales potentials, parcialmente valorados." In *La Argentina: Geografia general y los marcos regionales*, edited by Juan A. Roccatagliata, 203–30. Buenos Aires, Argentina: Planeta, 1992.

Bortagaray, Lucia L., and Alberto H. Pelaez. "El sistema productive." In *Geografia Economica Argentina*, edited by Juan A. Roccatagliata, 93–125. Buenos Aires, Argentina: El Atereo, 1993.

Brieva, Susana S., et al. "Proceso de privatizacion, organizacion y competencia interportuaria en Argentina: El caso del Puerto Quequen." *Investigaciones Geograficas* 54 (2004): 93–113.

Conte, Analia S. "Historia del pensamiento en geografia economica." In *Geografia Economica Argentina*, edited by Juan A. Roccatagliata, 5–16. Buenos Aires, Argentina: El Atereo, 1993.

Erbiti, Cecilia. "El sistema especial: Estructura de los asentamientos." In *Geografia Economica Argentina*, edited by Juan A. Roccatagliata, 44–69. Buenos Aires, Argentina: El Atereo, 1993.

Gaignard, Romain. "L'economie de la Republique Argentine." *Les Cahiers d'Outre Mer* 13 (1960): 59–103, 265–339.

Jofre, Ana. "Poblacion y economia." In *Geografia Economica Argentina*, edited by Juan A. Roccatagliata, 70–92. Buenos Aires, Argentina: El Atereo, 1993.

Lara, Albania L. "Sistema ecologica, medio ambiente y recursos naturales." In *Geografia Economica Argentina*, edited by Juan A. Roccatagliata, 17–43. Buenos Aires, Argentina: El Ateneo, 1993.

Lara, Albina L., and Diana Duran. "Estructura regional y organizacion territorial." In *Geografia Economica Argentina*, edited by Juan A. Roccatagliata, 267–83. Buenos Aires, Argentina: El Ateneo, 1993.

Morris, Arthur S. "The Regional Problem in Argentine Economic Development." *Geography* 57, no. 4 (1972): 289–306.

Padula, Cristina C. "El proceso de integracion." In *Geografia Economica Argentina*, edited by Juan A. Roccatagliata, 331–61. Buenos Aires, Argentina: El Atereo, 1993.

Pelaez, Alberto H. "Breve resena de la evolucion de la economia Argentina." In *La Argentina: Geografia general y los marcos regionales*, edited by Juan A. Roccatagliata, 169–88. Buenos Aires, Argentina: Planeta, 1992.

Propin Frejomil, Enrique, et al. "Potencial territorial e integracion economica en America del Sur: El caso del noroeste argentine." *Investigaciones Geograficas* 41 (2000): 139–49.

Roccatagliata, Juan A., ed. *Geografia Economica Argentina*. Buenos Aires, Argentina: El Ateneo, 1993.

———. *Region Nuevo Cuyo: Una region asociativa en el Centro Oeste del Territorio Argentino*. Buenos Aires, Argentina: Editorial Docencia, 2010.

———. *Region pampeana bonaerense: Desarrollo sustenable con equidad*. Buenos Aires, Argentina: Editorial Docencia, 2005.

Santillan de Andres, Selva, and Teodoro R. Ricci. "La region del Noreste Argentino: Paisajes heterogeneous con economia mixta." In *La Argentina: Geografia general y los marcos regionales*, edited by Juan A. Roccatagliata, 579–612. Buenos Aires, Argentina: Planeta, 1992.

Smith, J. Russell. "The Economic Geography of the Argentine Republic." *Bulletin of the American Geographical Society* 35, no. 2 (1903): 130–43.

Zamorano, Mariano. "Region de las nucleos economicos fragmentados de las sierras pampeanas, con oasis pobres y economias de subsistencia." In *La Argentina: Geografia general y los marcos regionales*, edited by Juan A. Roccatagliata, 661–96. Buenos Aires, Argentina: Planeta, 1992.

Agriculture and Land Use

Bergmann, John. "Soil Salinization and Welsh Settlement in Chubut, Argentina." *Cahiers de Geographie du Quebec* 15, no. 35 (1971): 361–69.

Bruniard, Enrique, and Alfredo S. Bolsi. "Region agro-silvo ganadera con frentes pioneros de ocupacion del Nordeste." In *La Argentina: Geografia general y los marcos regionales*, edited by Juan A. Roccatagliata, 527–78. Buenos Aires, Argentina: Planeta, 1992.

Canibano, A., et al. "Relacion entre variables de intenes agropecuario en Azul, Argentina." *Investigaciones Geograficas*, Especial (1996): 35–42.

Carbone, Maria E., et al. "Valuation of the Surface Waters of the Claromeco River Basin for Supplementary Irrigation." *Revista Geografica* 137 (2005): 97–116.

Cepparo de Grosso, Maria E. "Desarrollo de un proyecto Agricola en la region marginal de la Patagonia Meridional Argentina. El case de Governadar Gregores." *Investigaciones Geograficas* 61 (2006): 58–74.

Daus, Federico A., and Ana del C. Yeannes. "La macroregion pampeana agro ganadera, con industrias urbanos y portuarias." In *La Argentina: Geografia general y los marcos regionales*, edited by Juan A. Roccatagliata, 477–526. Buenos Aires, Argentina: Planeta, 1992.

Deasy, George F. "Distribution of Flax Production in Argentina." *Economic Geography* 19, no. 1 (1943): 45–54.

Falasca, Silvia, and Maria A. Bernabe. "Aptitud Agroclimatica Argentina para la implantacion de bosques energeticas de Paulownia ssp." *Revista Geografica* 148 (2010): 151–64.

Falasca, Silvia, et al. "Analisis de algunas variables que inciden sobre la disponibilidad de agua util para los cultivos en un sector del partido de Tandil, provincial de Buenos Aires, Argentina." *Revista Geografica* 132 (2002): 141–55.

———. "Los potenciales areas de production de anis estrellado (*Illicium verum*) en Argentina." *Revista Geografica* 148 (2010): 83–96.

Falasca, Silvia, and Carina Rodriguez. "Eficiencia de riego complementario sobre el cultivo de papa (Solamene tuberosum) en el partido de Tandil, Argentina." *Revista Geografica* 139 (2006): 131–49.

Falasca, Silvia, and Ana C. Ulberich. "El agua disponible de los suelas del sudeste boraerense, Republica Argentina." *Revista Geografica* 140 (2006): 7–18.

———. "Cultivos alternativos par alas produccion de biocombustibles en el area semiarida Argentina." *Revista Geografica* 140 (2006): 135–50.

———. "Efecto de cambio en el conterido de humedad del aire en el centro-sud banaerense, Republica Argentina." *Revista Geografica* 135 (2004): 21–32.

———. "La production de mostazaetiope (Brassica carinata) en Argentina como cultivo energetico." *Revista Geografica* 148 (2010): 7–22.

Fernandez, Ricardo S., and Ignacio de Los Rios Carmenado. "Agricultural Transformations and Depopulation in Rural Communities of the Pampas Argentina [in Spanish]." *Estudios Geograficos* 71, no. 268 (2010): 235–66.

Furlani de Civit, Maria E., et al. "La Campetencia por el Suelo Guaymallen-Mendoza." *Revista Geografica* 78 (1973): 55–101.

Fulani de Civit, Maria, and Maria G. G. de Manchan. "Dinamica Agraria en un Oasis de Espacializacion Viticola." *Revista Geografica* 115 (1992): 85–137.

Gonzalez, Myriam S. "Economia agroganadera de vocacion exportadora." In *La Argentina: Geografia general y los marcos regionales*, edited by Juan A. Roccatagliata, 231–64. Buenos Aires, Argentina: Planeta, 1992.

Grosso-Cepparo, Maria V. "Uso, gestion e impactos de uso Agricola de aguas residuals en zonas aridas Departamento de Lavalle, Mendoza, Argentina." *Revista Geografica* 144 (2008): 47–96.

Hansis, Richard A. "Land Tenure Hazards and the Ecology: Viticulture in the Mendoza Oasis, Argentina." *Tijdschrift voor Economische en Sociale Geographie* 76, no. 2 (1985): 121–32.

Kentner Carby, Julia. "For Members and Markets: Neoliberalism and Cooperativism in Mendoza's Wine Industry." *Journal of Latin American Geography* 9, no. 2 (2010): 27–47.

Miller, Elbert E. "Eight Cows and a Bull: The Influence of Cattle on Argentine Culture." *Great Plains-Rocky Mountain Geographical Journal* 8 (1979): 20–30.

Morris, Arthur S. "Dairying in Argentina." *Revista Geografica* 76 (1972): 102–20.

———. "Development of the Irrigation Economy of Mendoza, Argentina." *Annals of the Association of American Geographers* 59 (1969): 97–115.

———. "Diffusion, Deformation or Epigenesis? Growth Processes of Colonia Agricola in Argentina." In *New Dimensions in Agricultural Geography: Human Perception and Technological Change in Agriculture*, edited by Noor Mohammad, 6:93–106. New Delhi, India: Concept, 1992.

———. "Globalization and Regional Differentiation: The Mendoza Wine Region." *Journal of Wine Research* 11, no. 2 (2000): 145–54.

Schellmann, G. *Jungkanozoische Landschaftsgeschichte Patagoniens (Argentinien). Andine Vorlandvergletscherungen, Talentwicklung und marine Terrassen.* Essen, Germany: Essener Geographische Arbeiten, Band 29, 1998.

Snyder, David E. "An Estancia in Argentina: Commercial Livestock Raising." In *Focus on Geographic Activity: A Collection of Original Studies*, edited by Richard Thoman and Donald Patton, 55–59. New York: McGraw-Hill, 1964.

Stewart, Norman R. "Tea: A New Agricultural Industry for Argentina." *Economic Geography* 36, no. 2 (1960): 267–76.

Tadeo, N. "La filiere agrumesdoux dans la province d'Entre Rios, Argentina." *Norois*, 210 (2009): 69–80.

Ulberich, Ana C. "La production de mostazetiape (Brassica carinata) en Argentina como cultivo energetico." *Revista Geografica* 148 (2010): 7–22.

Valenzuela de Mari, Cristina O. "La desarticulacion de logicas territoriales: El sector Agricola del Chaco entre 1994 y 2004." *Revista Geografica* 139 (2006): 103–29.

Valenzuela de Mari, Cristina O., and Alejandra H. Torre Geraldi. "Problematica de la articulacion entre distribucion, tenencia y usos del espacio agrario en dos provincias del norte argentine a fines del siglo XX." *Revista Geografica* 136 (2004): 5–47.

White, C. Langdon. "The Argentine Meat Question." *Geographical Review* 35, no. 4 (1945): 634–46.

Winsberg, Morton D. "The Introduction and Diffusion of the Aberdeen Angus in Argentina." *Geography* 55, no. 2 (1970): 187–95.

———. *Modern Cattle Breeds in Argentina: Origins, Diffusion, and Change*. Lawrence: University of Kansas, Center for Latin American Studies, 1968.

———. "Una Regionalizacion estadistica de la agricultura en la Pampa Argentina." *Revista Geografica* 72 (1970): 45–60.

Zamorano, Gloria L. "La Liberdad, Departimento de Rivadavia, un tipico districo agriario Mendocino." *Revista Geografica* 85 (1977): 11–61.

Zamorano, Mariano. "Region cuyana de los oasis agroindustriales." In *La Argentina: Geografia general y los marcos regionales*, edited by Juan A. Roccatagliata, 613–60. Buenos Aires, Argentina: Planeta, 1992.

Commerce and Trade

Beaujeu-Garnier, J. "Transformation et problemes du commerce argentin." *Les Cahiers d'Outre Mer* 2 (1949): 256–66.

Bortagaray, Lucia L. "Comercio e intercambio." In *La Argentina: Geografia general y los marcos regionales*, edited by Juan A. Roccatagliata, 385–423. Buenos Aires, Argentina: Planeta, 1992.

Crespo, Ricardo. "Las exportaciones no tradicionales en la Argentina." In *Geografia Economica Argentina*, edited by Juan A. Roccatagliata, 126–54. Buenos Aires, Argentina: El Atereo, 1993.

Di Nucci, Josefina, and Diana Lan. "Globalizacion, hiper concentracion y transnacionalizacion del comercio alimentico en Argentina: El territorio usado par las empresas supermercadistas." *Anales de Geografia de al Universidad Computense* 29, no. 1 (2009): 9–33.

Rudolph, William E. "Argentine Trade under Wartime Conditions." *Geographical Review* 34 (1944): 311–16.

Manufacturing and Labor Studies

Borello, Jose A. "The Territorial Dimension of Manufacturing Restructuring: The Transformation of Places in Contemporary Argentina." *Yearbook, Conference of Latin Americanist Geographers* 16 (1990): 41–55.

Dozo, Servando. "Industria no integrada y concentrada con desarrollo interrumpido." In *La Argentina: Geografia general y los marcos regionales*, edited by Juan A. Roccatagliata, 309–28. Buenos Aires, Argentina: Planeta, 1992.

Johns, Michael. "Industrial Capital and Economic Development in Turn of the Century Argentina." *Economic Geography* 68, no. 2 (1992): 188–204.

Kirchner, John A. *Sugar and Seasonal Labor Migration: The Case of Tucuman, Argentina.* Chicago: University of Chicago, Department of Geography, Research Paper 192, 1980.

Morris, Arthur S. "Regional Industrial Promotion in Argentine Patagonia: End of an Era." *Tijdschrift voor Economische en Sociale Geographie* 87, no. 5 (1996): 399–406.

Tomadoni, C. "Flexible Production and Labour Precariousness in a Glocally-Dependent: The Case of Renault Argentina [in French]." *Annales de Geographie* 113, no. 4 (2004): 3–28.

Energy, Mining, Fishing, and Lumbering

De Jorge, Carlos, and Maria Fioriti. "El sistema energetico." In *Geografia Economica Argentina*, edited by Juan A. Roccatagliata, 155–92. Buenos Aires, Argentina: El Ateneo, 1993.

Dozo, Servando. "Los recursos energeticas del territorio argentine." In *La Argentina: Geografia general y los marcos regionales*, edited by Juan A. Roccatagliata, 265–308. Buenos Aires, Argentina: Planeta, 1992.

Durland, W. D. "The Quebracho Region of Argentina." *Geographical Review* 14, no. 2 (1924): 227–41.

Lavie, Emilie. "La resource hydroelectrique dans la Province de Mendoza (Argentine)." *Les Cahiers d'Outre Mer* 60 (2007): 319–38.

Transportation and Communications

Ballistrier, Carlos A. "Geografia del transporte aereo en la Argentina. Historia de su evolucion; funcionamiento y expansion; crisis y perspectiva." In *Geografia Economica Argentina*, edited by Juan A. Roccatagliata, 241–66. Buenos Aires, Argentina: El Ateneo, 1993.

Keeling, David J. "Transport and Regional Development in Argentina: Structural Deficiencies and Patterns of Network Evolution." *Yearbook, Conference of Latin Americanist Geographers* 19 (1993): 25–34.

Miller, Fred. "Highway Improvements and Agricultural Production: An Argentine Case Study." In *Transportation and Development*, edited by Brian S. Hoyle, 104–24. New York: Harper & Row, 1973.

Roccatagliata, Juan A. "Las transportes y las comunicaciones." In *La Argentina: Geografia general y los marcos regionalies*, edited by Juan A. Roccatagliata, 353–84. Buenos Aires, Argentina: Planeta, 1992.

———. "Transporte. Algunas consideraciones actuales." In *Geografia Economica Argentina*, edited by Juan A. Roccatagliata, 193–240. Buenos Aires, Argentina: El Atereo, 1993.

Sagawe, Thorsten. "Privatisation of the Transport Sector in Argentina." *Geography* 79, no. 2 (1994): 183–86.

Theses and Dissertations

Baldridge, John. "Cooperative Institutions and Socio-spatial Transformation in Argentina's Recovered Businesses?" PhD diss., University of Arizona, Tucson, 2010.

Cornell, Thomas E. "Energy Resources in Argentina: Types, Regional Patterns, and Significance to the Economy." Master's thesis, Columbia University, New York, 1962.

Dambaugh, Luella M. "Industrialization in Argentina: A Geographical Analysis." PhD diss., University of Maryland, College Park, 1947.

Danna, Estaban. "Air Passenger Transport in Argentina, 1998–2001." Master's thesis, California State University, Northridge, 2006.

Gomez Insausti, Jose R. "The Industrial Restructuring of Argentina in the 1980s: Regional Decentralization and Economic Concentration." PhD diss., University of Toronto, Toronto, Ontario, 1996.

Hansis, Richard A. "Ethnogeography and Science: Viticulture in Argentina." PhD diss., Pennsylvania State University, University Park, 1976.

Hollman, Veronica C. "The Economy of Communion in Argentina: Exploring the Relationship between Culture and Economy." Master's thesis, University of British Columbia, Vancouver, 2001.

Jones, David M. "Shifting Patterns of Sugarcane Production in Northwest Argentina." PhD diss., Michigan State University, East Lansing, 1975.

HISTORICAL GEOGRAPHY

General Works

Bird, Junius. "Antiquity and Migrations of the Early Inhabitants of Patagonia." *Geographical Review* 28, no. 2 (1938): 250–75.

Bowen, E. G. "The Welsh Colony in Patagonia, 1865–1885: A Study in Historical Geography." *Geographical Journal* 132, no. 1 (1966): 16–31.

Carter, Eric D. "Development Narratives and the Uses of Ecology: Malaria Control in Northwest Argentina, 1890–1940." *Journal of Historical Geography* 33, no. 3 (2007): 619–50.

———. "State Visions, Landscape, and Disease: Discovering Malaria in Argentina, 1890–1920." *Geoforum* 39, no. 1 (2008): 278–93.

Daus, Federico A. "Resena sobre Historia de las Canocimientos Geograficos de la Argentina." *Revista Geografica*, 31–36 (1951–52): 1–5.

Jaime, Luis D., and Alicia I. Garcia. "La sismicidad en el NOA (Republica Argentina): Una Vision Historico-Geografica." *Revista Geografica* 146 (2009): 205–32.

Johnson, Lyman, and Susan Socolow. "Population and Space in Eighteenth-Century Buenos Aires." In Social *Fabric and Spatial Structure in Colonial Latin America*, edited by David J. Robinson, 339–68. Syracuse, NY: Syracuse University, Department of Geography, 1979.

Miller, Elbert E. "The Frontier and the Development of Argentina." *Revista Geografica* 90 (1979): 183–98.

Pyle, L. "Indigenous and Colonial Population of Argentina." *Geographical Survey* 1, no. 3 (1972): 55–61.

Quintero-Palacios, Silvina. "State Promotion of Mackinder's New Geography in the Argentine: 1900–10." In *Text and Image: Social and Construction of Regional Knowledges*, edited by Anne Buttiner, Stanley Brunn, and Ute Wardenga, 155–60. Leipzig, Germany: Institut für Landerkunde, Beitrage zur Regionalen Geographie, Heft, 49, 1999.

Rey Balmaceda, R. C. *Geografia historica de la Patagonia (1870–1960)*. Buenos Aires, Argentina: Ediciones Cervantes, 1976.

Robinson, David J. "Trade and Trading Links in Western Argentina during the Viceroyalty." *Geographical Journal* 136, no. 1 (1970): 24–41.

Robinson, David J., and Teresa Thomas. "New Towns in Eighteenth Century Northwest Argentina." *Journal of Latin American Studies* 6, no. 1 (1974): 1–33.

Royle, Stephen A. "The Falkland Islands, 1833–1876: The Establishment of a Colony." *Geographical Journal* 151, no. 2 (1985): 206–14.

Schmeider, Oscar. "Alteration of the Argentine Pampa in the Colonial Period." *University of California Publications in Geography* 2, no. 10 (1919–1920): 303–22.

———. "The Historical Geography of Tucuman." *University of California Publications in Geography* 2, no. 12 (1928): 359–86.

Takagi, Tadashi. "The Development of Railway Businesses and the Regionalization in Argentina (1850–1914) [in Japanese]." *Annals, Japan Association of Economic Geographers* 28, no. 3 (1982): 1–18.

Temberg, Rolf. "Occupance of the Humid Pampa, 1856–1914." *Revista Geografica* 76 (1972): 61–102.

PHYSICAL GEOGRAPHY

General Works

Bosch, Beatriz. "Les Tajamanes de Entre Rios (Argentina)." *Revista Geografica* 41 (1954): 67–80.

Capitanelli, Ricardo G. "Los ambientes naturales del territorio argentino." In *La Argentina: Geografia general y los marcos regionales*, edited by Juan A. Roccatagliata, 71–144. Buenos Aires, Argentina: Planeta, 1992.

Collado, A., et al. "Satellite Remote Sensing Analysis to Monitor Desertification Processes in the Crop-Rangelands Boundary of Argentina." *Journal of Arid Environments* 52, no. 1 (2002): 121–33.

Daus, Federico A. *Fisonomia regional de la Republica Argentina*. Rio de Janeiro, Brazil: Instituto Panamericano de Geografia e Historia, 1956.

Rey Balmaceda, Raul C. "La porcion oceania y la porcion antarctica, dos espacios en cuestion." In *La Argentina: Geografia general y los regionales*, edited by Juan A. Roccatagliata, 747–82. Buenos Aires, Argentina: Planeta, 1992.

Schneider, Thomas. "Argentina's Dry West: The Cuyo Oases on the Eastern Border of the Andes [in German]." *Mitteilungen zur Osterreichischen Geographische Gesellschaft* 141 (1999): 187–206.

Sili, Marcelo. "La Pampa argentine: un siècle d'evolution et restructuration." *Les Cahiers d'Outre Mer* 51 (1998): 35–64.

Villalba, Richardo, et al. "Climate, Tree-Ring, and Glacial Fluctuations in the Rio Frias Valley, Rio Negro, Argentina." *Arctic and Alpine Research* 22, no. 3 (1990): 215–32.

Vincenti, Rita D. "Les cambios ambientales en el Nordeste Argentino durante el Holocene." *Revista Geografica* 141 (2007): 53–110.

Zarate, M. A. "Late Pleistocene-Holocene History of Pampa Interserrena, Argentina." In *Landschaftsentwicklung, Palaookologie, und Klimageschichta der Ariden Diagonale Sudamerikas im Jungquarter*, edited by K. Garleff, 101–8. Bamberg, Germany: Bamberger Geographische Schriften, Heft 15, 1998.

Biogeography

Adamoli, Jorge, et al. "Stress and Disturbance: Vegetation Dynamics in the Dry Chaco Region of Argentina." *Journal of Biogeography* 17, nos. 4–5 (1990): 491–502.

Amat, Francisco, et al. "Further Evidence and Characteristics of Artemia from Ciscana Populations in Argentina." *Journal of Biogeography* 31, no. 11 (2004): 1735–50.

Blackhall, M., et al. "Cattle Affect Early Post-Fire Regeneration in a *Nothofagus dameyi–Austrocedrus chilensis* Mixed Forest in Northern Patagonia, Argentina." *Biological Conservation* 141 (2008): 2251–61.

Blair, W. F., et al. "Origins and Affinities of Vertebrates of the North American Sonoran Desert and the Monte Desert of Northwestern Argentina." *Journal of Biogeography* 3, no. 1 (1976): 1–18.

Brener, Alejandro G. F., and Adrianna Ruggiero. "Leaf-Cutting Ants (Atta and Acromyrmex) Inhabiting Argentina: Patterns in Species Richness and Geographical Range Size." *Journal of Biogeography* 21, no. 4 (1994): 391–400.

Cabido, Marcelo, et al. "Photosynthetic Pathway Variation among C4 Grasses along a Precipitation Gradient in Argentina." *Journal of Biogeography* 35, no. 1 (2008): 131–40.

Deasy, George F. "A Tentative Growing-Season Map of Argentina." *Journal of Geography* 42, no. 6 (1943): 225–29.

Faggi, A., et al. "Biodiversity in the Argentinian Rolling Pampa Ecoregion: Changes Caused by Agriculture and Urbanization." *Erdkunde* 60, no. 2 (2006): 127–38.

Fontana, S. L. "Coastal Dune Vegetation and Pollen Representation in South Buenos Aires Province, Argentina." *Journal of Biogeography* 32, no. 4 (2005): 719–35.

Gonzalez Roglich, Mariano, et al. "The Role of Private Lands for Conservation: Land Cover Change Analysis in the Caldenal Savanna Ecosystem, Argentina." *Applied Geography* 34 (2012): 281–88.

Grau, Hector R. "Regional-Scale Spatial Patterns of Fire in Relation to Rainfall Gradients in Sub-tropical Mountains, NW Argentina." *Global Ecology and Biogeography* 10, no. 2 (2001): 133–46.

Grau, Hector R., and Thomas T. Veblen. "Rainfall Variability, Fire, and Vegetation Dynamics in Neotropical Montane Ecosytems in Northwestern Argentina." *Journal of Biogeography* 27, no. 5 (2000): 1107–22.

Kitzberger, Thomas, et al. "Climatic Influences on Fire Regions along a Rain Forest-to-Xeric Woodland Gradient in Northern Patagonia, Argentina." *Journal of Biogeography* 24, no. 1 (1997): 35–48.

———. "Effects of Climatic Variability on Facilitation of Tree Establishment in Northern Patagonia." *Ecology* 81, no. 7 (2000): 1914–24.

———. "Variable Community Responses to Herbivory in Fire-Altered Landscapes of Northern Patagonia, Argentina." *African Journal of Range and Forage Science* 22 (2005): 85–91.

Kitzberger, Thomas, and Thomas T. Veblen. "Fire-Induced Changes in Northern Patagonian Landscapes." *Landscape Ecology* 14, no. 1 (1999): 1–15.

———. "Influences of Climate on Fire in Northern Patagonia, Argentina." In *Fire and Climate Change in Temperate Ecosystems of the Western Americas*, edited by Thomas T. Veblen et al., 290–315. New York: Springer Verlag, 2003.

———. "Influences of Humans on ENSO on Fire History of Austrocedruschilensis Woodlands in Northern Patagonia, Argentina." *Ecoscience* 4 (1997): 508–20.

Lopez, Hugo, et al. "Biogeographical Revision of Argentina (Andean and Neotropical Regions): An Analysis Using Freshwater Fishes." *Journal of Biogeography* 35, no. 9 (2008): 1564–79.

Menghi, M., et al. "Grassland Heterogeneity in Relation to Lithology and Geomorphology in the Cordoba Mountains, Argentina." *Vegetatio* 84, no. 2 (1989): 133–42.

Mermoz, M., et al. "Landscape Influences on Occurrence and Spread of Wildfires in Patagonian Forests and Shrublands." *Ecology* 86 (2005): 2705–15.

Naumann, M. "Vegetationsdynamic und Landschaftsdegradationseit dem 17. Jahrhundert im norplatag onishcen Seengebiet Nahuel-Huapi (Argentinien)." In *Landschaftsentwicklung, Palaookologie, und Klimageschichta der Ariden Diagonale Sudamerkicas im Jungquarter*, edited by K. Garleff, 321–34. Bamberg, Germany: Bamberger Geographische Schriften, Heft 15, 1998.

Nores, Manuel. "Insular Biogeography of Birds on Mountain-Tops in North Western Argentina." *Journal of Biogeography* 22, no. 1 (1995): 1–70.

Otte, Daniel. "Species Richness Patterns of New World Desert Grasshoppers in Relation to Plant Diversity." *Journal of Biogeography* 3, no. 3 (1976): 197–209.

Paez, M. M., et al. "Modern Pollen-Vegetation and Isopoll Maps in Southern Argentina." *Journal of Biogeography* 28, no. 8 (2001): 997–1022.

Paritsis, J., et al. "Assessing Dendro-ecological Methods to Reconstruct Defoliator Outbreaks in Nothofagus pumilio in Northwestern Patagonia, Argentina." *Canadian Journal of Forest Research* 39 (2009): 1617–29.

———. "Spatial Prediction of Caterpillar (Ormiscodes) Defoliation in Patagonian Nothofagus Forests." *Landscape Ecology* 26, no. 6 (2001): 791–803.

———. "Vegetation Disturbance by Fire Affects Plant Reproductive Phenology in a Shrubland Community in Northwestern Patagonia, Argentina." *New Zealand Journal of Ecology* 30 (2006): 387–95.

Paritsis, J., and Thomas T. Veblen. "Dendro-ecological Analysis of Defoliator Outbreaks on Nothofagus pumilio and Their Relation to Climate Variability in the Patagonian Andes." *Global Change Biology* 17, no. 1 (2011): 239–53.

———. "Temperature and Foliage Quality Affect Performance of the Outbreak Defoliator Ormiscodes amphimone (F.) (Lepidoptera: Saturniidae) in Northwestern Patagonia, Argentina." *Revista Chilena de Historia Natural* 83 (2008): 593–603.

Rabinovich, Jorge, and E. H. Rapoport. "Geographical Variation of Diversity in Argentine Passerine Birds." *Journal of Biogeography* 2, no. 3 (1975): 141–58.

Raffade, E., et al. "Interactive Effects of Introduced Herbivores and Post-Flowering Die-Off of Bamboos on Tree Regeneration and the Understory in a Nothofagus Forest, Argentina." *Journal of Vegetation Science* 18 (2007): 371–78.

———. "Synergistic Influences of Introduced Herbivorers and Fire on Vegetation Change in Northern Patagonia, Argentina." *Journal of Vegetation Science* 22 (2011): 59–61.

Raffade, E., and Thomas T. Veblen. "Effects of Cattle Grazing on Early Postfire Regeneration of Matorral in Northwest Patagonia, Argentina." *Natural Areas Journal* 21 (2001): 243–49.

Real, Raimundo, et al. "Relative Importance of Environment, Human Activity, and Spatial Situation in Determining the Distribution of Terrestrial Mammal Diversity in Argentina." *Journal of Biogeography* 30, no. 6 (2003): 939–48.

Rebertus, A. J., et al. "Blowdown History and Landscape Patterns in the Andes of Tierra del Fuego, Argentina." *Ecology* 78 (1997): 678–92.

Relva, M. A., and Thomas T. Veblen. "Impacts of Introduced Large Herbivores on Astrocedrus chilensis Forests in Northern Patagonia, Argentina." *Forest Ecology and Management* 108, nos. 1–2 (1998): 27–40.

Renison, Daniel, et al. "Cover and Growth Habitat of Polylepis Woodlands and Shrublands in the Mountains of Central Argentia: Human or Environmental Influence?" *Journal of Biogeography* 35, no. 5 (2008): 876–87.

Sarasola, Jose, and Juan Negro. "Role of Exotic Tree Stands on the Current Distribution and Social Behavior of Swainson's Hawk, Buteo swainsoni in the Argentine Pampas." *Journal of Biogeography* 33, no. 6 (2006): 1096–101.

Schmeider, Oscar. "The Pampa: A Natural or Culturally Induced Grassland?" *California Publications in Geography* 2 (1927): 255–70.

Tercero-Bucardo, N., et al. "A Field Experiment on Climatic and Herbivore Impacts on Postfire Tree Regeneration in Northwestern Patagonia." *Journal of Ecology* 95 (2007): 771–79.

Tonello, M. S., and A. R. Prieto. "Modern Vegetation-Pollen-Climate Relationships for the Pampa Grasslands of Argentina." *Journal of Biogeography* 35, no. 4 (2008): 926–33.

Torrusio, S., et al. "Grasshopper (Orthopter: Acridoidea) and Plant Community Relationships in the Argentine Pampas." *Journal of Biogeography* 29, no. 2 (2002): 221–30.

Veblen, Thomas T., et al. "Ecological Impacts of Introduced Animals in Nahuel Huapi National Park, Argentina." Conservation Biology 6 (1992): 71–83.

———. "Fire History in Northern Patagonia: The Roles of Humus and Climatic Variation." *Ecological Monographs* 69, no. 1 (1999): 47–67.

———. "Fire History and Vegetation Change in Northern Patagonia, Argentina." In *Fire Regimes and Climatic Change in Temperate Ecosystems of the Western Americas*, edited by Thomas T. Veblen et al., 259–89. Wiesbaden, Germany: Springer-Verlag, 2003.

Villalba, Ricardo, et al. "Spatial Patterns of Climate and Tree Growth Variations in Subtropical Northwestern Argentina." *Journal of Biogeography* 19, no. 6 (1992): 631–50.

Villalba, Ricardo, and Thomas T. Veblen. "Influences of Large-Scale Climatic Variability on Episodic Tree Mortality at the Forest-Steppe Ecotone in Northern Patagonia." *Ecology* 79, no. 8 (1998): 2624–40.

———. "Regional Patterns of Tree Population Age Structures in Northern Patagonia: Climatic and Disturbance Influences." *Journal of Ecology* 85, no. 2 (1997): 113–24.

———. "Spatial and Temporal Variation in the Growth along the Forest-Steppe Ecotone in Northern Patagonia." *Canadian Journal of Forest Research* 27 (1997): 580–97.

Climatology and Weather

Adams, S. B., and K. A. Crews-Meyer. "Aridity and Desertification: Exploring Environmental Hazards in Jachal, Argentina." *Applied Geography* 26, no. 1 (2006): 61–85.

Bruniard, Enrique. "La diagonal arida argentina: Un limite climatico real." *Revista Geografica* 95 (1982): 5–20.

Burgoy, J. J., and A. L. Vidalo. "The Climates of the Argentine Republic According to the New Thornthwaite Classification." *Annals of the Association of American Geographers* 41, no. 2 (1951): 231–63.

Deasy, George F. "A Tentative Growing Season Map of Argentina." *Journal of Geography* 42 (1943): 225–29.

Endlicher, Wilfried, et al. "Weather Types, Local Winds, and Air Pollution Problems in Mendoza, Argentina." *Yearbook, Conference of Latin Americanist Geographers* 25 (1999): 61–76.

Eriksen, W. "Ariditat und Trockengrenzenin Argentinien-Ein Beitrag zur Klimageographie der Trockendiagonale Sudamerikas." *Colloquium Geographicum* 16 (1983): 43–68.

———. "Fohnprozesse und fohnortige Winde in Argentinien." In *Frogen geographischer Forschung: Festschrift des Instituts fur Geographiszum 60*, edited by Peter Maimayer et al., 63–78. Innsbruck, Austria: Innsbrucker Geographische Studien, Band 5, 1979.

Falasca, Silvia, and Maria A. Bernabe. "Aptitud Agroclimatica Argentina para la implantacion de bosques energeticas de Paulownia ssp." *Revista Geografica* 148 (2010): 151–64.

Falasca, Silvia, and Juan Forte-Lay. "Cambio en la Evapotranspiracion potencial de la pradera pamperana (Republica Argentina) inducido par un periodo humedo." *Revista Geografica* 134 (2003): 125–62.

Falasca, Silvia, and Ana C. Ulberich. "El agroclima de Stevia Rebaudiana Bertoni en Argentina." *Revista Geografica* 137 (2005): 35–47.

Geraldi, A. M., et al. "Anamalias de Precipitacion y Temperatura en las Encadenadas del Oeste, Buenos Aires, Argentina." *Revista Geografica* 148 (2010): 137–50.

Huamantinco-Cisneros, Maria, and Maria C. Piccolo. "Characterization of the Sea Breeze in Monte Hermoso, Argentina [in Spanish]." *Estudios Geograficos* 72, no. 271 (2011): 461–75.

Krepper, Carlos, et al. "Time and Space Variability of Rainfall in Central-East Argentina." *Journal of Climate* 2, no. 1 (1989): 39–47.

Merlotto, Alejandra, et al. "Caracteristicas del oleaje y vientos del sector costero del partido de Necochea, Buenos Aires, Argentina." *Revista Geografica* 147 (2010): 113–32.

Minetti, J. L., and E. M. Sierra. "The Influence of General Circulation Patterns on Humid and Dry Years in the Cuyo Andean Region of Argentina." *International Journal of Climatology* 9, no. 1 (1989): 55–68.

Polimeni, Claudia M. "Las regions de la zona templada, calida, arida y semiarida argentina." *Revista Geografica* 95 (1982): 21–29.

Geomorphology, Landforms, and Volcanism

Anastasi, Atilio B. "Propuesta para una clasificacion de regiones aridas y en proceso de desertizacion en Argentina." *Revista Geografica* 95 (1982): 185–94.

Andre, M. F., et al. "Stone Runs in the Falklands: Periglacial or Tropical?" *Geomorphology* 95 (2008): 524–43.

Angeles, Guillermo, et al. "Fractal Analysis of Tidal Channels in the Bahia Blanca Estuary (Argentina)." *Geomorphology* 57, no. 3 (2004): 263–74.

Campo, Alicia M., et al. "Morfometria fluvial aplicada a una cuenca urbanan en Ingeniero White, Republica Argentina." *Investigaciones Geograficas* 77 (2012): 7–17.

Capitanelli, Ricardo G. "Aspectos Nivologicas, Glaciologices y Morphologicas del Valle del Atuel." *Revista Geografica* 57 (1962): 5–28.

Clark, R. J., and P. Wilson. "Occurrence and Significance of Ventifacts in the Falkland Islands, South Atlantic." *Geografiska Annaler* 74A, no. 1 (1992): 35–46.

Conway, Dennis, and Robert Holtz. "The Use of Near-Infrared Photography in the Analysis of Surface Morphology of an Argentine Alluvial Floodplain." *Remote Sensing of the Environment* 2, no. 4 (1973): 235–42.

Diaz, Ramon J. "El Valle de Sanagasta." *Revista Geografica* 95 (1982): 110–20.

Falasca, Silvia, and Maria A. Bernabe. "Indices de la Erosividad de la lluvia en el Partido de Tandil, Provincia de Buenos Aires." *Revista Geografica* 134 (2003): 117–24.

Gomez, Juan, and Lucia Magnin. "Cartografia geomorfologica aplicada un sector de interes arqueologico en el Nacizo del Deseado, Santa Cruz, (Patalonia Argentina)." *Investigaciones Geograficas* 65 (2008): 22–37.

Inbar, Moske, et al. "The Morphological Development of a Young Lava Flow in the South Western Andes-Neuquen, Argentina." *Zeitschrift fur Geomorphologie* 38, no. 4 (1995): 479–88.

Inbar, Moske, and C. Risso. "Holocene Yardangs in Volcanic Terrains in the Southern Andes, Argentina." *Earth Surface Processes and Landforms* 26, no. 6 (2001): 657–66.

Mamani, Manuel, et al. "Reconocimiento y ubicacion par teledeteccion de estrados partadores de fosiles paleontologicas: Cuenca Triasica la Rioja-San Juan (Argentina)." *Revista Geografica* 139 (2006): 151–61.

Molinillo, Marcelo F. "Is Traditional Pastoralism the Cause of Erosion Processes in Mountain Environments? The Case of the Cumbres Calchaques in Argentina." *Mountain Research and Development* 13, no. 2 (1993): 189–202.

Ogilvie, Alan G. "Argentine Physiographical Studies: A Review." *Geographical Review* 12 (1923): 112–21.

Parsons, D. D., et al. "Form Roughness and the Absence of Secondary Flow in a Large Confluence-Differences, Rio Parana, Argentina." *Earth Surface Processes and Landforms* 32, no. 1 (2007): 155–62.

Patron, Luis R. "The Altitude of Acancagua." *Geographical Review* 19 (1928): 485–88.

Regairaz, Alberto, and Alberto Bozzolo. "Redacion entre Tectonia y Morfologia en la Cuenca del Golfo San Jorge (Camodaro Rivadavia (Patagonia)-Republica Argentina)." *Revista Geografica* 61 (1964): 71–86.

Rivas, V., et al. "Geomorphic Consequences of Urban Development and Mining Activities: An Analysis of Study Areas in Spain and Argentina." *Geomorphology* 73, nos. 3–4 (2006): 185–206.

Sancho, Carlos, et al. "Morphological and Speleothemic Development in Brujas Cave (Southern Andean Range, Argentina: Palaeoenvironmental Significance." *Geomorphology* 57, nos. 3–4 (2004): 367–84.

Schellmann, G. *Jungkanozoische Landschaftsgeschichte Patagoniens (Argentinien). Andine Vorland vergletscheungen, Talentwicklung, und marine Terrassen.* Essen, Germany: Essener Geographische Arbeiten, Band 29, 1998.

So, C. L., et al. "Sand Waves in the Gulf of San Matias, Argentina." *Geografiska Annaler* 56A, nos. 3–4 (1974): 227–36.

Stigl, H., and K. Garleff. "Teriane une pleistozare Reliefenwicklung an der interozeanischen Wasserscheide in Sudpatogonien (Gebiet von Rio Turbio, Argentinien)." In *Regionale Beitrage zur Geomorphologie*, edited by G. Stablein, 113–18. Berlin: Berliner Geographische Abhandlungen, Band 36, 1984.

Thorne, Colin R. "Geomorphic Analysis of Large Alluvial Rivers." *Geomorphology* 44, nos. 3–4 (2002): 203–19.

Vich, Alberto. "Estimacion del factor de erosividad en areas de Piedmonte, proximo a Gran Mendoza (Argentina)." *Revista Geografica* 141 (2007): 145–64.

Vincenti, Rita D. "La Incidencia de los Factores Litologicas el Escurrimiento Fluvial." *Revista Geografica* 135 (2004): 63–78.

Hydrology and Glaciology

Alconada Magliaro, Margarita, et al. "El bio-drenaje para el control del exceso hidrico en Pampa Arenosa, Buenos Aires, Argentina." *Investigaciones Geograficas* 68 (2009): 50–72.

Arcuri, C. "Glacial Features in Huaca Huasi (Cumbres Caichaquies, Tucuman, Northwestern Argentina)." In *Landschaftsentwicklung Palaookologie und Klimageschichata der Ariden Diagonale Sudamerikas im Jungquarter*, edited by K. Garleff, 389–99. Bamberg, Germany: Bamberger Geographische Schriften, Heft 15, 1998.

Carbone, Maria, and Maria C. Piccolo. "Zonas de des bordes y anegamiento a traves de cartografia hidrogeomorfologia: Caso de estudio cuenca del Arroyo Claromeco, Argentina." *Revista Geografica* 148 (2010): 23–42.

Clapperton, Chalmers M., and D. E. Roberts. "Quaternary Sea Level Changes in the Falkland Islands." In *Quaternary of South America and Antarctic Peninsula*, vol. 4, edited by J. Rabassa. Rotterdam, Netherlands: Balkema, 1986.

Espinoza, Lydia. "Fluctuations of the Rio del Plomo Glaciers." *Geografiska Annaler* 68A, no. 4 (1986): 317–27.

Espinoza, Lydia, and J. D. Bengochea. "Surge of Grande del Nevado Glacier (Mendoza, Argentina) in 1984: Its Evolution through Satellite Images." *Geografiska Annaler* 72A, nos. 3–4 (1990): 255–60.

Falasca, Silvia, and Maria A. Bernabe. "Evolucion de variables hidrologicas en el Centro-Sud Boraerense, Republica Argentina." *Revista Geografica* 139 (2006): 71–82.

Falasca, Silvia, et al. "Constantes hidrologicas edaficas en la region subhumeda-humeda y humeda oriental de la Republica Argentina." *Revista Geografica* 142 (2007): 89–110.

Genchi, Sibila A., et al. "Hydrolic Response of the Drainage Basins That Intersect Las Crutas Town, Argentina." *Investigaciones Geograficas* 75 (2011): 23–36.

Hunzinger, Harald. "Hydrology of Montane Forests in the Sierra de San Javier, Tucuman, Argentina." *Mountain Research and Development* 17, no. 4 (1997): 299–308.

Lonfredi, Nestor, et al. "Sea-Level Rise and Related Potential Hazards on the Argentine Coast." *Journal of Coastal Research* 14, no. 1 (1998): 47–60.

Masiokas, M. H., et al. "Twentieth Century Glacier Recession and Regional Hydroclimatic Changes in Northeastern Patagonia." *Global and Planetary Change* 60, nos. 1–2 (2008): 85–100.

Mercer, John H., and M. A. Ager. "Glacial and Floral Changes in Southern Argentina since 14,000 Years Ago." *National Geographical Society Reports* 13 (1983): 457–77.

Nichols, Robert, and Maynard Miller. "Glacial Geology of Ameghino Valley, Lago Argentina." *Geographical Review* 41, no. 2 (1951): 274–94.

Radtke, Ulrich, et al. "Untersuchungen zum marinen Quarter Patagoniens (Argentiniens)." In *Neue Ergebnisse zur Kustenforschung-Vorgrage der Jahrestagung Wilhelmshaven 18. und 19. Mai 1989*, edited by D. Kelletat, 267–89. Essen, Germany: Essener Geographische Arbeiten, Band 17, 1989.

Torrero, M. P., and Alicia M. Campo. "Hydrological Dynamics of the Sauce Chico River, Argentina: Flow and Runoff." *Revista Geografica* 147 (2010): 95–112.

Valenzuela de Mari, Cristina O. "Una Aproximacion al Impacto Geografico de las Inundaciones en el Sector Agropecuario Chaqueno en la Segunda Mitad del Siglo XX." *Revista Geografica* 130 (2001): 33–59.

Vincenti, Rita D. "Las dos caras del agua en el litoral Chaqueno Durante los Ultimas Veinticinco Anos de Siglo XX." *Revista Geografica* 131 (2002): 117–64.

Waylen, Peter R., et al. "Interannual and Interdecadel Variability in Stream Flow from the Argentine Andes." *Physical Geography* 20, no. 5 (2000): 452–65.

Wayne, W. J., and A. E. Carte. "Multiple Glaciations of the Cordon del Plata, Mendoza, Argentina." *Paleogeography, Paleoclimatology, and Paleoecology* 42, nos. 1–2 (1983): 185–209.

Wenzens, Gerd. "Fluctuations of Outlet and Valley Glaciers in the Southern Andes (Argentina) during the Past 13,000 Years." *Quaternary Research* 51, no. 3 (1999): 238–47.

———. "Glacier Advances East of the Southern Andes between the Last Glacial Maximum and 5000BP Compared with Lake Terraces of the Endorrheic Lago Cardiel (49 Degrees S, Patagonia, Argentina)." *Zeitschrift fur Geomorphologie* 49, no. 4 (2005): 433–54.

———. "Pliocene Piedmont Glaciation in the Rio Shehuen Valley, Southeast Patagonia, Argentina." *Arctic, Antarctic, and Alpine Research* 32, no. 1 (2000): 46–54.

———. "Terminal Moraines, Outwash Plains, and Lake Terraces in the Vicinity of Lago Cardiel (40 Degrees South, Patagonia, Argentina)-Evidence for Miocene Andean Foreland Glaciations." *Arctic, Antarctic, and Alpine Research* 38, no. 2 (2006): 276–91.

Soils

Derbyshire, E. *Loess and the Argentine Pampa.* Leicester, England: University of Leicester, Department of Geography, Occasional Paper 23, 1992.

Frederiksen, Peter. "Climatic Influence on Morphology and Chemistry of Nine Loess-Profiles, Argentina." *Geografisk Tidsskrift* 82 (1982): 16–24.

———. "LANDSAT, Aerial Photography, and State Factors in Soil Survey of Arid and Humid Patagonia, Argentina." *Geografisk Tidsskrift* 81 (1981): 39–48.

Kemp, R. A., and M. A. Zarate. "Pliocene Pedosedimentary Cycles in the Southern Pampas, Argentina." *Sedimentology* 47, no. 1 (2000): 3–14.

Scarpati, O., et al. "Impacts of ENSO Events in Soil Water Moisture in Pampen Region (Argentina)." *Revista Geografica* 141 (2007): 39–52.

Schrott, Lothar. "Global Solar Radiation, Soil Temperature, and Permafrost in the Central Andes, Argentina: A Progress Report." *Permafrost and Periglacial Processes* 2, no. 1 (1991): 59–66.

———. "The Hydrological Significance of High Mountain Permafrost and Its Relation to Solar Radiation. A Case Study in the High Andes of San Juan, Argentina." In *Landschaftsentwicklung, Palaookologie, und Klimageschichta der Ariden Diagonale Sudamerikas im Jungquarter*, edited by K. Garleff, 71–84. Bamberg, Germany: Bamberger Geographische Schriften, Heft 15, 1998.

Trombotto, D., et al. "Monitoring of Mountain Permafrost in the Central Andes, Cordon del Plata, Mendoza, Argentina." *Permafrost and Periglacial Processes* 8, no. 1 (1997): 123–29.

Vogt, Thea, and H. F. Del Valle. "Calcretes and Cryogenic Structures in the Area of Puerto Madryre (Chubut, Patagonia, Argentina)." *Geografiska Annaler* 76A, nos. 1–2 (1994): 57–76.

Wilson, P. "Forms of Unusual Patterned Ground: Examples from the Falkland Islands, South Atlantic." *Geografiska Annaler* 77A, no. 3 (1995): 159–66.

Wilson, P., et al. "Soil Erosion in the Falkland Islands: An Assessment." *Applied Geography* 13, no. 4 (1993): 329–52.

Zarate, M. A. "Late Pleistocene and Holocene Loess Deposits in the Southeastern Buenos Aires Province, Argentina." *GeoJournal* 24, no. 2 (1991): 211–20.

Theses and Dissertations

Santimano, Tasca. "Kinematics and Mechanisms of Upper-Crustal Deformation in the Eastern Cordillera, Southern Central Andes, Northwest Argentina." Master's thesis, McMaster University, Hamilton, Ontario, 2010.

POLITICAL GEOGRAPHY

General Works

Ackerman, William V. "Development Strategy for Cuyo, Argentina." *Annals of the Association of American Geographers* 65, no. 1 (1975): 36–47.

Benwell, Matthew, and Klaus Dodds. "Argentine Territorial Nationalism Revisited: The Malvisas/Falklands Geographies of Everyday Nationalism." *Geoforum* 30, no. 8 (2011): 441–49.

Bosco, Fernando J. "Emotions That Build Networks: Geographies of Human Rights Movements in Argentina and Beyond." *Tijdschrift voor Economische en Sociale Geographie* 98, no. 5 (2007): 545–63.

———. "Human Rights Politics and Sealed Performances of Memory: Conflicts among the Madres de Plaza de Mayo in Argentina." *Social and Cultural Geography* 5, no. 3 (2004): 381–402.

———. "The Madres de Plaza de Mayo and Three Decades of Human Rights Activism: Embeddedness, Emotions, and Social Movements." *Annals of the Association of American Geographers* 96, no. 2 (2006): 342–65.

———. "Place, Space, Networks, and the Sustainability of Collective Action: The Madres de Plaza de Mayo." *Global Networks* 1, no. 4 (2001): 307–29.

Bruniard, Enrique, et al. "La Region Funcional de Resistencia (Provincia del Chaco-Republica Argentina)." *Revista Geografica* 81 (1974): 7–46.

Cara, Robert B. "Los Sistemas Territoriales: Etapas de Estructuracian y Destruccion en Argentina." *Anales de Geografia de la Universidad Complutense* 22 (2002): 113–30.

Carter, Eric D. "Misiones Province, Argentina: How Borders Shape Political Identity." In *Borderlines and Borderlands: Political Oddities at the Edge of the Nation-State*, edited by Alexander Diener and Joshua Hagen, 155–72. Lanham, MD: Rowman & Littlefield, 2010.

Chatterton, P. "Making Autonomous Geographies: Argentina's Popular Uprising and the Movimiento de Trabajodores Desocupados (Unemployed Workers Movement)." *Geoforum* 36, no. 5 (2005): 545–61.

Child, Jack. "Antarctica and Argentine Geopolitical Thinking." *Yearbook, Conference of Latin Americanist Geographers* 12 (1986): 12–16.

Dodds, Klaus. "Geography, Identity, and the Creation of the Argentine State." *Bulletin of Latin American Research* 12, no. 3 (1993): 311–33.

———. "Unfinished Business in the South Atlantic: The Falklands/Malvinas in the Late 1990s." *Political Geography* 17 (1998): 623–26.

Escobar, Marcelo, et al. "Geographical Identity and Patriotic Representation in Argentina." In *Geography and National Identity*, edited by David Hooson, 346–66. Cambridge, England: Blackwell, 1994.

Fritschy, Blanca A. "Apartes de la geografia para la mitigacion de los delitos: Case de estudion: Seccireal 8va de Policia, Santa Fe, Argentina, ano 2005." *Revista Geografica* 145 (2009): 7–30.

Gomez, Cesar. "Conflictos de tierras en la provincia de Chaco (Argentina): Una aproximacion a las organizacion indigenas y sus estrategias territoriales." *Revista Geografica* 146 (2009): 171–204.

Hepple, Leslie W. "The Geopolitics of the Falkland/Malvinas and the South Atlantic: British and Argentine Perceptions, Misperceptions, and Rivalries." In *Geopolitics of the Southern Cone and Antarctica*, edited by Philip Kelly and Jack Child, 223–36. Boulder, CO: Lynne Rienner, 1988.

De Hoyos, Ruben. "Malvinas/Falklands, 1982–1988: The New Gibraltar in the South Atlantic?" In *Geopolitics of the Southern Cone and Antarctica*, edited by Philip Kelly and Jack Child, 237–49. Boulder, CO: Lynne Rienner, 1988.

Jackiewicz, Edward, and Jeffrey Boggs. "Apples and Oranges? A Cross-National Comparison of Privatization in Argentina and Germany." *Middle States Geographer* 29 (1996): 57–66.

Kanai, J. M. "Barrio Resurgence in Buenos Aires: Local Autonomy Claims amid State-Sponsored Transnationalism." *Political Geography* 30, no. 4 (2011): 225–35.

Keeling, David J. "Regional Development and Transport Policies in Argentina: An Appraisal." *Journal of Developing Areas* 28, no. 4 (1994): 487–502.

Kentner Carby, Julia. "For Members and Markets: Neoliberalism and Cooperativism in Mendoza's Wine Industry." *Journal of Latin American Geography* 9, no. 2 (2010): 27–47.

Laymond, Philippe. "Mise en valeur et organization d'un territoire: La province de Tucuman." *Les Cahiers d'Outre Mer* 60 (2007): 339–62.

Morris, Arthur S. "The Argentine Colorado-Interprovincial Rivalries over Water Resources." *Scottish Geographical Journal* 94, no. 3 (1978): 169–79.

———. *Regional Disparities and Policy in Modern Argentina*. Glasgow, Scotland: Institute of Latin American Studies, 1975.

Penning-Rowsell, E. C. "Flood-Hazard Response in Argentina." *Geographical Review* 86, no. 1 (1996): 72–90.

Reboratti, Carlos E. "Politicas Publicas y Resdistribucion de la Poblacion en una Frontera Agraria (Argentina)." *Revista Geografica* 97 (1983): 104–12.

———. "Socio-environmental Conflict in Argentina." *Journal of Latin American Geography* 11, no. 2 (2012): 3–20.

Rey Balmaceda, Raul C., and Graciela M. DeMarco. "Conformacion del sistema politicoterritorial." In *La Argentina: Geografia general y los marcos regionales*, edited by Juan A. Roccatagliata, 25–70. Buenos Aires, Argentina: Planeta, 1992.

Roccatagliata, Juan A. *Argentina: Hacia un Nuevo ordenamiento territorial.* Buenos Aires, Argentina: Editorial Pleamar, 1986.

———. "Crisis, cambio economic y politico de ordenacion territorial." In *La Argentina: Geografia general y los marcos regionales,* edited by Juan A. Roccatagliata, 783–96. Buenos Aires, Argentina: Planeta, 1992.

———. *Geografia y politicas territorials: La ordenacion del espacio.* Buenos Aires, Argentina: Editorial Ceyne, 1994.

Roucek, Joseph S. "The Geopolitics of Antarctica and the Falkland Islands." *World Affairs Interpreter* 22 (1951): 44–56.

Rydant, Alan L. "Adaptation in Argentina: Effects of Inflation and Democratization." *Proceedings of the New England-St. Lawrence Valley Division, Association of American Geographers* 20 (1990): 86–91.

Salmenkori, Taru. "Geography of Protest: Places of Demonstration in Buenos Aires and Seoul." *Urban Geography* 30, no. 3 (2009): 239–60.

Salomon, Jean-Noel. "Le Rio Atuel, un exemple d'amenagement en milieu naturel subaride (Andes de Mendoza, Argentine)." *Les Cahiers d'Outre Mer* 60 (2007): 301–18.

Seone, Jose. "Rebellion, Dignity, Autonomy, and Democracy: Shared Voices from the South." *Antipode* 36, no. 3 (2004): 383–91.

Shaw, Earl B. "Geographic Aspects of United States–Argentine Relations." *Journal of Geography* 46, no. 4 (1947): 136–46.

Starck, Emilie. "Le precarite paysanne argentine en milieu semiaride: Exemple de Santiago del Estero." *Les Cahiers d'Outre Mer* 60 (2007): 363–94.

Velut, S. "Argentina: National Identity and Globalization [in French]." *Annales de Geographie* 113, nos. 4–5 (2004): 489–510.

URBAN GEOGRAPHY

General Works

Abaleron, Carlos A. "The Pros and Cons of Peri-urban Management in a Tourism City." *Geographer, Aligarh* 49, no. 2 (2002): 15–25.

Ackermann, William V. "Testing Central Place Concepts in Western Argentina." *Professional Geographer* 30, no. 4 (1978): 377–88.

Armstrong, Warwick. "Militarism and Malnutrition: The Impact of a Monetarist Strategy, Buenos Aires, 1976–1983." In *Proceedings, Fourteenth New Zealand Geography Conference,* edited by Richard LeHeron et al., 43–51. Palmerston North, New Zealand: New Zealand Geographical Society, 1987.

Auyero, J. "The Geography of Popular Contention: An Urban Protest in Argentina." *Canadian Journal of Latin American and Caribbean Studies* 28 (2003): 37–70.

Bouvet, Y., et al. "Mar de Plata, archetype de la station balneaire au service d'une metropole." *Les Cahiers d'Outre Mer* 56, no. 223 (2003): 281–300.

Celemin, Juan P., and Guillermo A. Velazquez. "Estimacion de un indice de ciudad ambiental para la ciudad y provincia de Buenos Aires." *Journal of Latin American Geography* 10, no. 1 (2011): 71–84.

De Civit, Maria, et al. "Un ejemplo de las relaciones Ciudad-Campo: Las heras y sus vinculaciones con el nucleo de la aglomeracion mendocina." *Revista Geografica* 76 (1972): 5–59.

Denis, Paul-Yves. "Developpement urbain et developpement regional: rapports Ville-Compagne, l'exemple de la Republique Argentina." *Canadian Geographer* 14 (1970): 217–28.

Difrieri, Horacio. *Buenos Aires: Geohistoria de una metropolis.* Buenos Aires, Argentina: Universidad de Buenos Aires, 1981.

Esteia Ares, Sofia, and Claudia A. Mikkelsen. "Dimedondevives y sobre por que llegaste. Movilidad territorial y poblamiento de localidades pequenos del partido de General Pueyrredon (Buenos Aires)." *Invesitgaciones Geograficas* 72 (2010): 101–19.

Fournier, J. M. "Orders and Disorders in Argentina's Cities: The Example of Alto Comedero, San Salvador de Jujuy [in French]." *Annales de Geographie* 111, no. 2 (2002): 179–97.

Fugier, A. "Buenos Aires et ses problems de croissance." *Les Cahiers d'Outre Mer* 2 (1949): 97–111.

Gomez Insausti, Ricardo. "La region metropolitan de Buenos Aires, una desproporcionada concentracion." In *La Argentina: Geografia general y los marcos regionales*, edited by Juan A. Roccatagliata, 451–76. Buenos Aires, Argentina: Planeta, 1992.

Gray de Cerdan, Nelly A. "La Ciudad de San Juan su influencia regional y su proyeccion en la red de Ciudades de Cuyo." *Revista Geografica* 81 (1974): 47–79.

———. "El suburbio mendocino su organizacion actuel y sus perspectives en una planificacion." *Revista Geografica* 90 (1979): 157–81.

Keeling, David J. *Buenos Aires: Global Dreams, Local Crises.* New York: Wiley, 1996.

Lacarrieu, Monica, and Guy Thuillier. "Une utopie de l'ardre et de la fermeture: 'quartiers prives' et 'countries' a Buenos Aires." *L'Espace Geographique* 37, no. 2 (2004): 149–64.

Leveau, Carlos. "Contraurbanizacion en Argentina? Una aproximacion a varias escalas con base en datos censales del periodo 1991–2001." *Investigaciones Geograficas* 69 (2009): 85–95.

Roccatagliata, Juan A., and Susana Beguiristain. "Urbanizacion sistema urbano." In *La Argentina: Geografia general y los marcos regionales*, edited by Juan A. Roccatagliata, 329–52. Buenos Aires, Argentina: Planeta, 1992.

Rofman, A. B. "Argentina: A Mature Urbanization Pattern." *Cities* 2, no. 1 (1985): 47–54.

Sargent, Charles S. "Argentina." In *Latin American Urbanization: Historical Profiles of Major Cities*, edited by Gerald M. Greenfield, 1–38. Westport, CT: Greenwood, 1994.

Sassone, Susana. "Subsistema urbanos policentricas en los sistemas nacionales de ciudades: Un case en la Argentina." *Revista Geografica* 116 (1992): 85–111.

———. "Transformacion de un Espacio Urbano: El Caso del Mercado de Abasto de Buenos Aires." *Anales de Geografia de la Universidad Complutense* 21 (2001): 99–118.

Takahashi, N. "The Urban System in Argentina [in Japanese]." *Tsukuba Studies in Human Geography* 14 (1990): 41–80.

Velazquez, Guillermo A. "Ranking del bienestar segun categorias urbanas en la Argentina 2001." *Revista Geografica de America Central* 46 (2011): 185–210.

Veron, Eleonora, and Alejandra Menlotto. "Urbanizaciones cerradas en la ciudad Intereses y conflictos en Mar del Plata, Argentina." *Revista Geografica* 146 (2009): 19–36.

Wilhelmy, Herbert. "Urban Changes in Argentina: Historical Roots and Modern Trends." In *World Patterns of Modern Urban Change: Essays in Honor of Chauncy D. Harries*, edited by Michael P. Conzen, 273–92. Chicago: University of Chicago, Department of Geography, Research Paper 217–218, 1986.

Urban Social Geography

Bouvet, Y., et al. "Mar del Plata, archetype de la station balneaire au service d'un metropole." *Les Cahiers d'Outre Mer* 56, no. 223 (2003): 281–300.

Buzai, Gustavo, and Mariana Marcos. "The Social Map of Greater Buenos Aires as Empirical Evidence of Urban Models." *Journal of Latin American Geography* 11, no. 1 (2012): 67–78.

Capron, G. "Les cafes a Buenos Aires. Une analyse historique de la contraction sociale et culturelle de l'espace public et de l'urbanite." *Geographie et Cultures* 24 (1997): 29–50.

Foschiatti de Dell'Orto, Ann M. "El desarrollo urbano y las particularidades demograficas del Chaco y su Capital entre 1960 y 1990." *Revista Geografica* 115 (1992): 37–54.

Hall, G. B., et al. "Integration of Remote Sensing and GIS to Detect Pockets of Urban Poverty: The Case of Rosario, Argentina." *Transactions in GIS* 5, no. 3 (2001): 235–53.

Keeling, David J. "Waterfront Development and the Puerto Madeo Project in Buenos Aires, Argentina." In *Cities and Urban Geography in Latin America*, edited by Vincent O. Chabrera et al., 113–36. Castellan de la Plana, Spain: Publicaciones Universitat Uaume I. Serie Coleccion Americas, 2005.

Lombardo, Juan, and Mercedes Di Virgilio. "Vinculas y relaciones entre espacio urbano y reproduccion social en el contexto de economia emergentes: El Caso de cinco municipios de la region Metropolitan de Buenos Aires." *Revista Geografica* 136 (2004): 49–83.

Poppe, Nicolas. "Siteseeing Buenos Aires in the Early Argentine Sound Film Los Tres Berne-
tines." *Journal of Cultural Geography* 26, no. 1 (2009): 49–69.
Tecco, Claudio A. "A cerca de las desafios socio-economicas y ambientales en la genstion
urbana moderna: El case del municipio de la Ciudad de Cordoba, Argentina." *Revista
Geografica* 117 (1993): 75–92.

Morphology and Neighborhoods

Aguirre, P. "How the Very Poor Survive: The Impact of Hyper-inflationary Crises on Low-
Income Households in Buenos Aires, Argentina." *GeoJournal* 34, no. 3 (1994): 295–304.
Alberto, Jorge A., and Juan A. Alberto. "Procesos de occupacion formal e informal del suelo
con fines urbanos del area Metropolitana del Gran Resistencia (AMRR), Republica Argenti-
na." *Revista Geografica* 142 (2007): 7–37.
Annessi, Gustavo J. "Espacio rural, turismo y desarrollo local en Tandil, (Buenos Aires, Argen-
tina)." *Revista Geografica* 133 (2003): 27–52.
Barksy, Andres. "Problems of Access to Land for Bolivian Horticultural Producers in the
Transitional Zone of Western Greater Buenos Aires." *Journal of Latin American Geography*
5, no. 2 (2006): 127–32.
Bockelman, Brian. "Along the Waterfront: Alejandro Malaspina, Fernando Brambila, and the
Invention of the Buenos Aires Cityscape, 1789–1809." *Journal of Latin American Geogra-
phy*, 11 (Special Issue) (2012): 61–88.
Braumann, Vernonika, and Christoph Stadel. "Boom Town in Transition: Development Process and Urban Structure in Ushauaia, Tierra del Fuego, Argentina." *Yearbook, Conference
of Latin Americanist Geographers* 25 (1999): 33–44.
Ford, Larry. "The Highrise in City Structure and Urban Image: A Cross-Cultural Comparison
between Argentina and the United States." *Geographical Survey* 1, no. 2 (1972): 1–23.
Formiga, Nidia. "El deredeo a la ciudad y la cuestion del espacio publico: Experiencias en la
ciudad de Bahier Blanca." *Journal of Latin American Geography* 6, no. 1 (2007): 173–96.
Gaffney, Christopher T. "Stadiums and Society in Twenty-First Century Buenos Aireas." *Soc-
cer and Society* 10, no. 2 (2009): 160–82.
Gilbert, Alan. "Moving the Capital of Argentina: A Further Example of Utopian Planning."
Cities 6 (1989): 234–42.
Grace, Jacqueline. "El Patrimonio Urbano-Arquitectonico Como Factor de Desarrollo para la
Ciudad de San Miguel de Tucuman, Argentina." *Revista Geografica* 132 (2002): 61–78.
Gray de Cerdon, Nelly A. "Internal Structure of the City of San Juan, Republic of Argentina."
Geographical Survey 2, no. 1 (1973): 41–61.
Howell, David. "A Model of Argentine City Structure." *Revista Geografica* 109 (1989):
129–40.
Keeling, David J. "Neoliberal Reform and Landscape Change in Buenos Aires." *Yearbook,
Conference of Latin Americanist Geographers* 25 (1999): 15–32.
Monti, Alejandro, and Ana Maria Escofet. "Ocupacion urbana de espacios litorales: Gestion del
riesgo e iniciativas de manejo en une comunidad patagonica automotivada (Playa Magagna,
Chubut, Argentina)." *Investigaciones Geograficas* 67 (2008): 113–29.
Sargent, Charles S. "Elements of Urban Platt Development: Great Buenos Aires, Argentina."
Revista Geografica 74 (1971): 7–32.
———. *The Spatial Evolution of Greater Buenos Aires, Argentina, 1870–1930*. Tempe: Arizo-
na State University, 1974.
Torres, H. A. "La Aglomeracion de Buenos Aires: Centralidady Suburbanizacion
(1940–1990)." *Estudios Geograficos* 54 (2011): 301–22.

Urban Economic Geography

Balino, Luciana, and Ana C. Ulberich. "La actividad minera en le entarno urbano: Estudio de
caso Cerro Leonas, Tandil, Argentina." *Revista Geografica* 140 (2006): 151–63.

Capron, G. "La centralite commerciale dans une municipalite peripherique de l'aire metropolitaine de Buenos Aires (Pilar). Un role de recomposition." *Bulletin de l'Association de Geographes Francais* 78, no. 4 (2001): 350–62.

Johns, Michael. "The Urbanization of Peripheral Capitalism: Buenos Aires, 1880–1920." *International Journal of Urban and Regional Research* 16, no. 2 (1992): 352–74.

Luduena, M. A. "Region metropolitan Buenos Aires: Estructuracion, problematica, y aspect de cambio." In *Geografia Economica Argentina*, edited by Juan A. Roccatagliata, 284–330. Buenos Aires, Argentina: El Atereo, 1993.

Manzonal, Mabel. "Argentina's Urban System and the Economic Crisis." *Cities* 3, no. 3 (1988): 260–67.

Whitson, Risa. "Beyond the Crisis: Economic Globalization and Informal Work in Urban Argentina." *Journal of Latin American Geography* 6, no. 2 (2007): 121–36.

Wrigley, Gladys M. "Salta, an Early Commercial Center of Argentina." *Geographical Review* 2 (1916): 116–33.

Zamorano, Mariano. "The City of San Luis, Argentina: Administrative Center and Hub of Communications." *Geographical Survey* 2, no. 4 (1973): 204–20.

Urban Environments

Alfredo, Alberto. "Urban Occupations at Fluviolacustrine Environments: Related Environmental Problems. A Case Study: The Area Metropolitana de Gran Resistencia." *Revista Geografica* 138 (2005): 109–27.

Bolton, Sarah, and Bernardo de Gouvello. "Water and Sanitation in the Buenos Aires Metropolitan Region: Fragmented Markets, Splintered Effects?" *Geoforum* 39, no. 6 (2008): 1859–70.

Colemin, Juan P., and Guillermo A. Velazquez. "Proposal and Application of an Environmental Quality Index for the Metropolitan Area of Buenos Aires, Argentina." *Geografisk Tiddskrift* 112, no. 1 (2012): 15–26.

Geraldi, A. M., et al. "Anomalies de precipitacion y temperature en las Encadenados del Oeste. Buenos Aires, Argentina." *Revista Geografica* 148 (2010): 137–50.

Verón, Eleonora Marta. "Estimacionde la isla de calar en Santa Teresita, Partido de la costa, Provicia de Buenos Aires, Argentina." *Revista Geografica de America Central* 45 (2010): 129–48.

Theses and Dissertations

Johns, Michael. "The Production of Primacy in Latin America: An Enquiry into the Urbanization of Argentina, 1870–1930." PhD diss., Johns Hopkins University, Baltimore, 1989.

Loftus, Alexander. "The Wet Dream of Capital and Its Class: A Study of Private Sector Water Provision, Uneven Development, and Nature in Buenos Aires." Master's thesis, Queen's University, Kingston, Ontario, 2001.

Miller, Jacob. "Malls without Stores: Political Effects in Spaces of Consumption, Buenos Aires, Argentina." Master's thesis, University of Arizona, Tucson, 2011.

Sargent, Charles S. "Urban Dynamics and the Changing Pattern of Residential Development: Buenos Aires, 1870–1930." PhD diss., University of California, Berkeley, 1971.

Steurer, Erin. "A Private Commodity or Public Good? A Comparative Case Study of Water and Sanitation Privatization in Buenos Aires, Argentina, 1993–2006." Master's thesis, University of South Florida, Tampa, 2008.

Whitson, Risa. "Reworking Place, Gender, and Power: Informal Work in Urban Argentina." PhD diss., Pennsylvania State University, University Park, 2004.

Chapter Three

Bolivia

GENERAL WORKS

Atlases and Graphic Presentations

Keller, Frank L. "Aerial Mapping in the Santa Cruz Region of Bolivia." *Journal of Geography* 52, no. 7 (1953): 299–304.

Books, Monographs, and Texts

Fifer, J. Valerie. *Bolivia*. Santa Barbara, CA: ABC-CLIO Press, 2000.
———. *Bolivia: Land, Location, and Politics since 1825*. New York: Cambridge University Press, 1972.
Schmeider, Oscar. *The East Bolivian Andes South of the Rio Grande at Guapay*. Berkeley: University of California, Publications in Geography 2, no. 5, 1926.

Articles and Book Chapters

Bingham, Hiram. "Potosi." *Bulletin of the American Geographical Society* 43, no. 11 (1911): 1–13.
Crist, Raymond E. "Bolivia-Land of Contrast." *American Journal of Economics and Sociology* 5, no. 3 (1946): 297–325.
Mather, K. F. "Along the Andean Front in Southeastern Bolivia." *Geographical Review* 12 (1922): 358–74.
———. "Eastern Bolivia: A Land of Opportunity." *Bulletin of the Geographical Society of Philadelphia* 18, no. 3 (1920): 49–55.
———. "Explorations in the Land of the Yuracares, Eastern Bolivia." *Geographical Review* 12 (1922): 42–56.
Miller, Leo E. "Across the Bolivian Highlands from Cochabamba to the Chapone." *Geographical Review* 4 (1917): 267–83.
Tata, Robert J. "Bolivia." In *Focus on South America*, edited by Alice Taylor, 185–204. New York: Praeger, 1973.
Walcott, Frederic C. "An Expedition to the Laguna Colorada, Southern Bolivia." *Geographical Review* 15 (1925): 345–66.

CULTURAL AND SOCIAL GEOGRAPHY

General Works

Bebbington, Anthony J. "Sustaining the Andes? Social Capital and Policies for Rural Regeneration in Bolivia." *Mountain Research and Development* 18, no. 2 (1998): 173–81.
Preston, David A. "Restructuring Bolivian Rurality: Batalla in the 1990s." *Journal of Rural Studies* 8, no. 3 (1992): 323–33.

The Built Environment

Hiraoka, Mario. "Structural Variations among Dwellings in the Japanese Colony of San Juan de Yapacani, Bolivia." *Yearbook, Association of Pacific Coast Geographers* 34 (1972): 137–52.

Medical Geography

Perry, B., and W. Gesler. "Physical Access to Primary Health Care in Andean Bolivia." *Social Science and Medicine* 50, no. 9 (2000): 1177–88.
Weil, Connie. "The Diffusion of Oral Rehydration Therapy and Obstacles to Its Acceptance in Cochabamba, Bolivia." *Papers of the Applied Geography Conferences* 10 (1987): 181–88.
———. "Morbidity, Mortality, and Diet as Indicators of Physical and Economic Adaptation among Bolivian Migrants." *Social Science and Medicine* 13D, no. 4 (1979): 215–22.
Wrigley, Gladys M. "The Traveling Doctors of the Andes: The Caulahuayas of Bolivia." *Geographical Review* 4 (1917): 183–96.

Ethnic, Social, and Population Geography

Bee, Beth. "Gender, Solidarity, and the Paradox of Microfinance: Reflections from Bolivia." *Gender, Place, and Culture* 18, no. 1 (2011): 23–43.
Bowman, Isaiah. "The Distribution of Population in Bolivia." *Bulletin of the Geographical Society of Philadelphia* 7 (1909): 28, 46, 74–93.
———. "The Highland Dwellers of Bolivia." *Bulletin of the Geographical Society of Philadelphia* 7, no. 4 (1909): 1–26, 159–184.
———. "The Valley People of Eastern Bolivia." *Journal of Geography* 11 (1912): 114–19.
Denevan, William M. *The Aboriginal Cultural Geography of the Llanos de Mojos of Bolivia.* Berkeley: University of California Press, Ibero-Americana 48, 1966.
Jones, Richard C. "The Local Economic Imprint of Return Migrants in Bolivia." *Population, Space, and Place* 17, no. 5 (2011): 435–53.
Jones, Richard C., and Leonardo de la Torre. "Diminished Tradition of Return? Transnational Migration in Bolivia's Valle Alto." *Global Networks* 11, no. 2 (2011): 180–202.
———. "Endurance of Transnationalism in Bolivia's Valle Alto." *Migration Letters* 6, no. 1 (2009): 63–74.
Miller, Leo E. "The Yuracare Indians of Eastern Bolivia." *Geographical Review* 4 (1917): 450–64.
O'Hare, Greg, and Sara Rivas. "Changing Poverty Distribution in Bolivia: The Role and Rural-Urban Migration and Urban Services." *GeoJournal* 68, no. 4 (2007): 307–26.
Stearman, Allyn M. *Camba and Kolla: Migration and Development in Santa Cruz, Bolivia.* Orlando: University of Central Florida Press, 1987.
Thomas, Robert N., and Robert Wittick. "Longitudinal Analysis of International Migration to La Paz, Bolivia." *Revista Geografica* 99 (1984): 57–67.
Weil, Connie. "Migration among Landholdings by Bolivian Campesinos." *Geographical Review* 73, no. 2 (1983): 182–97.

————. "Survey Research among Third World Mothers: Bolivian Puzzle Pieces." *Papers of the Applied Geography Conferences* 11 (1988): 56–60.

Yarnall, Kaitlin, and Marie D. Price. "Migration, Development, and a New Rurality in the Valle Alto, Bolivia." *Journal of Latin American Geography* 9, no. 1 (2010): 107–24.

Community and Settlement Studies

Eastwood, David A., and H. J. Pollard. "Colonization and Coca in the Chapare, Bolivia: A Development Paradox for Colonization Theory." *Tijdschrift voor Economische en Sociale Geographie* 77, no. 4 (1986): 258–68.

————. "The Development of Bolivia's Rurrenbaque Land Colonization Scheme: The Colonists' Perception." *Malaysian Journal of Tropical Geography* 15 (1987): 13–25.

————. "Rurrenabaque-Secure: The State of the Art in Bolivian Land Settlement Planning." *Revista Geografica* 101 (1985): 141–52.

Fifer, J. Valerie. "Bolivia's Pioneer Fringe." *Geographical Review* 57, no. 1 (1967): 1–23.

————. "The Search for a Series of Small Successes: Frontiers of Settlement in Eastern Bolivia." *Journal of Latin American Studies* 14, no. 2 (1982): 407–32.

Hiraoka, Mario. "Colonization and Economic Development in the Bolivian Lowlands [in Japanese]." *Jimbun Chiri/Human Geography* 23, no. 5 (1971): 88–104.

————. *Japanese Agricultural Settlement in the Bolivian Upper Amazon: A Study in Regional Ecology.* Ibaraki, Japan: University of Tsukuba, Sakuramura, Latin American Studies 1, Special Research Project on Latin America, 1980.

————. "Settlement and Development of the Upper Amazon: The East Bolivia Example." *Journal of Developing Areas* 14, no. 3 (1980): 327–48.

Preston, David A. "Freeholding Communities and Rural Development: The Case of Bolivia." *Revista Geografica* 73 (1972): 29–42.

————. "New Towns: A Major Change in the Rural Settlement Pattern in Highland Bolivia." *Journal of Latin American Studies* 2, no. 1 (1970): 1–27.

Washington-Allen, R. A., et al. "Change Detection of the Effect of Severe Drought on Subsistence Agropastoral Communities on the Bolivian Altiplano." *International Journal of Remote Sensing* 19, no. 7 (1998): 1319–34.

Works, Martha A. "An Evaluation of Planned Settlement in the Bolivian Tropics: The 1974 Plans." *Geographical Bulletin* 19, no. 2 (1980): 7–13.

Theses and Dissertations

Anthamatten, Peter. "Childhood Malnutrition in Bolivia: An Examination of Associations with Individual, Household, and Contextual Variables." PhD diss., University of Minnesota, Minneapolis, 2007.

Denevan, William M. "The Aboriginal Geography of the Llanos de Mojos: A Seasonally Inundated Savanna in Northeastern Bolivia." PhD diss., University of California, Berkeley, 1963.

Hiraoka, Mario. "Pioneer Settlement in Eastern Bolivia." PhD diss., University of Wisconsin, Madison, 1974.

Keller, Frank L. "Geography of the Lake Titicaca Basin of Bolivia: A Comparative Study of Great Landed Estates and Highland Indian Communities." PhD diss., University of Maryland, College Park, 1949.

Weil, Connie, "The Adaptiveness of Tropical Settlement in the Chapare of Bolivia." PhD diss., Columbia University, New York, 1980.

ECONOMIC GEOGRAPHY

General Works

Conway-Gomez, Kristen. "Market Integration, Perceived Wealth, and Household Consumption of River Turtles (*Podocnemis* spp.) in Eastern Lowland Bolivia." *Journal of Latin American Geography* 7, no. 1 (2008): 85–108.

Crossley, J. Colin. "Santa Cruz at the Crossroads: A Study of Development in Eastern Bolivia: 1–2." *Tijdschrift voor Economische en Sociale Geographie* 52, no. 8 (1961): 197–206; 52, no. 9 (1961): 230–40.

Gonzales Tapia, Ismael. "De l'enclavement a la globalization: Une Ouverture risqué pour la Bolivia." *Les Cahiers d'Outre Mer* 53 (2000): 317–42.

Keller, Frank L. "Institutional Barriers to Economic Development: Some Examples from Bolivia." *Economic Geography* 31, no. 4 (1955): 351–63.

Schroeder, Kathleen. "Economic Globalization and Bolivia's Regional Divide." *Journal of Latin American Geography* 6, no. 2 (2007): 99–120.

South, Robert B. "The Economic Organization of Bolivia: An Analysis of Commodity Flows." *Southeastern Geographer* 16, no. 1 (1976): 9–25.

Stadel, Christoph. "Development Needs and Mobilization of Rural Resources in Highland Bolivia." *Yearbook, Conference of Latin Americanist Geographers* 21 (1995): 37–48.

Stouse, Pierre. "Regional Specialization in Developing Areas: The Altiplano of Bolivia." *Revista Geografica* 74 (1971): 51–70.

Zimmerer, Karl S. "Soil Erosion and Labor Shortages in the Andes with Special Reference to Highland Bolivia, 1950–1991: Implications for Conservation-With-Development." *World Development* 21 (1993): 1659–75.

Agriculture and Land Use

Clark, R. J. "Land-Holding Structure and Land Conflicts in Bolivia's Lowland Cattle Regions." *Inter-American Economic Affairs* 28, no. 2 (1974): 15–38.

Denevan, William M. "Cattle Ranching in the Mojos Savannas of Northeastern Bolivia." *Yearbook, Association of Pacific Coast Geographers* 25 (1963): 37–44.

Kessler, C. A. "Decisive Key-Factors Influencing Farm Households' Soil and Water Conservation Investments." *Applied Geography* 26, no. 1 (2006): 40–60.

Laurie, Nina, and S. Marvin. "Globalization, Neoliberalism, and Negotiated Development in the Andes: Water Projects and Regional Identity in Cochabamba, Bolivia." *Environment and Planning A* 31, no. 12 (1999): 1401–16.

Pacheco, Pablo. "Agricultural Expansion and Deforestation in Lowland Bolivia: The Import Substitution versus the Structural Adjustment Model." *Land Use Policy* 23, no. 3 (2006): 205–25.

Preston, David A. *Agriculture in a Highland Desert: The Central Altiplano of Bolivia.* Leeds, England: University of Leeds, Department of Geography, Paper 18, 1972.

———. "L'agriculture dans un desert d'altitude: L'altiplano central de Bolivie." *Les Cahiers d'Outre Mer* 26 (1973): 113–28.

———. "Land Tenure and Agricultural Development in the Central Altiplano, Bolivia." In *Spatial Aspects of Development*, edited by Brian S. Hoyle, 231–52. London: Wiley, 1974.

———. "Post-peasant Capitalist Graziers: The Twenty First Century in Southern Bolivia." *Mountain Research and Development* 18, no. 2 (1998): 151–58.

Preston, David A., et al. "Fewer People, Less Erosion: The Twentieth Century in Southern Bolivia." *Geographical Journal* 163, no. 2 (1997): 198–205.

———. "Grazing and Environmental Change on the Tarija Altiplano, Bolivia." *Mountain Research and Development* 23, no. 2 (2003): 141–48.

Redo, Daniel J., et al. "The Relative Importance of Socioeconomic and Environmental Variables in Explaining Land Change in Bolivia, 2001–2010." *Annals of the Association of American Geographers* 102, no. 4 (2012): 778–807.

Romecin, Eduardo. "Agricultural Adaptation in Bolivia." *Geographical Review* 19 (1929): 248–55.

South, Robert B. "Coca in Bolivia." *Geographical Review* 67, no. 1 (1977): 22–33.

Stouse, Pierre. "The Distribution of Llamas in Bolivia." *Proceedings, Association of American Geographers* 2 (1970): 136–40.

Vellard, J. "L'experience agraire en Bolivie." *Les Cahiers d'Outre Mer* 16 (1963): 201–13.

Weeks, David. "Bolivia's Agricultural Frontier." *Geographical Review* 36, no. 4 (1946): 546–67.

Zimmerer, Karl S. "Woodlands and Agrobiodiversity in Irrigation Landscapes Amidst Global Changes: Bolivia, 1990–2002." *Professional Geographer* 62, no. 3 (2010): 335–56.

Commerce and Trade

Bee, Beth. "Gender, Solidarity, and the Paradox of Microfinance: Reflections from Bolivia." *Gender, Place, and Culture* 18, no. 1 (2011): 23–43.

Bowman, Isaiah. "Trade Routes in the Economic Geography of Bolivia." *Bulletin of the American Geographical Society* 42, nos. 1–3 (1910): 22–37, 90–104, 180–192.

Manufacturing and Labor Studies

Sage, Colin. "Small-Scale Brewing and Rural Livelihoods: The Case of Chicha in Bolivia." In *Rural Enterprise: Shifting Perspectives on Small-Scale Production*, edited by Sarah Whatmore et al., 58–77. London: David Fulton, 1991.

Schroeder, Kathleen. "Spatial Constraints on Women's Work in Tarija, Bolivia." *Geographical Review* 90, no. 2 (2000): 191–205.

Energy, Mining, Fishing, and Lumbering

Arnous de Riviere, H. "Explorations in the Rubber Districts of Bolivia." *Journal of the American Geographical Society of New York* 32, no. 5 (1900): 432–40.

Auty, R. M. "Bolivia: Accelerating Weakness despite Positive External Stocks." In *Sustaining Development in Mineral Economies: The Resource Curse Thesis*, edited by R. M. Auty, 73–90. London: Routledge, 1993.

———. "The Resource Curse Thesis: Minerals in Bolivian Development, 1970–90." *Singapore Journal of Tropical Geography* 15, no. 2 (1994): 95–111.

Craig, Alan K. "Mining Ordenanzas and Silver Production at Potosi: The Toledo Reforms." In *Precious Metals, Coinage and the Changes in Monetary Structures in Latin America, Europe, and Asia*, edited by Eddy Van Cauwenberghe, 159–83. Brussels, Belgium: Leuven University Press, 1989.

Hindery, Derrick. "Social and Environmental Impacts of World Bank/IMF-Funded Economic Restructuring in Bolivia: An Analysis of Enron and Shell's Hydrocarbon Projects." *Singapore Journal of Tropical Geography* 25, no. 3 (2004): 281–303.

Marsik, Matthew, et al. "Amazon Deforestation: Rates and Patterns of Land Cover Change, Northern Bolivia." *Progress in Physical Geography* 35, no. 3 (2011): 353–74.

Milstead, Harley P. "Bolivia as a Source of Tin." *Economic Geography* 2 (1927): 354–60.

Murphy, James T., and Seth Schindler. "Globalizing Development in Bolivia? Alternative Networks and Value-Capture Challenges in the Wood Products Industry." *Journal of Economic Geography* 11, no. 1 (2011): 61–85.

Transportation and Communications

Bowman, Isaiah. "Geographical Aspects of the New Madeira-Mamore Railroad." *Bulletin of the American Geographical Society* 45 (1913): 275–81.

West, Max J. "Transportation Networks as Determinants of Effective National Territory: The Case of Bolivia." *Ohio Geographer* 10 (1982): 31–42.

Theses and Dissertations

Bee, Beth. "Financing Solidarity, Peddling Paradox: The Possibilities and Perils of a Feminist Geographic Approach to Microfinance in Bolivia." Master's thesis, Pennsylvania State University, University Park, 2006.
Conway, K. "Human Use of River Turtles (*Podochemis* sp.) in Lowland Eastern Bolivia." PhD diss., University of Florida, Gainesville, 2004.
Griess, Phyllis R. "The Tin Industry of Bolivia: A Geographical Interpretation." PhD diss., Pennsylvania State University, University Park, 1948.
Henkel, Ray. "The Chapone of Bolivia: A Study of Tropical Agriculture in Transition." PhD diss., University of Wisconsin, Madison, 1971.
Jenkins, Abigail. "Landscape Change in San Christobal, Bolivia: Quinoa, Mining, Water, and Migration." Master's thesis, California State University, Chico, 2011.
Redo, Daniel J. "Understanding and Mapping Land-Use and Land-Cover Change along Bolivia's Cornedar Bioceanico." PhD diss., Texas A&M University, College Station, 2010.
Schiller, William E. "The Mining Industry of Bolivia's Highlands." Master's thesis, University of Oklahoma, Norman, 1966.
South, Robert B. "An Analysis of the Spatial Structure of the Bolivian Economy." PhD diss., University of Maryland, College Park, 1972.

HISTORICAL GEOGRAPHY

General Works

Craig, Alan K. "The Indigenous Ingenios: Spanish Colonial Watermills at Potosi." In *Culture, Form, and Place: Essays in Cultural and Historical Geography*, edited by Kent Mathewson, 125–56. Baton Rouge: Louisiana State University, Geoscience and Man 32, 1993.
Denevan, William M. "Additional Comments on the Earthworks of Mojos in Northeastern Bolivia." *American Antiquity* 28 (1963): 540–45.
———. "Pre-Spanish Earthworks in the Llanos de Mojos of Northeastern Bolivia." *Revista Geografica* 60 (1964): 17–26.
Evans, Brian M. "The Indian Population of Alta Peru, 1520–1780." *Bulletin of the Association of North Dakota Geographers* 36 (1986): 62–77.
———. "Migration Processes in Upper Peru in the Seventeenth Century." In *Migration in Colonial Spanish America*, edited by David J. Robinson, 62–85. Cambridge, England: Cambridge University Press, 1990.
———. "The Structure and Distribution of the Indian Population of Alta Peru in the Late Seventeenth Century." *Yearbook, Conference of Latin Americanist Geographers* 11 (1985): 33–40.
Grieshaber, Edwin P. "Survival of Indian Communities in Nineteenth-Century Bolivia: A Regional Comparison." *Journal of Latin American Studies* 12, no. 2 (1980): 223–69.
Keller, Frank L. "Finca Ingavi: A Medieval Survival on the Bolivian Altiplano." *Economic Geography* 26, no. 1 (1950): 37–50.
Klein, Herbert S. "The Impact of the Crisis in Nineteenth-Century Mining on Regional Economics: The Example of the Bolivian Yungas, 1786–1838." In *Social Fabric and Spatial Structure in Colonial Latin America*, edited by David J. Robinson, 315–38. Syracuse, NY: Syracuse University, Department of Geography, 1979.
Lombardo, U., et al. "Eco-archaeological Regions in the Bolivian Amazon. An Overview of Pre-Columbian Earthworks Linking Them to Their Environmental Settings." *Geografica Helvetica* 66, no. 3 (2011): 173–82.

Preston, David A. "The Revolutionary Landscape of Highland Bolivia." *Geographical Journal* 135, no. 1 (1969): 1–16.

PHYSICAL GEOGRAPHY

General Works

Gerold, Gerhard, et al. "Hydro-meteorologic, Pedologic, and Vegetation Patterns along an Elevational Transect in the Montane Forest of the Bolivian Yungas." *Die Erde* 139, nos. 1–2 (2008): 141–68.

Hardy, D., et al. "Near-Surface Faceted Crystals, Avalanches, and Climate in High-Elevation Tropical Mountains of Bolivia." *Cold Regions Science and Technology* 3 (2001): 291–302.

Pancel, Laslo, and Claus Wiebecke. "Problems in Evaluating Forestry Development Projects: A Case Study Based on the Rehabilitation of Eroded Soils in the Tarija Valley, Bolivia." *Applied Geography and Development* 22 (1983): 94–107.

Valdivia, Corinne, et al. "Adapting to Climate Change in Andean Ecosystems: Landscapes, Capitals, and Perceptions Shaping Rural Livelihood Strategies and Linking Knowledge Systems." *Annals of the Association of American Geographers* 100, no. 4 (2010): 818–34.

Biogeography

Langstoth Plotkin, R. "Biogeography of the Llanos de Moxos: Natural and Anthropogenic Determinants." *Geografica Helvetica* 66, no. 3 (2011): 183–92.

Lopez, R. P. "Phytogeographical Relations of the Andean Dry Valleys of Bolivia." *Journal of Biogeography* 30, no. 11 (2003): 1659–68.

Marsik, Matthew, et al. "Amazon Deforestation: Rates and Patterns of Land Cover Change and Fragmentation in Pando, Northern Bolivia, 1986 to 2005." *Progress in Physical Geography* 35, no. 3 (2011): 353–74.

Mizuno, Kazubaru. "Upper Limit of Plant Distribution in Response to Lithology and Rubble Size of Land Surfaces in Tropical High Mountains in Bolivia." *Geographical Reports of Tokyo Metropolitan University* 37 (2001): 67–74.

Pacheo, L. F., et al. "Conservation of Bolivian Flora: Representation of Phytogeographic Zones in the National System of Protected Areas." *Biodiversity and Conservation* 3 (1994): 751–56.

Reese, Carl A., et al. "Pollen Dispersal and Deposition on the Ice Cap of Volcan Parinacota, Southwestern Bolivia." *Arctic, Antarctic, and Alpine Research* 35, no. 4 (2003): 469–74.

Resse, Carl A., and Kam Bui Liu. "Interannual Variability in Pollen Dispersal and Deposition on the Tropical Quelreaya Ice Cap." *Professional Geographer* 57, no. 2 (2005): 185–97.

Steininger, Marc K., et al. "Tropical Deforestation in the Bolivian Amazon." *Environmental Conservation* 28 (2001): 127–34.

Climatology and Weather

McGlade, Michael, et al. "The Impact of Rainfall Frequency on Coca (*Etythrozylum coca*) Production in the Chapore Region of Bolivia." *Yearbook, Conference of Latin Americanist Geographers* 20 (1994): 97–106.

Thompson, L. G., et al. "A 25,000 Year Tropical Climate History from Bolivian Ice Cores." *Science* 282 (1998): 1858–64.

Vuille, M. "Atmospheric Circulation over the Bolivian Altiplano during Dry and Wet Periods and Extreme Phases of the Southern Oscillation." *International Journal of Climatology* 19, no. 14 (1999): 1579–1600.

Geomorphology, Landforms, and Volcanism

Coppus, R., et al. "Identification, Distribution, and Characteristics of Erosion Sensitive Areas in Three Different Central Andean Ecosystems." *Catena* 51, nos. 3–4 (2003): 315–28.
Coppus, R., and A. C. Imeson. "Extreme Events Controlling Erosion and Sediment Transport in a Semi-arid and Sub-Andean Valley." *Earth Surface Processes and Landforms* 27, no. 13 (2002): 1365–76.
Gautier, E., et al. "Temporal Relations, Meander Deformation, Water Discharge, and Sediment Fluxes in the Floodplain of the Rio Besi (Bolivian Amazonia)." *Earth Surface Processes and Landforms* 32, no. 2 (2007): 230–48.
May, J. H. "The Rio Parapeti-Holocene Megafan Formation in the Southernmost Amazon Basin." *Geographica Helvetica* 66, no. 3 (2011): 193–201.
Plotzki, A., et al. "Review of Past and Recent Fluvial Dynamics in the Beni Lowlands, Northeast Bolivia." *Geographica Helvetica* 66, no. 3 (2011): 164–72.

Hydrology and Glaciology

Abbott, M. B., et al. "Holocene Hydrological Reconstruction from Stable Isotopes and Paleolimnology, Cordillera Real, Bolivia." *Quaternary Science Review* 19 (2000): 1801–20.
Martin, Shelby. "Glacial Lakes in the Bolivian Andes." *Geographical Journal* 131, no. 4 (1965): 519–25.

Soils

Barber, R. G. "Soil Degradation in the Tropical Lowlands of Santa Cruz, Eastern Bolivia." *Land Degradation and Rehabilitation* 6, no. 2 (1995): 95–109.
Barber, R. G., and F. Navarro. "The Rehabilitation of Degraded Soils in Eastern Bolivia by Subsoiling and the Incorporation of Cover Crops." *Land Degradation and Rehabilitation* 5, no. 4 (1994): 247–60.
Zimmerer, Karl S. "Discourses on Soil Loss in Bolivia." In *Liberation Ecologies, Environment, Development, and Social Movements*, edited by Richard Peet and Michael Watts, 110–24. London: Routledge, 1996.
———. "Local Soil Knowledge: Answering Basic Questions in Highland Bolivia." *Journal of Soil and Water Conservation* 49, no. 1 (1994): 29–34.
———. "Soil Erosion and Labor Shortages in the Andes, with Special Reference to Bolivia, 1953–91." *World Development* 21, no. 10 (1993): 1659–75.
———. "Soil Erosion and Social (Dis)Course in Cochabamba, Bolivia: Perusing the Nature of Environmental Degradation." *Economic Geography* 69, no. 3 (1993): 312–27.

Theses and Dissertations

Covarrubias-Rocha, Jose H. "Multi-temporal Remote Sensing Evaluations of Vegetal Coverage in the Bolivian Andean Plate." Master's thesis, State University of New York, Buffalo, 2007.
Fariman, Jonathan. "Investigating Paleoclimatic Conditions in the Tropical Andes Using a 2D Model of Glacial Mass Balance and Ice Flow." Master's thesis, Ohio State University, Columbus, 2006.
Glassner, Martin I. "Bolivia and an Access to the Sea." Master's thesis, California State University, Fullerton, 1965.
Langstoth, Robert P. "Forest Islands in an Amazonian Savanna of Northeastern Bolivia." PhD diss., University of Wisconsin, Madison, 1996.
Zermoglio, F. "An Empirical Analysis of the Relationship between Biodiversity and Landscape Pattern in the Savanna Mosaics of Parque Nacional, Noel Kempff, Mercado, Bolivia." Master's thesis, University of Florida, Gainesville, 2000.

POLITICAL GEOGRAPHY

General Works

Andolina, Robert, et al. "Development and Culture: Transnational Identity Making in Bolivia." *Political Geography* 24, no. 6 (2005): 678–702.

Arbona, Juan M. "Neo-liberal Ruptures: Local Political Entities and Neighborhood Networks in El Alto, Bolivia." *Geoforum* 38, no. 1 (2007): 127–37.

Bottazzi, Patrick, and Hy Dao. "On the Road through the Bolivian Amazon: A Multi-level Land Governance Analysis of Deforestation." *Land Use Policy* 30, no. 1 (2013): 137–46.

Brusle, Lactitia P. "The Front and the Line: The Paradox of South American Frontiers Applied to the Bolivian Case." *Geopolitics* 12, no. 1 (2007): 57–77.

Eastwood, David A., and H. J. Pollard. "Rurrenabaque-Secure: The 'State of the Art' in Bolivian Land Settlement Planning." *Revista Geografica* 101 (1985): 141–52.

Gade, Daniel. "Spatial Displacement of Latin American Seats of Governance: From Sucre to La Paz as the National Capital of Bolivia." *Revista Geografica* 73 (1970): 43–57.

Glassner, Martin I. "Bolivia's Orientation: Towards the Atlantic or the Pacific?" In *Geopolitics of the Southern Cone and Antarctica*, edited by Philip Kelly and Jack Child, 154–72. Boulder, CO: Lynne Rienner, 1988.

Haarsted, H. O. "FDI Policy and Political Spaces for Labour: The Disarticulation of the Bolivian Petroleros." *Geoforum* 40, no. 2 (2009): 239–48.

———. "Maneuvering the Spaces of Globalization: The Rearticulation of the Bolivian Labor Movement." *Norsk Geografisk Tidsskrift* 64, no. 1 (2010): 9–20.

Hagevik, Cheryl, and Christopher Bodurek. "Visualizing the Social and Political Divide of Bolivia with Multivariate Maps." *Pennsylvania Geographer* 49, no. 1 (2011): 3–17.

Henkel, Ray. "Government Policy Regarding the Disposal and Utilization of the Public Domain in Eastern Bolivia." *Papers of the Applied Geography Conferences* 7 (1984): 45–54.

Hinojosa, Leonith, and Karl Hennermann. "A GIS Approach to Ecosystem Services and Rural Territorial Dynamics Applied to the Case of The Bolivia Gas Industry." *Applied Geography* 34 (2012): 487–97.

Humphreys Begginton, Denise. "Consultation, Compensation, and Conflict: Natural Gas Extraction in Weenhayak Territory, Bolivia." *Journal of Latin American Geography* 11, no. 2 (2012): 49–71.

Kaup, Brent Z. "Negotiating through Nature: The Resistant Materiality and Materiality of Resistance in Bolivia's Natural Gas Sector." *Geoforum* 39, no. 5 (2008): 1734–42.

Kohl, Benjamin. "Restructuring Citizenship in Bolivia: El Plan de Todos." *International Journal of Urban and Regional Research* 27, no. 2 (2003): 337–50.

Kohl, Benjamin, and Linda Farthing. "Material Constraints to Popular Imaginaries: The Extractive Economy and Resource Nationalism in Bolivia." *Political Geography* 31, no. 4 (2012): 225–35.

Laurie, Nina, and Carlos Crespo. "Deconstructing the Best Case Scenario: Lessons from Water Politics in La Paz-El Alto, Bolivia." *Geoforum* 38, no. 5 (2007): 841–54.

Morree Dirkje, Johanna de. *Cooperacion campesina en los Andes. Un studio sobre estrategias de organizacion para el desarrollo rural en Bolivia*. Utrecht, Netherlands: Netherlands Geographical Studies 298, 2002.

Nijenhuis, G. *Decentralization and Popular Participation in Bolivia: The Link between Local Governance and Local Development*. Utrecht, Netherlands: Netherlands Geographical Studies, Band 299, 2002.

Pellegrini, Lorenzo, and Marco O. Ribera Arismendi. "Consulation, Compensation, and Extraction in Bolivia after the Left Turn: The Case of Oil Exploration in the North of La Paz Department." *Journal of Latin American Geography* 11, no. 2 (2012): 103–20.

Perrault, Thomas A. "Conflictos del gas y se gobernanza: El caso de los Guarani de Tarija, Bolivia." *Anthropologia*, 28, suppl. 1 (2010): 139–62.

————. "Custom and Contradiction: Rural Water Governance and the Politics of Usos y Costumbres in Bolivia's Irrigators' Movement." *Annals of the Association of American Geographers* 98, no. 4 (2008): 834–54.

————. "From the Guerra de Agua to the Guerra de Gas: Resource Governance, Neoliberalism, and Popular Protest in Bolivia." *Antipode* 38, no. 1 (2006): 150–72.

————. "Popular Protest and Unpopular Policies: State Restructuring, Resource Conflict, and Social Justice in Bolivia." In *Environmental Justice in Latin America*, edited by David Carruthers, 239–62. Cambridge, MA: MIT Press, 2008.

————. "State Restructuring and the Scale Politics of Rural Water Governance in Bolivia." *Environment and Planning A* 37, no. 2 (2005): 263–84.

Perrier-Brush, L., and J. C. Roux. "Natural Gas and Geopolitical Issues in Bolivia: Globalization and Lose Sovereignty [in French]." *Annales de Geographie* 112, no. 2 (2003): 167–87.

Radhuber, Isabella. "Indigenous Struggles for a Plurinational State: An Analysis of Indigenous Rights and Competences in Bolivia." *Journal of Latin American Geography* 11, no. 2 (2012): 167–93.

Redo, Daniel J., et al. "Deforestation Dynamics and Policy Changes in Bolivia's Post-neoliberal Era." *Land Use Policy* 28, no. 1 (2011): 227–41.

Reyes-Garcia, Victoria, et al. "Does Participatory Mapping Increase Conflicts? A Randomized Evaluation in the Bolivian Amazon." *Applied Geography* 34 (2012): 650–58.

Stadel, Christoph. "The Mobilization of Human Resources by Non-governmental Organizations in the Bolivian Andes." *Mountain Research and Development* 17, no. 3 (1997): 213–28.

————. "The Role of NGOs for the Promotion of Children in the Highlands of Bolivia." In *Worlds of Pain and Hunger: Geographical Perspectives on Vulnerability and Food Security*, edited by H. G. Bohle, 187–208. Saarbrucken, Germany: Breitenbach, 1993.

Valdivia, Gabriela. "Coca's Haunting Presence in the Agrarian Politics of the Bolivian Lowlands." *GeoJournal* 77, no. 5 (2012): 615–31.

West, Max J. "Transportation Networks as Determinants of Effective National Territory: The Case of Bolivia." *Ohio Geographer* 10 (1982): 31–42.

Theses and Dissertations

Cortina de Cardenas, Susana. "Does Private Management Lead to Improvement of Water Services? Lessons Learned from the Experiences of Bolivia and Puerto Rico." PhD diss., University of Iowa, Iowa City, 2011.

Green, Barbara. "Capitalism in a Poncho: The Articulation of Meaning in Struggles over Gas in Bolivia." Master's thesis, Syracuse University, Syracuse, NY, 2010.

URBAN GEOGRAPHY

General Works

Franqueville, Andre. "Villes et reseau urbain de Bolivie." *Les Cahiers d'Outre Mer* 43 (1998): 273–88.

Gade, Daniel. "Sucre, Bolivia, and the Quality of Place." *Journal of Latin American Geography* 9, no. 2 (2010): 99–117.

Kranenburg, Ronald H. *Buurtconsolidatie en urbane transformative in El Alto. Een longitudinal onderzoek naar veronderingsprocessen in de voor malige periferie van La Paz, Bolivia.* Utrecht, Netherlands: Netherlands Geographical Studies 295, 2002.

Preston, David A. *Post-revolutionary Rural-Urban Interaction of the Bolivian Altiplano.* Leeds, England: University of Leeds, Department of Geography, Paper 18, 1972.

Urban Social Geography

Schroeder, Kathleen. "Urban Squatters as Agricultural Migrants: The Case of Tarija, Bolivia." *Southwestern Latin Americanist* 41, nos. 3–4 (1998): 33–43.
Thomas, Robert N., and Robert Wittick. "Migrant Flows to La Paz, Bolivia, as Related to the Internal Structure of the City: A Methodological Treatment." *Revista Geografica* 94 (1981): 41–51.
Van Lindert, Paul. "Moving Up or Staying Down? Migrant-Native Differential Mobility in La Paz." *Urban Studies* 28, no. 3 (1991): 433–64.

Morphology and Neighborhoods

Kent, Robert B. "The Municipal Development Institute and Local Institution Building: Recent Bolivian Experience." *International Review of Administrative Sciences* 49 (1983): 279–87.
Sambrook, Richard, and Robert N. Thomas. "A Locational Strategy for Family Planning Centers in La Paz, Bolivia." *Revista Geografica* 109 (1989): 73–86.

Urban Economic Geography

Coen, Stephanie E., et al. "Without Tiendras It's a Dead Neighborhood: The Socio-economic Importance of Small Trade Stores in Cochabamba, Bolivia." *Cities* 25, no. 6 (2008): 327–39.

Urban Environments

O'Hare, Greg, and Sara Rivas. "The Landslide Hazard and Human Vulnerability in La Paz City, Bolivia." *Geographical Journal* 171, no. 3 (2005): 239–58.

Theses and Dissertations

Coen, Stephanie E. "Local Small-Trader Sites and Neighborhood Social Organization in Cochabamba, Bolivia." Master's thesis, McGill University, Montreal, Quebec, 2007.

Chapter Four

Brazil

GENERAL WORKS

Atlases and Graphic Presentations

Almeida Prado, Fernanda, et al. "Fuzzy Logic and Neural Network Approaches for Land Cover Mapping of the Brazilian Territory." *Journal of Geography and Regional Planning* 4, no. 15 (2011): 741–50.

Alves da Silva Neves, Sandra M., et al. "Aplicacao de imagens do radar interferometrico (SRTM) na avaliacao da fragilidade da bacia do corrego cachoeirinha, nos municipios de caceres e parto estrela, Mato Grosso." *Revista Geografica Academica* 2, no. 2 (2008): 124–37.

Barbosa, Reinaldo, and Ciro Campos. "Detection and Geographical Distributions of Clearing Areas in the Savannas (Lavrada) of Roraima Using Google Earth Webtood." *Journal of Geography and Regional Planning* 4, no. 3 (2011): 122–36.

Barbosa do Nascimento, Barbara K., et al. "Avaliacao de tecnicas de classificacao de imagens SARR 99B para al mapeamento do des florestamento." *Revista Geografica Academica* 6, no. 2 (2012): 15–24.

Boari, Mukesh, and Venerands E. Amaro. "Natural and Eco-environmental Vulnerability Assessment through Multi-temporal Satellite Data Sets in Apodi Valley Region, Northeast Brazil." *Journal of Geography and Regional Planning* 4, no. 4 (2011): 216–30.

Campbell, J. B., and John O. Browder. "Field Data Collection for Remote Sensing Analysis: SPOT Data, Rondonia, Brazil." *International Journal of Remote Sensing* 16, no. 2 (1995): 333–50.

Campos Nogueira, Thiago, et al. "Compartimentacao morfologica com base em dados SRTM: Estudo de caso bacio do Rio Uberabinha, Uberlandia-Minas Gerais." *Revista Geografica Academica* 2, no. 2 (2008): 154–69.

Cardoso Medeiros, Levindo, et al. "Sistema de disponibilizacao de informacoes geograficos do estado de Goias na Internet." *Revista Geografica Academica* 1, no. 1 (2007): 37–43.

Corves, C., and C. l. Place. "Mapping the Reliability of Satellite-Derived Landcover Maps: An Example from the Central Brazilian Amazon Basin." *Bulletin of the Pan American Union* 46 (1942): 132–44.

DeMorrson Valeriano, Marico, and D. de F. Rossett. "Topodata: Brazilian Full Coverage Refinement of SRTM Data." *Applied Geography* 32, no. 2 (2012): 300–309.

Dubreuil, Vincent, et al. "Paysages et fronts pionniers amazoniens sous le regard des satellites: L'exemple du Mato Grosso." *L'Espace Geographique* 37, no. 1 (2008): 57–74.

Hugueney de Mattos, A. "A Cartografiano Brasil." *Revista Geografica* 7–9 (1943): 133–44.
Jefferson, Mark. "Pictures from Southern Brazil." *Geographical Review* 16, no. 4 (1926): 521–49.
Martins, Luciana. "Geographical Exploration and the Elusive Mapping of Amazonia." *Geographical Review* 102, no. 2 (2012): 225–44.
Martins Dias, Josimora, et al. "Aplicacao de imagens Ikonos II e TM/Landsat T-5 na elaboracao de uma base cartografica para a reserve de desenvolvimento sustentavel Mamiraua-Amazonas." *Revista Geografica Academica* 3, no. 2 (2009): 15–27.
Morato de Carvalho, Thiago. "Metodos de sensoriamento remote aplicadosa geomorfologia." *Revista Geografica Academica* 1, no. 1 (2007): 44–54.
Moreira Alves, Thais, and Thiago Morato de Carvalho. "Tecnicas de sensoriamento remote para a classificacao e quantificacao do sistema lacustre do Rio Araguaia entre barra do Garcas e foz do Rio Cirstalino." *Revista Geografica Academica* 1, no. 1 (2007): 79–94.
Nascentes Coelho, Andre L. "Uso de dados SRTM como ferramenta de apoio aomapeamento geomorfologico de bacia de medio Grande Porte." *Revista Geografica Academica* 2, no. 2 (2008): 138–53.
————. "Uso de productos de sensoriamento remote para delimitacao de area efetivamente inundavel: Estudio de caso do Bixo Curso do Rio Benevente Anchicta-Espirito Santo." *Revista Geografica Academica* 4, no. 2 (2010): 53–63.
Pacheco do Amaral Filho, Zebino. "Importania das imagens de radar no mapeamento de solos da Folha Sergipe-22 Goiana." *Revista Brasileira de Geografia* 59, no. 1 (2005): 61–74.
Silva, Aguinaldo, et al. "Sensoriamento remote aplicado ao estudo da erosao marginal da Rio Paraguai: Bairro San Miguel em caceres/Mato Grosso-Brasil." *Revista Geografica Academica* 2, no. 3 (2008): 19–27.
Silva, Daltan R. V., and Venerands E. Amaro. "Integracao entre dados opticos e radar (SRTM) para a caracterizacao geoambiental da costa setentrionel do Rio Grande do Norte." *Revista Geografica Academica* 2, no. 2 (2008): 111–23.
Zeilhofer, Peter, and Vladimir Topanotti. "GIS and Ordination Techniques for Evaluation of Environmental Impacts in Informal Settlements: A Land Study from Cuiaba, Central Brazil." *Applied Geography* 28, no. 1 (2008): 1–15.

Books, Monographs, and Texts

Becker, Bertha, et al. *Fronteira Amazonica: Questoes Sobre a Gestao do Territorio.* Brasilia, Brazil: Edumb, 1990.
Brazil: A Country Study. 4th ed. Washington, DC: American University, Foreign Area Studies Series, 1983.
Brazilian Embassy, London. *Brazil: A Geography.* Northampton, England: Belmont Press, 1976.
De Andrate, M. Correia. *The Land and People of Northeast Brazil.* Albuquerque: University of New Mexico Press, 1980.
Dickenson, John P. *Brazil.* Boulder, CO: Westview, 1978; London: Longman, 1982.
James, Preston E. *Brazil.* New York: Odyssey Press, 1946, 1950.
Jobin, A. *Monografia geografia do Estado do Amazones.* Manaus, Brazil: Poperlaria Velho Lino, 1949.
Kleinpenning, J. M. G. *An Evaluation of the Brazilian Policy for the Integration of the Amazon Basin, 1964–1976.* Nijmegen, Netherlands: Geografisch en Planologisch Institut, Publicatie 9, 1979.
Mahar, Dennis J. *Frontier Development Policy in Brazil: A Study of Amazonia.* New York: Praeger, 1979.
Momsen, Richard P. *Brazil: A Giant Stirs.* Princeton, NJ: Van Nostrand, 1968.
Oliveira, A. V. *A Geografia dos Lutas no Campo.* Sao Paulo, Brazil: Conexto, 1988.
Schafter, Alois. *Die kustensunsud-brasiliens.* Stuttgart, Germany: Franz Steiner, 1989.
Sternberg, Hilgard O. *The Amazon River of Brazil.* Wiesbaden, Germany: Steiner, 1975.
Webb, Kempton E. *Brazil.* Boston: Ginn, 1970.
————. *The Changing Face of Northeast Brazil.* New York: Columbia University Press, 1974.

Wesche, Rolf, and Thomas Bruneau. *Integration and Change in Brazil's Middle Amazonia.* Ottawa, Ontario: University of Ottawa Press, 1990.

Articles and Book Chapters

Ajara, Cesar, et al. "O Estado do Tocantins: Reinterpretacao de um Espaco de Frontier." *Revista Brasileira de Geografie* 53, no. 4 (1991): 5–48.

Almeida dos Santos, Milton. "Sao Paulo, uma evolucao contradictoria." *Anales de Geografia de la Universidad Complutense* 16 (1996): 101–23.

Augel, Johannas. "O Lago de barragem de Sobradinho/Bahier (Brasil): Implicacoes economicas e sociais de um projecto de desenvolvimento." *Revista Geografica* 98 (1983): 30–43.

Bak, L. "Brazil and the Netherlands: Some Statistical Facts." *Revista Geografica* 49 (1958): 111–18.

Barber, Maude E. "A Project on Brazil." *Journal of Geography* 21 (1922): 159–74.

Cardoso, Maria E. "Organizaco e Reorganizacao do Espaco no Vale do Paraiba do Sul; Uma Analise Geografica Ate 1940." *Revista Brasileira de Geographia* 53, no. 1 (1991): 81–136.

Chamberlain, G. W. "The Conditions and Prospects of Brazil." *Journal of the American Geogreaphical Society of New York* 22 (1890): 537–54.

Correia de Andrade, Manuel. "Le nordeste bresilien." *Les Cahiers d'Outre Mer* 49, no. 1 (1996): 3–30.

Dickenson, John P. "Innovation for Regional Development in Northeast Brazil: A Century of Failures." *Third World Planning Review* 2 (1980): 57–74.

Galloway, J. H. "Geography in Brazil during the 1970s: Debates and Research." *Luso-Brazilian Review* 19, no. 1 (1982): 1–21.

Geiger, Pedro P. "Regional Organization of Brazil." *Revista Geografica* 61 (1964): 25–68.

Godfrey, Brian J. "Brazil." *Focus* 45, no. 4 (1999): 1–29.

Grabois, Jose, et al. "A Organizacao do Espaco tro Baixo Vale do Taperoa: Uma Ocupacao Extensiva em Mundanca." *Revista Brasileira de Geografia* 53, no. 4 (1991): 81–114.

Haas, William H. "Studies in the Geography of Brazil." *Journal of Geography* 22 (1923): 285–317; 23 (1924): 161–80; 24 (1925): 83–93; 24 (1925): 165–83.

Jefferson, Mark. "Pictures from Southern Brazil." *Geographical Review* 16, no. 4 (1926): 521–47.

Latl, S. V., and Z. Shalizi. "Location and Growth in the Brazilian Northeast." *Journal of Regional Science* 43 (2003): 1–19.

Lasserre, G. "Le Nord-Est du Bresil." *Les Cahiers d'Outre Mer* 1 (1948): 40–46.

Lim Ramirez, Julio C., et al. "Triangulo Mineiro: Regiao e regionalism." *Revista Brasileira de Geografia* 59, no. 1 (2005): 37–60.

Lovell, Peggy A. "The Geography of Economic Development and Racial Discrimination in Brazil." *Development and Change* 24, no. 1 (1993): 83–102.

Mandell, Paul I. "The Rise and Decline of Geography in Brazilian Development Planning: Some Lessons to Be Learned." *Luso-Brazilian Review* 10, no. 2 (1973): 187–96.

Momsen, Janet H. "Projeto Radan: A Better Look at the Brazilian Tropics." *GeoJournal* 3, no. 1 (1979): 3–14.

Nogueira-Mareira, Amelia A. "Nordeste Brasileiro: Problemas e Tentativas de Desenvolvimento." *Revista Geografica* 70 (1969): 29–56.

Pebayla, R. "Une typologie de l'innovation rurale au Bresil." *Les Cahiers d'Outre Mer* 27 (1974): 338–55.

Prost, G. "Apropos du Nord-Est du Brersil." *Les Cahiers d'Outre Mer* 19 (1966): 82–90.

———. "Dans le Nord-Est du Bresil." *Les Cahiers d'Outre Mer* 21 (1968): 78–102.

Shaw, E. W., and J. J. Darnell. "A Frontier Region in Brazil." *Geographical Review* 16, no. 2 (1926): 177–95.

Snethlage, Emile. "Nature and Man in Eastern Para, Brazil." *Geographical Review* 4 (1917): 41–50.

Sternberg, Hilgard O. "Brazilian Amazonia: A Metamorphosis in Progress." *Revista Geografica* 125 (1999): 5–48.

————. "Geographic Thought and Development in Brazil." *Professional Geographer* 11, no. 4 (1959): 12–17.
————. "The Status of Geography in Brazil." *Professional Geographer* 3, no. 2 (1951): 23–29.
Taylor, Harry W. "Sao Paulo: Hallow Frontier." *Revista Geografica* 79 (1973): 149–56.
Ward, Robert. "The Southern Campos of Brazil." *Bulletin of the American Geographical Society* 40, no. 11 (1908): 652–62.
Webb, Kempton E. "Brazil." In *Focus on South America*, edited by Alice Taylor, 162–84. New York: Praeger, 1973.
Zepp, Josef, and Kempton E. Webb. "Brazil." In *Latin America: A Geographical Commentary*, edited by I. Pohl, 29–72. London: John Murray, 1946.

Theses and Dissertations

Cohen, Jennifer K. "Emerald of the Amazon: Journalistic Interpretation of the Brazilian Rainforest, 1969–1990." Master's thesis, University of Texas, Austin, 1990.
Ketteringham, William J. "The Road to Belem." PhD diss., University of California, Los Angeles, 1972.
Koeppe, Matthew T. "Environment, Urbanization, and the Rise of Mechanized Agriculture in Southeastern Rondonia, Brazil." PhD diss., University of Kansas, Lawrence, 2005.
Nissly, Charles M. "Acre: An Amazonian Frontier of Brazil." PhD diss., University of Florida, Gainesville, 1966.
Rothwell, Stuart C. "The Caxias Area, Rio Grande do Sul (Brazil)." PhD diss., Syracuse University, Syracuse, NY, 1956.

CULTURAL AND SOCIAL GEOGRAPHY

General Works

Brooks, M. Y. "Drought and Adjustment Dynamics in Northeastern Brazil." *GeoJournal* 6, no. 2 (1982): 121–28.
Brooks, Reuben. "Differential Perceptions of Drought in Northeastern Brazil." *Proceedings of the Association of American Geographers* 5 (1973): 31–34.
————. "Human Response to Recurrent Drought in Northeastern Brazil." *Professional Geographer* 23, no. 1 (1971): 30–44.
Brostolin, Marta, and Heitor Marques. "Education in the Context of Brazilian Development: Perspective from Local Development." *Anales de Geografia de la Universidad Complutense* 31, no. 2 (2011): 11–29.
Candido Oscar, Sergio. "Educacao ambiental no estado do Rio de Janeiro." *Revista Geografica Academica* 1, no. 1 (2007): 65–78.
Crist, Raymond E. "Cultural Crosscurrents in the Valley of the Rio Sao Francisco." *Geographical Review* 34, no. 4 (1944): 587–612.
Delson, Robert M., and John P. Dickenson. "Perspectives on Landscape Changes in Brazil." *Journal of Latin American Studies* 16, no. 1 (1984): 101–25.
Fearnside, Philip M. *Human Carrying Capacity of the Brazilian Rainforest*. New York: Columbia University Press, 1986.
Hoefle, Scott W. "Enchanted (and Disenchanted) Amazonia: Environmental Ethics and Cultural Identity in Northern Brazil." *Ethics, Place, and Environment* 12, no. 1 (2009): 107–30.
James, Preston E. "The Cultural Regions of Brazil." In *Brazil: Portrait of Half a Continent*, edited by T. Lynn Smith and Alexander Marchant, 86–103. New York: Dryden, 1951.
LeCointe, P. *L'Amazonie bresilienne: le payses habitants, ses resources, notes et statistiques jusquen 1920*. 2 vols. Paris, France: Augustin Challamel, 1922.
Lobato Correa, Roberto, and Zeny Rosendahl. "Brazilian Studies in Cultural Geography." *Social and Cultural Geography* 5, no. 4 (2004): 651–62.

Magnanini, Ruth. "Human Potential in Northeastern Brazil." *Revista Geografica* 56 (1962): 43–50.

Pacheco, Pablo. "Actor and Frontier Types in the Brazilian Amazon: Assessing Interactions and Outcomes Associated with Frontier Expansion." *Geoforum* 43, no. 4 (2012): 864–74.

Rodrigues, S. C. "Impacts of Human Activity on Landscapes in Central Brazil: A Case Study in the Arraguari Watershed." *Australian Geographical Studies* 40, no. 2 (2002): 167–78.

Rosendahl, Zeny. "Le pouvoir du sacre sur l'espace: Muquem et Santa Cruz dos Milagres au Bresil." *Geographie et Cultures* 12 (1994): 71–86.

Sacco dos Anjos, Flavio, et al. "La seguridad alimentaria bajo otra Mirada: Analisis sobre la evolucion de la poblacion brasileira ocupada en actividades de autoconsumo." *Investigaciones Geograficas* 73 (2010): 103–18.

Schmeider, Oscar. "The Brazilian Culture Hearth." *University of California Publications in Geography* 3, no. 3 (1929–30): 159–98.

Smith, Nigel J. H. *Amazon Sweet Sea: Land, Life, and Water at the River's Mouth.* Austin: University of Texas Press, 2002.

———. "Utilization of Game along Brazil's Transamazon Highway." *Acta Amazonia* 6 (1976): 455–66.

Walker, Robert, et al. "The Amazonian Theater of Cruelty." *Annals of the Association of American Geographers* 101, no. 5 (2011): 1156–70.

The Built Environment

DeCastro, Ina E. "Housing Projects: Widening the Controversy Surrounding the Removal of Favelas." *Revista Geografica* 97 (1983): 56–69.

Nash, Roy. "The Houses of Rural Brazil." *Geographical Review* 13 (1923): 329–44.

O'Neill, Maria H. "Segragacao residencial: Os condominios exclusivos." *Revista Geografica*, 97 (1987): 36–43.

Medical Geography

Angulo, Juan J., et al. "Diffusion Dynamics in a Small Brazilian Town: An Epidemiological Reconstruction." *Proceedings, Conference of Latin Americanist Geographer* 5 (1976): 24–34.

———. "Variola Minor in Braganca Paulista County, 1956: A Trend-Surface Analysis." *American Journal of Epidemiology* 105 (1979): 272–80.

Arden, William B., and M. Leitner. *The New Medical Geography of Public Health and Tropical Medicine: Case Studies from Brazil.* Saarbrucken, Germany: VDM, 2008.

Baracho Eduardo, Anna R., et al. "Geograficas e topografias medicas: Os primeiros estudos ambientals da cidade concreta." *Investigaciones Geograficas* 52 (2003): 83–98.

Brooks, Reuben, and Wayne T. Enders. "The Role of Migrants in the Spread of Schistosomiasis Mansone or What Nine Million Brazilians Have and You Don't." *Proceedings, Conference of Latin Americanist Geographers* 5 (1976): 89–96.

Caldas de Castro, Marcia, et al. "Spatial Patterns of Malaria in the Amazon: Implications for Surveillance and Targeted Intervention." *Health and Place* 13, no. 2 (2007): 368–80.

Guimares de Souza, Michael, and Eduardo da Silva Pinheiro. "Incidencia e distribucao da tuberculose na cidade de Manaus/Amazonas, Brasil." *Revista Geografica Academica* 3, no. 2 (2009): 35–43.

Morrill, Richard L., and Juan J. Angulo. "Spatial Aspects of a Smallpox Epidemic in a Small Brazilian City." *Geographical Review* 69, no. 3 (1979): 319–30.

Shimoda, Akio, et al. "Regional Differences of Death from Chronic Diseases in Rio Grande do Sul, Brazil, from 1970 to 1976." *Social Science and Medicine* 15D, no. 1 (1981): 187–98.

Ethnic, Social, and Population Geography

Beaujeu-Garnier, J. "Les migrations vers Salvador (Bresil)." *Les Cahiers d'Outre Mer* 15 (1962): 291–300.

Bonaswiez, A. "Population Pressure in the State of Sao Paulo, Brazil." In *Geography and a Crowding World: Symposium on Population Pressures upon Physical and Social Resources in the Developing Lands,* edited by Wilbur Zelinsky, Leszak Kosinski, and R. M. Prothero, 556–73. Oxford, England: Oxford University Press, 1970.

Bret, B. "Faits de population et types d'organisation regionale dans le Sud de Bresil." *Les Cahiers d'Outre Mer* 29 (1976): 225–50.

Brooks, Edwain. "Twilight of Brazilian Tribes." *Geographical Magazine* 45, no. 4 (1973): 304–10.

Carney, Judith, and Robert A. Voeks. "Landscape Legacies of the African Diaspora in Brazil." *Progress in Human Geography* 27, no. 2 (2003): 139–52.

Cavalconti de Oliveira, Zuleica L. "Aspectos socio-demograficas do trabalho feminine nas areas urbanas do estado de Sao Paulo, 1970–1976." *Revista Geografica* 97 (1983): 70–82.

Chardon, Roland E. "Changes in the Geographic Distribution of Population in Brazil, 1950–1960." In *New Perspectives of Brazil,* edited by Eric Baklanoff, 155–78. Nashville, TN: Vanderbilt University Press, 1966.

Chase, Jacqueline. "In the Valley of the Sweet Mother: Gendered Metaphors, Domestic Lives, and Reproduction under a Brazilian State Mining Company." *Gender, Space, and Culture* 8, no. 2 (2001): 169–88.

Coy, Martin, and Reinhold Lucker. "Mutation dans un espace peripherique en cours de modernization: Espaces sociaux dans le milieu rural du Centro-Oeste bresilien." *Les Cahiers d'Outre Mer* 46 (1993): 153–74.

Crist, Raymond E. "German Colonization in Rio Grande do Sul." *Geographical Review* 56, no. 2 (1966): 187–208.

Da Cunha, Jose, et al. "Social Segregation and Academic Achievement in State-Run Elementary Schools in the Municipality of Campinas, Brazil." *Geoforum* 40, no. 5 (2009): 873–83.

Dawsey, Cyrus B. "Migration in Brazil: Research during the 1980s." *Yearbook, Conference of Latin Americanist Geographers* 17–18 (1990): 109–16.

———. "Push Factors and Pre-1970 Migration to Southwest Parana, Brazil." *Revista Geografica* 98 (1983): 54–58.

———. "Source of Settlers in Southwest Parana." *Ecumene* 11 (1979): 6–11.

Dyer, Donald R. "Growth of Brazil's Population." *Journal of Geography* 65, no. 9 (1966): 417–28.

Fischlowitz, E. "Internal Migration in Brazil." *International Migration Review* 3, no. 3 (1969): 36–44.

Gauld, Charles A. "Brazil Takes a Census." *Journal of Geography* 40, no. 4 (1941): 138–44.

Gauthier, Howard L. "Migration Theory and the Brazilian Experience." *Revista Geografica* 82 (1975): 51–62.

Geiger, Pedro P., and Susan Oxnard. "Aspects of Population Growth in Brazil." *Revista Geografica* 70 (1969): 7–28.

Geraiges de Lemos, Amalia. "La diversidad de la geografia brasilena: Una Mirada a la educacion basica y superior en la actualidad." *Anales de Geografia de la Universidad Complutense* 29, no. 2 (2009): 207–30.

Godfrey, Brian J. "Migration to the Gold-Mining Frontier in Brazilian Amazonia." *Geographical Review* 82, no. 4 (1992): 458–69.

Gutberlet, Jutta. "Rural Development and Social Exclusion: A Case Study of Sustainability and Distributive Issues in Brazil." *Australian Geographer* 30, no. 2 (1999): 221–38.

Haller, Archibold O. "A Socio-economic Regionalization of Brazil." *Geographical Review* 72, no. 4 (1982): 450–64.

Haller, Archibold O., et al. "Migration and Socio-economic Status in Brazil: Interregional and Rural-Urban Variations in Education, Occupational Status, and Income." *Luso-Brazilian Review* 18, no. 1 (1981): 117–38.

James, Preston E. "The Changing Patterns of Population in Sao Paulo State, Brazil." *Geographical Review* 28, no. 3 (1938): 353–62.

Kleinpenning, J. M. G., and S. Volbeda. "Recent Changes in Population Size and Distribution in the Amazon Region of Brazil." In *Change in the Amazon Basin: The Frontier after a Decade of Colonization*, edited by John Hemming, 2:6–36. Manchester, England: Manchester University Press, 1985.

Kohlhepp, G. "Raumwirksame Tatigkeit ethnosozialer Gruppen in Brasilien: Am Beispiel donauschwabischer Siedler in Entre Rios/Parana." In *Europische Ethnien in Landlichen Raum der neuen Welt*, edited by Behandelle Thenen, 31–46. Passau, Germany: Passauer Schriften zur Geographie, Band 7, 1989.

Lewis, Clare, and Steven Pike. "Women, Body, and Space: Rio Carnaval and the Politics of Performance." *Gender, Place, and Culture* 3, no. 1 (1996): 23–42.

Lucker, Reinhold. "Raumlicher Wandeleuropische Kolonisationsgebiete in Sudbrasilen: Innovation und Dynamik durch ethnische Gruppen?" In *Europaische Ethnien im Landlichen Raum der Neuen Welt*, edited by Behandelte Thenen, 47–59. Passau, Germany: Passauer Schriften zur Geographie, Band 7, 1989.

Ludwig, Armin K. "Early Population Movements to Brasilia: Substance and Model." *Proceedings of the New England-St. Lawrence Valley Division, Association of American Geographers* 2 (1972): 21–27.

Magnanini, Ruth. "Geografia da Populacao: Potencial Human do Nordeste Brasileiro." *Revista Geografica* 56 (1962): 43–50.

Marcus, Alan. "Brazilian Immigration to the United States and the Geographical Imagination." *Geographical Review* 99, no. 4 (2009): 481–98.

———. "(Re)Creating Places and Spaces in Two Countries: Brazilian Transnational Migration Processes." *Journal of Cultural Geography* 26, no. 2 (2009): 173–98.

Menezes de Rodriguez, N. "A segregacao socioespacial e acoes do plano director: Os casos do bairro Maria Eugenia e a area central de Vicosa, Minas Gerais." *Revista Geografica Academica* 4, no. 1 (2010): 56–70.

Momsen, Janet H. "Are Women Reservoirs of Traditional Plant Knowledge? Gender, Ethnobotany, and Globalization in Northeast Brazil." *Singapore Journal of Tropical Geography* 28, no. 1 (2007): 1–6.

Mougeot, Luc J. A. *Alternative Migration Targets and Brazilian Amazonia Closing Frontiers.* Norwich, England: GeoBooks, 1984.

———. "Frontier Population Absorption and Migrant Socio-economic Mobility: Evidence from Brazilian Amazonia." *Proceedings, Conference of Latin Americanist Geographers* 8 (1981): 150–61.

Muller, Keith D., and M. D. C. Oliveira. "Gender and the Criminal Justice System: Northeast Brazil." *Papers of the Applied Geography Conferences* 20 (1997): 293–96.

Pooler, James, and Joao F. de Abreu. "Income Fronts and Migration Winds in Brazil: A Graphical Analysis." *Ontario Geographer* 13 (1979): 25–40.

Preston, Dennis R. "Mental Maps of Language Distribution in Rio Grande do Sul (Brazil)." *Geographical Bulletin* 27 (1985): 46–64.

Reis, Elisa, and Simon Schwarzman. "Spatial Dislocation and Social Identity in Brazil." *International Social Science Journal* 30, no. 1 (1978): 98–115.

Rodrigues Has, Dora. "Os povos da floresta, os imigrantes e os madelos de occupacao territorial: Impactos a alternativos." *Revista Brasileira de Geografia* 52, no. 3 (1990): 83–90.

Rosa Souza, Romario. "Analise preliminar da preferencia das chuvas na Amazonia Mato-Grossense no period de 2004 a 2007 (Janeiro, Fevereiro e Marco)." *Revista Geografica Academica* 2, no. 1 (2008): 56–72.

Rothwell, Stuart C. "The Old Italian Colonial Zone of Rio Grande do Sul, Brazil." *Revista Geografica* 46 (1957): 22–54; 47 (1958): 1–21; 49 (1958): 89–101; 50 (1959): 67–108.

Sahr, W. "Religion und Szientismus in Brasilien: Versuch (Essy) uber eine dekonstruktive Regionalgeographie des wissens." *Geografische Zeitschraft* 94, no. 2 (2006): 27–42.

Schurz, W. L. "The Distribution of Population in the Amazon Valley." *Geographical Review* 15, no. 2 (1925): 206–25.

Sims, Harold D. "Japanese Post-war Migration to Brazil: An Analysis of Data Presently Available." *International Migration Review* 6, no. 3 (1972): 246–65.

Sobelberg, E. "Landnahme und Eigentumsveranderungen im italienischen Kolonisationgebiet von Ibiracu/Espirito Santo (Brasilien)." In *Eurpaische Ethnien im Landlichen Raum der neuen Welt*, edited by Behandelte Thenen, 75–86. Passau, Germany: Passaurer Schriften zur Geographie, Band 7, 1989.

Takagi, Hideki. "Socio-economic Changes of Bastos, a Japanese Colony in Brazil [in Japanese]." *Jimbun Chiri/Human Geography* 29, no. 1 (1977): 96–112.

Teulieres, R. "Les noirs du Minas." *Les Cahiers d'Outre Mer* 11 (1958): 47–61.

Urban, Greg. "Developments in the Situation of Brazilian Tribal Populations from 1976 to 1982." *Latin American Research Review* 20, no. 1 (1985): 7–25.

Webb, Kempton E. "Landscape Evolution: A Tool for Analyzing Population-Resource Relationships: Northeast Brazil as a Test Area." In *Geography and a Crowding World*, edited by Wilbur Zelinsky et al., 218–34. New York: Oxford University Press, 1970.

Wesche, Rolf. "The Transformation of Rural Caboclo Society upon Integration into Brazil's Amazon Frontier: A Study of Itacoatiara." *Studies in Third World Society* 32 (1985): 115–42.

WinklerPrins, Antoinette M. "Seasonal Floodplain-Upland Migration along the Lower Amazon River." *Geographical Review* 92, no. 3 (2002): 415–31.

Young, Drew M. "An Evaluation of Preliminary Age and Sex Data from the 1970 Brazilian Census." *Revista Geografica* 81 (1974): 91–104.

Community and Settlement Studies

Augel, Johannes. "Human Settlement Problems in Brazilian Development." *Ekistics* 49 (1982): 31–36.

Augelli, John P. "Cultural and Economic Changes of Bastos: A Japanese Colony on Brazil's Paulista Frontier." *Annals of the Association of American Geographers* 48, no. 1 (1958): 3–19.

———. "The Latvians of Varpu: A Foreign Colony on the Brazilian Frontier Fringe." *Geographical Review* 48, no. 3 (1958): 365–87.

Butland, Gilbert J. "The Colonization of Northern Parana." *Geography* 40, no. 2 (1955): 126–28.

Caldas, Marcellus, et al. "Settlement Formation and Land Cover and Land Use Change: A Case Study in the Brazilian Amazon." *Journal of Latin American Geography* 9, no. 1 (2010): 125–44.

Casetti, Emilio, and Howard L. Gauthier. "A Formalization and Test of the 'Hollow Frontier' Hypothesis." *Economic Geography* 53, no. 1 (1977): 70–78.

Chaloult, Yves. "Settlement Pattern and Peasant Organization in Northeastern Brazil." *International Journal of Urban and Regional Research* 4, no. 4 (1980): 536–48.

Chase, Jacqueline. "Their Space: Security and Service Workers in a Brazilian Gated Community." *Geographical Review* 98, no. 4 (2008): 476–95.

Coy, Martin. "Pioneer Front and Urban Development: Social and Economic Differentiation of Pioneer Towns in Northern Mato Grosso." *Applied Geography and Development* 39 (1992): 7–30.

Dawsey, Cyrus B. "Income and Residential Location in Piracicaba, Sao Paulo. Brazil." *Revista Geografica* 89 (1979): 185–90.

Deffontaines, Pierre. "Mountain Settlement in the Central Brazilian Plateau." *Geographical Review* 27, no. 3 (1937): 394–413.

Fearnside, Philip M. "Brazilian Amazon Settlement Schemes." *Habitat International* 8 (1989): 82–137.

Furley, Peter A., ed. *The Forest Frontier: Settlement and Change in Brazilian Roraima*. New York: Routledge, 1994.

Furley, Peter A., and Luc J. A. Mougeot. "Perspectives." In *The Forest Frontier: Settlement and Change in Brazilian Roraima*, edited by Peter A. Furley, 1–38. New York: Routledge, 1994.

Henshall, Janet D. "Agricultural Colonization in Rondonia, Brazil." *Luso-Brazilian Review* 19, no. 2 (1982): 169–85.

James, Preston E. "The Expanding Settlements of Southern Brazil." *Geographical Review* 30, no. 4 (1940): 601–26.

Jefferson, Mark. "An American Colony in Brazil." *Geographical Review* 18, no. 2 (1928): 226–31.

Kleinpenning, J. M. G. *The Integration and Colonization of the Brazilian Portion of the Amazon Basin*. Nijmegen, Netherlands: Nijmeegse Geografische Cahiers 4, 1975.

———. "Road Building and Agricultural Colonization in the Amazon Basin." *Tijdschrift voor Economische en Sociale Geographie* 62, no. 5 (1971): 285–89.

Kohlhepp, Gerd. "Types of Agricultural Colonization on Subtropical Brazilian Campas Limpos." *Revista Geografica* 70 (1969): 131–55.

Methey, Kosta. "The Rehabilitation Program of Alagados Squatter Settlement, Brazil." *Ekistics* 45 (1978): 257–61.

Muller, Keith D. "Colonizacao pioneira no Sul do Brasil: O caso de Todelo, Parana." *Revista Brasileira de Geografica* 48, no. 1 (1986): 83–139.

———. "The Future of the Long Lot in South Brazil: The Case of West Parana." *Papers of the Applied Geography Conferences* 17 (1994): 129–34.

———. "Pindorama: European Agricultural Settlement Scheme, Coastal Northeast Brazil." *Papers of the Applied Geography Conferences* 16 (1993): 1–7.

Muller, Keith D., and A. M. G. Lages. "Emerging Rural Settlement Patterns in the Sugar Cane Region of Alagos, Northeast Brazil." *Papers of the Applied Geography Conferences* 18 (1995): 151–54.

Neuburger, Martina. *Pionierfrontentwicklung im Hinterland von Careres (Mato Grosso, Brasilien): okologische Degradierung Verwundbarkeit und Kleinbauerliche Uberle bensstrategien*. Tubingen, Germany: Tubinger Geographische Studien, Heft 135, 2002.

Prost, G. "Dans Le Nord-Est du Bresil: Les pionniers du Cariris dans la Barborem a semi-aride." *Les Cahiers d'Outre Mer* 20 (1967): 367–93.

Roche, J. "Quelques aspects de la vie rurale dans les colonies allemandes du Rio Grande du Sol." *Les Cahiers d'Outre Mer* 12 (1959): 5–25.

Rothwell, Stuart C. "The Old Italian Colonial Zone of Rio Grande do Sul, Brazil: A Geographic Interpretation." *Revista Geografica* 46 (1957): 22–54.

———. "The Old Italian Colonial Zone of Rio Grande do Sul, Brazil: A Geographical Interpretation." *Revista Geografica* 47 (1957): 1–21.

———. "The Old Italian Colonial Zone of Rio Grande do Sul, Brazil: A Geographical Interpretation." *Revista Geografica* 50 (1959): 67–108.

Sader, Maria R. "Ocupacao de uma area de fronteira: bico do papagaio-extremo norte de goias brasil." *Revista Geografica* 97 (1983): 96–103.

Schadt, Siegfried. "Agricultural Colonization of the Zona da Mata of North-East Brazil: The Example of Pindorama." *Applied Geography and Development* 17 (1981): 71–90.

Struck, E. "Mitelpunktsiedlungen im deutschen Kolonisationsgebiet von Espirito Santo (Brasilien)." In *Europaische Ethnien im Landlichen raum der Neuen Welt*, edited by Behandelte Thenen, 61–74. Passau, Germany: Passauer Schriften zur Geographie, Band 7, 1989.

Thypin-Bermeo, Sam, and Brian J. Godfrey. "Envisioning Amazonian Frontiers: Place-Making in a Brazilian Boomtown." *Journal of Cultural Geography* 29, no. 2 (2012): 215–38.

Waibel, Leo. "European Colonization in Southern Brazil." *Geographical Review* 40, no. 4 (1950): 529–47.

Wesche, Rolf. "Transporte, organizacion especial y logras en las proyectos de colonization de la Region amazonica brasilien." *Revista Geografica* 98 (1983): 44–53.

Wolford, Wendy. "Producing Community: The MST and Land Reform Settlements in Brazil." *Journal of Agrarian Change* 3, no. 4 (2003): 500–525.

Tourism and Recreation

Braghini, Claudio R., et al. "Perspectivas de sutentabilidade ecologica do turismo em Xingo, Sergipe/Alagoas." *Revista Geografica Academica* 3, no. 1 (2009): 56–69.

Castilho da Costa, Nadja M., and Vivian Castilho da Costa. "Turismo ecologico e educacao ambiental no Parque Estadual da Pedra Branca (PEPB), municipio de Rio de Janeiro, Brasil." *Revista Geografica* 138 (2005): 71–87.

Castilho da Costa, Nadja M., et al. "Turismo ecologico e educacao ambiental no parquet estadual da Pedra Branca-Municipio do Rio de Janeiro, Brazil." *Revista Geografica* 138 (2005): 113–28.

Castilho da Costa, Vivian, and Josilda M. Rodrigues da Silva. "Analise do potential turistico nas regioes administrativas (RAs) de Campo Grande e Guaratiba zona destedo municipio do Rio de Janeiro (Brasil)." *Investigaciones Geograficas* 52 (2003): 137–52.

Corcia, Nilson. "O rejuvenescimento da regiao turista par expansao geografica e redistribuicao territorial das funcoes: A destinacao turistica de Pipa, Litoral do Nordeste do Brazil." *Revista Geografica* 133 (2003): 91–102.

Maruyama, Hiraoki, et al. "Ecotourism in the North Pantanal, Brazil: Regional Bases and Subjects for Sustainable Development." *Geographical Review of Japan* 78, no. 5 (2005): 43–64.

Medeiros de Araujo, L., and W. Bromwell. "Partnership and Regional Tourism in Brazil." *Annals of Tourism Research* 29, no. 4 (2002): 1138–64.

Moreita Bento, Lilian C., and Silvio C. Rodriguez. "Geoturismo geomorfisitios: Refletindo sobre o potential turistico de quedas d'agua-um estudo de caso do municipio de Indianopolis/Minas Gerais." *Revista Geografica Academica* 4, no. 2 (2010): 96–104.

Saito, Isao, et al. "Gravata, One of the Hill Resorts in Northeast Brazil [in Japanese]." *Tsukuba Studies in Human Geography* 22 (1998): 1–26.

Stillwell, H. Daniel. "National Parks in Brazil: A Study in Recreational Geography." *Annals of the Association of American Geographers* 53, no. 3 (1963): 391–406.

Xavier de Souza, Jose A. "A 'litoralizacao' de comocian (Ceara) e o teritorio usado da praia de Maceio-Ceara." *Revista Geografica Academica* 2, no. 1 (2008): 88–97.

Zanott, Laura, and Janet Chernela. "Conflicting Cultures of Nature: Ecostourism, Education, and the Kayapo of the Brazilian Amazon." *Tourism Geographies* 10, no. 4 (2008): 495–521.

Place Names

Marchand, Annie. "Brazilian Spelling and Geographic Names." *Journal of Geography* 42 (1943): 208–15.

Theses and Dissertations

Aragonvaca, Luis E. "Migration to Northern Goias: Geographical and Occupational Mobility in Southeastern Amazonia, Brazil." PhD diss., Michigan State University, East Lansing, 1978.

Arden, William B. "Medical Geography in Public Health and Tropical Medicine: Case Studies from Brazil." PhD diss., Louisiana State University, Baton Rouge, 2008.

Bein, Frederick. "Patterns of Pioneer Settlement in Southern Mato Grosso: Two Case Studies." PhD diss., University of Florida, Gainesville, 1974.

Borges, Marcos M. "Tourism on the Rio Araguaia, Brazil: Tourists' Perceptions and Motor Boat Erosion." Master's thesis, University of Wyoming, Laramie, 1995.

Brooks, Reuben. "Drought Perception as a Force in Migration from Northeast Brazil." PhD diss., University of Colorado, Boulder, 1972.

Caldas, Marcellus. "Settlement Formation and Land Cover and Land Use Change: A Case Study in the Brazilian Amazon." PhD diss., Michigan State University, East Lansing, 2008.

Cutrim, Ana C. "Rural Tourism as an Alternative to Rural Development in the State of Para, Brazil." Master's thesis, Western Michigan University, Kalamazoo, 2008.

Darnel, Bernard. "Land Settlement in Northeast Brazil: A Study of Seven Projects." PhD diss., McMaster University, Hamilton, Ontario, 1973.

Demoor, David A. "Japanese Immigration and Colonization in the Lower Amazon Basin, Brazil." Master's thesis, University of California, Los Angeles, 1964.

Diniz, Alexander. "Step and Chain Migration to Rondonia State, Brazil." Master's thesis, Kansas State University, Manhattan, 1994.

Dozier, Craig L. "Northern Parana, Brazil: Settlement and Development of a Recent Frontier Zone." PhD diss., Johns Hopkins University, Baltimore, 1954.

Enders, Wayne T. "The Spatial Behavior of Low Income Hospital Patients: A Case in Porto Alegre, Brazil." PhD diss., University of Texas, Austin, 1978.

Fish, Warren R. "Constraints on the Dietary Status of Rural Man in the Paraiba Valley Region, Sao Paulo, Brazil." PhD diss., University of Illinois, Urbana, 1972.

Garmany, Jeffrey. "Appropriated Representation of Space: A Socio-spatial Analysis of O Moyimento dos Trabalhadones Rurais Sem Terre in Ceara, Brazil." Master's thesis, University of Arizona, Tucson, 2006.

Hiraoka, Mario. "Tome-Acu and Guatapara: A Comparative Study of Two Japanese Colonies in Brazil." Master's thesis, University of California, Los Angeles, 1967.

Hulima, Stephanie M. "Spatial Patterns of Vector-Borne Disease: Cutaneous Leishmoniasis in Northeast Brazil." PhD diss., Clark University, Worcester, MA, 2001.

Kermath, Brian. "Ecotourism as Sustainable Development in Brazilian Amazonia." PhD diss., University of Tennessee, Knoxville, 1993.

Ladourceur, Micheline. "Bresil: espace pluri-culturel et geographie nationale, 1964–1985." PhD diss., Laval University, Ste. Foy, Quebec, 1990.

Maraus, Alan. "The Geography of Ethnic and Racial Identities and Complexities in Brazil." Master's thesis, University of Massachusetts, Amherst, 2004.

Martin, Robert M. "Integration in the Brazilian Amazon: Bringing Wishful Thinking in Line with Geographic Reality." Master's thesis, Pennsylvania State University, University Park, 1991.

Miranda, Marta. "In Search of the Promised Land: Land Tenure in Northeast Brazil and Migration to the Amazonian Frontier." Master's thesis, San Diego State University, San Diego, 1993.

Schwarz, Maria L. "Les representations des enfants sur la biodiversite de la Mata Atlantica-Bresil." PhD diss., University of Montreal, Montreal, Quebec, 2007.

Smith, Nigel J. H. "Transamazon Highway: A Cultural-Ecological Analysis of Settlement in the Humid Tropics." PhD diss., University of California, Berkeley, 1976.

Stern, Peter. "Sao Paulo, Brazil: Foreign-Settlement along Its Pioneer Fringe." Master's thesis, Columbia University, New York, 1948.

Weinert, Julie. "A Framework for Multileveled Gendered Analysis of Ecotourism: With Examples from Costa Rica, Brazil, and Indonesia." Master's thesis, Ohio State University, Columbus, 2000.

Wolford, Wendy. "This Land Is Our's Now: Social Mobilization and Production in Land Reform Settlements in Brazil." PhD diss., University of California, Berkeley, 2001.

ECONOMIC GEOGRAPHY

General Works

Arque, P. "Notes sur l'economie bresilienne." *Les Cahiers d'Outre Mer* 15 (1962): 418–24.

Auty, R. M. *Creating Comparative Advantage in HCI: Brazil, Korea, and Mexico.* Washington, DC: World Bank Report Research Project PRO 675–41, IENIN, 1990.

———. "Industrial Policy, Sectoral Maturation, and Postwar Economic Growth in Brazil: The Resource Curse Thesis." *Economic Geography* 71, no. 3 (1995): 257–72.

Azzoni, C. R., et al. *Geography and Income Convergence among Brazilian States.* New York: American Development Bank, Network Working Paper R-395, 2000.

Barberia, Lorena, and Ciro Biderman. "Local Economic Development: Theory, Evidence, and Implications for Policy in Brazil." *Geoforum* 41, no. 6 (2010): 951–62.

Becker, Bertha. "The Frontier at the End of the Twentieth Century: Eight Propositions for a Debate on Brazilian Amazonia." In *International Economic Restructuring and the Regional*

Community, edited by H. Muegge and Walter B. Stohr, 375–92. Aldershot, England: Avebury, 1987.

Becker, Bertha, and Claudio A. G. Egler. *Brazil: A New Regional Power in the World-Economy*. Cambridge, England: Cambridge University Press, 1992.

Castilho da Costa, Vivian. "O potencial eco-rural da zona oeste do municipion do Rio de Janeiro, Brazil." *Revista Geografica* 132 (2002): 79–89.

Correa, Robert L. A. "Corporacao e Organizacao Espacial: Um Estudio de Case." *Revista Brasileira de Geografia* 53, no. 3 (1991): 33–66.

Cunningham, Susan M. "Multinational Enterprises in Brazil: Locational Patterns and Implications for Regional Development." *Professional Geographer* 33, no. 1 (1981): 48–62.

Dillman, C. Daniel. "Land and Labor Patterns in Brazil during the 1960s." *American Journal of Economics and Sociology* 35, no. 1 (1975): 49–70.

Dozier, Craig L. "Northern Parana, Brazil: An Example of Organized Regional Development." *Geographical Review* 46, no. 3 (1956): 318–33.

Egler, Claudio A. G. "As Escadas da Economica: Uma Introducao a Dimensao Territorial da Crise." *Revista Brasileira de Geografia* 53, no. 3 (1991): 229–46.

Enders, Wayne T. "Regional Development in Brazil: Research Trends in the Seventies." *Proceedings, Conference of Latin Americanist Geographers* 8 (1981): 408–20.

Fearnside, Philip M. "Jari Development in the Brazilian Amazon." *Interciencia* 5 (1980): 145–56.

Fernandes, Ana C. *Stabilization, Exports, and Regional Development in Brazil: The North East in the 1980s*. Sussex, England: University of Sussex, Research Paper in Geography 8, 1993.

Guihote, J. J. M., et al. "Production Relations in the Northeast and the Rest-of-Brazil Regions in 1995: Decomposition and Synergy in Input-Output Systems." *Geographical Analysis* 34, no. 1 (2002): 62–75.

Haddad, Eduardo, et al. "The Spatial Formation of the Brazilian Economy: Historical Overview and Future Trends." *Geographia Polonica* 72, no. 1 (1999): 89–106.

Henshall, Janet D., and Richard P. Momsen. *Geography of Brazilian Development*. London: G. Bell, 1976.

———. "Regional Disparities and Economic Growth in Brazil." *Revista Geografica* 79 (1973): 167–76.

Kearns, Kevin C. "Northeast Brazil: A Study in Regional Economic Development." *Pennsylvania Geographer* 8, no. 2 (1970): 12–20.

Machado Oliveira, Sebastiao S. "O capital comunitatio e o PPC: A participacao publica, privada e comunitaria na construcao do desenvolviment local." *Revista Geografica Academica* 3, no. 2 (2009): 68–77.

Moran, Emilio. *Developing the Amazon: The Social and Ecological Consequences of Government-Directed Colonization along Brazil's Transamazonian Highway*. Bloomington: Indiana University Press, 1981.

Moreia, Amelia A. "Northeastern Brazil: Development Problems and Attempts to Solve Them." *Revista Geografica* 70 (1969): 29–56.

Pokshishevskiy, V. V. "The Major Economic Regions of Brazil." *Soviet Geography* 1 (1960): 48–67.

Pooler, James, and Joal de Abrecu. "Income Fronts and Migration Winds in Brazil: A Graphical Analysis." *Ontario Geography* 13 (1979): 25–39.

Ramieris, Julio. "As corporacoes multinacionais e a organizacao espacial: Ume introducao." *Revista Brasileira de Geografia* 51, no. 1 (1989): 103–12.

Rios-Villamizar, Eduardo A., and Ana C. F. Nogueira. "Silvinita, una nueva matriz de dessarrollo? El caso del municipio de Nova Olinda do Norte, Amazonas, Brasil." *Revista Geografica Academica* 5, no. 1 (2011): 68–78.

Rodriguez-Pose, Andres, et al. "Local Empowerment through Economic Restructuring in Brazil: The Case of the Greater ABC Region." *Geoforum* 32, no. 4 (2001): 459–69.

Semple, R. Keith, et al. "Growth Poles in Sao Paulo, Brazil." *Annals of the Association of American Geographers* 62, no. 4 (1972): 591–98.

Semple, R. Keith, and Howard L. Gauthier. "Spatial-Temporal Trends in Income Inequalities in Brazil." *Geographical Analysis* 4, no. 2 (1972): 169–79.

Skillings, Robert F. "Economic Development of the Brazilian Amazon: Opportunities and Constraints." *Geographical Journal* 150, no. 1 (1984): 48–54.

Sternberg, Hilgard O. "Agriculture and Industry in Brazil." *Geographical Journal* 121 (1955): 488–502.

Stevens, Ray, and Paolo Brandao. "Diversification of the Economy of the Cacao Coast of Bahia, Brazil." *Economic Geography* 37, no. 2 (1961): 231–53.

Struck, E. "Persistenz und Wandel des zentalortlichen Gefuges im brasilianischen Bundesstaat Espirito Santo." *Zeitschrift fur Wirtschaftsgeographie* 36, no. 4 (1992): 229–37.

Timmons-Roberts, J. "Trickling Down and Scrabling Up: The Informal Sector, Food Provisioning, and Local Benefits of the Carajat Mining Growth Pole in the Brazilian Amazon." *World Development* 23 (1995): 385–400.

Webb, Kempton E. "Origins and Development of a Food Economic in Central Minas Gerais." *Annals of the Association of American Geographers* 49 (1959): 409–19.

Agriculture and Land Use

Aldrich, Stephen P., et al. "Land-Cover and Land-Use Change in the Brazilian Amazon: Smallholders, Ranchers, and Frontier Stratification." *Economic Geography* 82, no. 3 (2006): 265–88.

Alves, D. "An Analysis of the Geographical Patterns of Deforestation in Brazilian Amazonia in the 1991–1996 Period." In *Patterns of Processes of Land Use and Forest Change on the Amazon*, edited by C. Wood and R. Porro, 95–106. Gainesville: University of Florida Press, 2003.

Andreoti, Claudio E. "Utilazacao do modelo sib 2 na avaliacao da dinamica da agua do solo entre o cultivo de cana-de-acucar e a floresta tropica atlantica." *Revista Geografica Academica* 5, no. 1 (2011): 79–84.

Arvor, Damien, et al. "Analyzing the Agricultural Transition in Mato Grosso, Brazil, Using Satellite-Derived Indices." *Applied Geography* 32, no. 2 (2012): 702–13.

Becker, Bertha. "Changing Land Use Patterns in Brazil." In *Selected Papers*. Vol. 2. New Delhi, India: International Geographical Congress, 1968.

———. "Changing Land Use Patterns in Brazil: The Spread of Cattle Ranching in Sao Paulo State." *Revista Geografica* 71 (1969): 35–63.

Bein, Frederick. "Changing Land Use in Brazil's Colonia Agricola de Terenos, 1924–1972." *Bulletin, Association of North Dakota Geographers* 34 (1984): 44–58.

Berg, M. *Land Use Systems Research on Strongly Weathered Soils in South and South-East Brazil*. Utrecht: Netherlands Geographical Studies, Band 271, 2000.

Blum, Winfried E. H. "Land Use in the Humid Tropics, Exemplified by the Amazon Region of Brazil." *Applied Geography and Development* 25 (1985): 71–87.

Blumenschein, Markus. *Landnutzungs veranderungen in dermodernisierten Landwirthschaft in Mato Grosso, Brasilien: Die Role von Netzwerken, institutionellen und okonomischen Faktoren fur agrarwirtschaftliche Innovationen auf der Capoda das Pareas*. Tubingen, Germany: Tubinger Geographische Studien, Heft 133, 2001.

Brandizio, Eduardo S., et al. "Land Use Change in the Amazon Estuary: Patterns of Cabodo Settlement and Land Management." *Human Ecology* 22, no. 3 (1994): 249–78.

Brannstrom, Christian. "Coffee Labor Regimes and Deforestation on a Brazilian Frontier, 1915–1965." *Economic Geography* 76, no. 4 (2000): 326–46.

———. "A Q-Method of Environmental Governance Discourses in Brazil's Northeastern Soy Frontier." *Professional Geographer* 63, no. 4 (2011): 531–49.

Brannstrom, Christian, and Paulo R. B. Brandao. "Two Hundred Hectares of Good Business: Brazilian Agriculture in a Themed Space." *Geographical Review* 102, no. 4 (2012): 465–85.

Brannstrom, Christian, et al. "Compliance and Market Exclusion in Brazilian Agriculture: Analysis and Implications for Soft Governance." *Land Use Policy* 29, no. 2 (2012): 357–66.

Brannstrom, Christian, and A. M. Filippi. "Classification of Cerrado (Savanna) and Agricultural Land Covers in North-Eastern Brazil's Agricultural Frontier." *Geocarto International* 23 (2008): 109–34.

Brito, Mario S. "O programa nacional de irrigacao: Uma avaliacao previados resultados." *Revista Brasileira de Geografia* 53, no. 2 (1991): 113–26.
Brito, Mario S., and S. T. Silva. "The Small Production Role in Different Brazilian Agrarian Spaces." *Rural Systems*, 1 (1983).
Broggio, C., et al. "L'Irrigation dans le Nordeste du Bresil." *L'Information Geographique* 63, no. 5 (1999): 223–31.
Brown, J. C. "Responding to Deforestation: Productive Conservation, the World Bank, and Beekeeping in Rondonia, Brazil." *Professional Geographer* 53, no. 1 (2001): 106–18.
Brown, J. C., et al. "Expansion of Mechanized Agriculture and Land-Cover Change in Southern Rondonia, Brazil." *Journal of Latin American Geography* 3, no. 1 (2004): 96–102.
Caetano de Souza, Jaimeval, et al. "Combinacoes agricolas no estado da Bahia, 1970–1980: Uma contricuicao metodologica." *Revista Brasileira de Geografia* 53, no. 2 (1991): 95–112.
Cairns, D. M. "Spatial Pattern Analysis of Witches Broom Disease of Cacao at a Landscape Scale in Rondonia, Brazil." *Tropical Agriculture* 71 (1994): 31–35.
Caldas, Marcellus, et al. "Theorizing Land Cover and Land Use Change: The Peasant Economy of Amazonian Deforestation." *Annals of the Association of American Geographers* 97, no. 1 (2007): 86–110.
Cameron, C. R. "Mate: An Important Brazilian Product." *Journal of Geography* 29, no. 2 (1930): 54–69.
Carrero, G. C., and Philip M. Fearnside. "Forest Clearing Dynamics and the Expansion of Landholdings in Apui, a Deforestation Hotspot on Brazil's Transamazaon Highway." *Ecology and Society* 16, no. 2 (2011): 26.
Casarin, Rosalia, et al. "Uso da terra qualidade: Da agua da bacia hidrografica Paraguai/Jauquara-Mato Grosso." *Revista Geografica Academica* 2, no. 1 (2008): 33–42.
Castillo, Ricardo. "Transporde e logistica de graneis solidos agricolas: Componentes estruturais do novo sistema de movimiendo do territorio brasileiro." *Investigaciones Geograficas* 55 (2004): 79–96.
Caviedes, Cesar N., and Keith D. Muller. "Fruiticulture and Uneven Development in Northeast Brazil." *Geographical Review* 84, no. 4 (1994): 380–93.
Ceron, Antonio, and Lucia Gerardi. "Modernizacao da agricultura brasileira: Transformacoes agrarios em um pais em desenvolvimento." *Revista Geografica* 101 (1985): 5–28.
Chase, Jacqueline. "Controlling Labor Commitment in Brazil's Global Agriculture: The Crisis of Competing Flexibilities." *Environment and Planning D* 15, no. 5 (1997): 587–610.
Comitz, K. M., and T. S. Thomas. *Geographic Patterns of Land Use and Intensity in the Brazilian Amazon.* Washington, DC: World Bank, Development Research Group, Infrastructure and Environment, 2001.
Corneia de Andrade, Manuel. "L'elevage dans le Nord-Est du Bresil." *Les Cahiers d'Outre Mer* 21 (1968): 56–77.
Dambaugh, Luella M. "Westward Migration of Coffee in Brazil." *Journal of Geography* 57, no. 4 (1958): 173–77.
De Espindola, Giovana, et al. "Agricultural Land Use Dynamics in the Brazilian Amazon Based on Remote Sensing and Census Data." *Applied Geography* 32, no. 2 (2012): 240–52.
De Silva, Maria, and D. J. T. Pinheiro. "Depicota a agrara: Las transformaciones de Pelourinho (Salvador, Bahia, Brasil)." *Anales de Geografia de la Universidad Complutense* 17 (1997): 69–97.
Dickenson, John P. "Viticulture in Pre-independence Brazil." *Journal of Wine Research* 6, no. 3 (1995): 195–200.
Dillman, C. Daniel. "Absentee Landlords and Farm Management in Brazil during the 1960s." *American Journal of Economics and Sociology* 34, no. 1 (1975): 1–7.
Diniz, Jose. "IGU's Suggestions and the Types of Agriculture: A Case Study." *Revista Geografica* 70 (1969): 91–108.
Dupon, J. F., and A. Vant. "Contrastes et Changements dans l'agriculture du Goias central." *Les Cahiers d'Outre Mer* 42 (1979): 217–52.
Eudamidas Bezerra, Juscelino, and Denise Elias. "Difusao do trabalho agricola formal no Brasil e sua dinamica multiescarlai." *Investigaciones Geograficas* 76 (2011): 103–17.

Fearnside, Philip M. "Deforestation and Agricultural Development in the Brazilian Amazon." *Interciencia* 14, no. 6 (1989): 291–97.

———. "Land-Tenure Issues as Factors in Environmental Destruction in Brazilian Amazonia: The Case of Southern Para." *World Development* 28, no. 8 (2001): 1361–72.

———. "Rethinking Continuous Cultivation in Amazonia." *Bioscience* 37, no. 3 (1987): 209–13.

———. "Soybean Cultivation as a Threat to the Environment in Brazil." *Environmental Conservation* 28, no. 1 (2001): 23–38.

Fernandes, Bernardo. "Bresil: Quelle reforme agraire?" *Les Cahiers d'Outre Mer* 53 (2000): 393–400.

Fernandes, Bernardo, and Clifford Welch. "Brazil's Experience with Agrarian Reform, 1995–2006." Challenges for Agrarian Geography." *Human Geography* 1, no. 1 (2008): 59–69.

Filho, Virgilio C. "A Proposito da Industria Pastoril no Brasil." *Revista Geografica* 46 (1957): 55–83.

Flowers, Nancy W., et al. "Variation in Swidden Practices in Four Central Brazilian Indian Societies." *Human Ecology* 10, no. 2 (1982): 203–17.

Gauche, E. "Family Farming and Sustainable Development in a Protected Amazonian Area: The Case of the Apa do Igarape Gelado (South-Eastern Para in Brazil) [in French]." *Annales de Geographie* 120, no. 5 (2011): 528–53.

Geiger, Pedro P., et al. "Distribuicao de Atividades Agropastoris em Torno da Metropole de Sao Paulo." *Revista Geografica* 80 (1974): 35–70.

Gervaise, Y. "La cultura de la pomme de terre a Ouro Branco (Minas Gerais): Un exemple de specialization agricole." *Les Cahiers d'Outre Mer* 26 (1973): 279–313.

Grabois, Jose, et al. "Reodenacao Espacial e Evolucao da Economia Agraria: O Caso das Terras Altas da Transicao Agreste-Mata do Norte de Pernambuco." *Revista Brasileira de Geografia* 54, no. 1 (1992): 121–78.

Haas, William H. "The Coffee Industry of Brazil." *Journal of Geography* 28 (1929): 41–57.

Haggett, Peter. "Land Use and Sediment Yield in an Old Plantation Tract of the Serra do Mar, Brazil." *Geographical Journal* 127, no. 1 (1961): 50–62.

Halperian, Daniel T. "The Jari Project: Large-Scale Land and Labor Utilization in the Amazon." *Geographical Survey* 9, no. 1 (1980): 13–21.

Hecht, Susanna B. "Environmental Development and Politics: Capital Accumulation and the Livestock Sector in Eastern Amazonia." *World Development* 13 (1985): 663–84.

———. "Valuing Land Uses in Amazonia: Colonist Agriculture, Cattle, and Petty Expansion in Comparative Perspective." In *Conservation of Neotropical Forests: Working from Traditional Resource Use*, edited by K. H. Redford and C. Padoch, 379–99. New York: Columbia University Press, 1992.

Hiraoka, Mario, and Shozo Yamamoto. "Changing Agricultural Land Use in the Agreste of Northeast Brazil." *Latin American Studies, Japan* 2 (1981): 81–124.

Huetz de Lemps, Alain. *La Canne a sucre au Bresil*. Talence, France: Travaux et documents de geographie tropicale 29, 1979.

James, Preston E. "The Coffee Lands of Southeastern Brazil." *Geographical Review* 22, no. 2 (1932): 225–44.

———. "Patterns of Land in Northeast Brazil." *Annals of the Association of American Geographers* 43, no. 1 (1953): 98–126.

———. "Trends in Brazilian Agricultural Development." *Geographical Review* 43, no. 2 (1953) 301–328.

Jepson, Wendy. "Producing a Modern Agricultural Frontier: Firms and Cooperatives in Eastern Mato Grosso, Brazil." *Economic Geography* 82, no. 3 (2006): 289–316.

Jepson, Wendy, et al. "Access Regimes and Regional Land Change in the Brazilian Cerrado, 1972–2002." *Annals of the Association of American Geographers* 100, no. 1 (2010): 87–111.

Johnson, Frederick I. "Sugar in Brazil: Policy and Production." *Journal of Developing Areas* 17, no. 2 (1983): 243–56.

Junior, Manuel D. "Propriedade e Uso da Terra na 'Plantation' Brasileira." *Revista Geografica* 48 (1958): 66–100.

Keithan, Elizabeth F. "The Cacao Industry of Brazil." *Economic Geography* 15, no. 2 (1939): 195–204.

Kerr, John A. "Cattle Raising in Brazil." *Revista Geographica* 83 (1975): 95–108.

Kleinpenning, J. M. G. "Losing Ground: Processes of Land Concentratiaon in the Cocoa Region of Southern Bahia (Brazil)." *Boletin de Estudios Latinoamericans y del Caribe* 33 (1982): 59–83.

Kuplich, T. M., et al. "The Study of ERS-1 SAR and Landsat TM Synergism for Land Use Classification." *International Journal of Remote Sensing* 21, no. 10 (2000): 2101–12.

Lima, Andre, et al. "Land Use and Land Cover Changes Determine the Spatial Relationship between Fire and Deforestation in the Brazilian Amazon." *Applied Geography* 34 (2012): 239–46.

Lorandi, Reinaldo, et al. "Geotechnical Investigation of Land in the Municipality of Descalvado (SP, Brazil) for Selection of Suitable Areas for Residency Water Treatment Lagoons." *Journal of Geography and Regional Planning* 3, no. 8 (2010): 200–207.

Ludwig, Armin K. "Brazil's New Gasohol Program: Its Sugar Cane Production and Land Requirements." *Proceedings of the Conference of Latin Americanist Geographers* 9 (1983): 174–90.

Machado, Lia. "A fronteira agricola na Amazonia Brasileira." *Revista Brasileira de Geografia* 54, no. 2 (1992): 27–56.

Marinlo, Mara, et al. "Bibliografia Brasileira de Lavantamento e de Interpretacao de Solos Para Fins Agricolas." *Revista Brasileira de Geografia* 53, no. 1 (1991): 172–76.

Marsden, T. K. "Reshaping Environments: Agriculture and Water Interactions and the Creation of Vulnerabililty." *Transactions, Institute of British Geographers* 22, no. 3 (1997): 321–37.

Maruyama, Hiraoki, et al. "Grazing Behavior of Cows in Dry Season Measured by Handheld GPS and BiteCounter Collar: A Case of Fazenda Baia Bonita in South Pantanal, Brazil [in Japanese]." *Tsukuba Studies in Human Geography* 32 (2008): 17–36.

Maruyama, Hiraoki, and Takaaki Nihei. "Grazing Behavior of Cows Measured by Handheld GPS and Bite Counter Collar: A Case of Fazenda Baia Bonita in South Pantanal, Brazil [in Japanese]." *Jimbun Chiri/Human Geography* 59, no. 1 (2007): 30–43.

Matzretter, J. "Derkakao-Aabau in Brasilien." *Wirtschaftsgeographische Studien* 7 (1980): 23–48.

Meaney-Leckie, Anne. "The Cashew Industry of Ceara, Brazil: A Development Alternative." *Proceedings, Middle States Division, Association of American Geographers* 22 (1984): 35–44.

———. "Large-Scale Irrigation Projects as a Means of Socio-economic Development in the Third World: A Brazilian Example." *Middle States Geographer* 20 (1987): 76–81.

Mesquita, Olindina, and Solange Silva. "A evolucao da agricultura Brasileira na decada de 70." *Revista Brasileira de Geografia*, 49, No 1 (1987): 3–10.

Messias de Almeida, Thiara, and Anom C. de Oliveira Teixeira. "Inter-relacoes entre fatores fisicos e socioeconomicos na dinamica de uso da terra no extreme sul da Bahia." *Revista Geografica Academica* 4, no. 2 (2010): 64–72.

Minzenberg, Eric, and Richard Wallace. "Amazonian Agriculturalists Bound by Subsistence Hunting." *Journal of Cultural Geography* 28, no. 1 (2011): 99–121.

Monbeig, Pierre. "Les petits cultivateurs de l'Etat de Sao Paulo." *Les Cahiers d'Outre Mer* 5 (1952): 278–82.

———. "Les structures agraires dans la frange pionniere de Sao Paulo." *Les Cahiers d'Outre Mer* 4 (1951): 1–22.

Montanher, Otavio C., et al. "A relacao entre o meio fisico e o avanco da cana-de-acucar no noroeste do Parana." *Revista Geografica Academica* 4, no. 1 (2010): 20–31.

Moran, Emilio, and E. Brondizio. "Land Use Change after Deforestation in Amazonia." In *People and Pixels: Linking Remote Sensing and Social Science*, edited by Diana Livermore et al., 94–120. Washington, DC: National Academy of Science Press, 1998.

Nishikawa, Daijiro. "Regional Development of Soybean Agriculture in Brazil: A Case Study of Braganca Paulista, Sao Paulo [in Japanese]." *Annals, Japan Association of Economic Geographers* 14, no. 1 (1968): 22–54.

Noble, Allen G. "Geographic Aspects of the Agriculture of Santa Catarina State, Brazil." *Ohio Journal of Science* 67, no. 5 (1967): 257–73.

Nogueira da Fonseca, Walmar. "Nocoes sobre aproveitamento agrossilviculture na Amazonia." *Revista Brasileira de Geografia* 59, no. 1 (2005): 75–92.

Oliveira Melo, Elizabete, and Manfred Feler. "O uso atual do solo e da agua na bacia do ribeirao Picarrao-Araguari-mg-Brasil." *Investigaciones Geograficas* 72 (2010): 39–48.

De Olivera, Leticia, and Paula Waquil. "Preforestation Land Use and Social Development in the Rio Grande do Sul State, Brazil." *Journal of Geography and Regional Planning* 3, no. 11 (2010): 297–305.

Pacheco, Pablo. "Agrarian Change, Cattle Ranching, and Deforestation: Assessing Their Linkages in Southern Para." *Environment and History* 15, no. 4 (2009): 493–520.

———. "Agrarian Reform in the Brazilian Amazon: Its Implications for Land Distribution and Deforestation." *World Development* 37, no. 8 (2007): 1337–47.

———. "Smallholder Livelihoods, Wealth, and Deforestation in Eastern Amazonia." *Human Ecology* 37, no. 1 (2009): 27–41.

Pacheco, Pablo, and Rene Poccard-Chapuis. "The Complex Evolution of Cattle Ranching Development Amid Market Intergration and Policy Shifts in the Brazilian Amazon." *Annals of the Association of American Geographers* 102, no. 6 (2012): 1366–90.

Pacheco do Amaral Filho, Zebino, and Isaias Oenning. "Apridao Agricola e zoneamento Agricola da Gleba Ol-Projecto assentamento Machadinho." *Revista Brasileira de Geografia* 59, no. 1 (2005): 135–48.

Papy, L. "Au pays de plantations cafeires de Sao Paulo." *Les Cahiers d'Outre Mer*, 7 (1954): 195–203.

———. "En marge de l'Empire du café: la façade atlantique de Sao Paulo." *Les Cahiers d'Outre Mer* 5 (1952): 357–98.

Pebayla, Raymond. *Aspects de l'agriculture commercial et de l'elevage au Bresil*. Talence, France: Travaux et documents de geographie tropicale 11, 1973.

———. "Aspects de l'elevage dans le Pantanal bresilien." *Les Cahiers d'Outre Mer* 27 (1974): 395–400.

———. "Fondements physiques de la geographie agrarie et economie de cueillette au Bresil, selon Orlando Valverde." *Les Cahiers d'Outre Mer* 21 (1968): 209–16.

Pices Guimares, Luiz, and N. R. Innocencio. "A Evolucao da Agricultura na Regiao Sudeste na decada de 70." *Revista Brasileira de Geografia* 49, no. 1 (1987): 107–58.

Pinedo-Vasquez, Miguel, et al. "Brazil (Amazonia)." In *Agrodiversity: Learning from Farmers across the World*, edited by Harold Brookfield et al., 43–78. Tokyo, Japan: UN University Press, 2003.

Pinto de Gusmao, Rivaldo, et al. "Difusao da Infraestrutura de Armazenagem e suas Vinculacoes com a atividade agrarian no noroeste do Rio Grande do Sul." *Revista Geografica* 88 (1978): 103–49.

Platt, Robert S. "Coffee Plantations of Brazil: A Comparison of Occupance Pattern in Established and Frontier Areas." *Geographical Review* 25, no. 2 (1935): 231–39.

Poccard-Chapuis, Rene, et al. "The Beef Value Chain: A Tool for Monitoring the Dynamics in Pioneer Frontiers in the Brazilian Amazon [in French]." *Cadernos de Ciencia y Tecnologia* 22, no. 1 (2005): 125–36.

Poelhekke, Fabio. "Fences in the Jungle: Cattle Raising and the Economic and Social Integration of the Amazon Region in Brazil." *Revista Geografica* 104 (1986): 33–48.

Rodrigues Mendonca, Marcelo, and Antonio Thomaz Junior. "A modernizacao da agricultura nas areas de cerrado em Goias (Brasil) e os impactos sobre a travalho." *Investigaciones Geograficas* 55 (2004): 91–121.

Rose, Newton. "Rainfall Reliability and Agriculture in the Semi-arid Tropics: The Case of the Jaguaribe River Basin, Ceara, Northeast Brazil." In *Spatial Aspects of Development*, edited by Brian S. Hoyle, 115–28. London: Wiley, 1974.

Rossini, Rosa E. "Women as Labor Force in Agriculture: The Case of the State of Sao Paulo, Brazil." *Revista Geografica* 97 (1983): 91–95.

Saito, Isao. "Agro-industries in the Middle Sao Francisco Valley, Northeast Brazil [in Japanese]." *Annals of the Japan Association of Economic Geographers* 37 (1991): 225–44.

———. "Irrigation Farming in Boqueirao in the Middle Paraiba Valley, Northeast Brazil." *Tsukuba Studies in Human Geography* 13 (1989): 23–52.

Saito, Isao, et al. "Agriculture and Land Tenure in Salgado de Sal Felix along the Middle Reaches of the Paraiba River in Northeast Brazil." *Latin American Studies, Japan* 8 (1986): 91–106.

———. "Drought, Canal Irrigation Project, and Family Farmers in the Icohima Compas Area of Ceara, Northeast Brazil [in Japanese]." *Tsukuba Studies in Human Geography* 21 (1997): 69–92.

———. "Irrigation Projects and Corporate Agriculture in the Middle Sao Francisco Valley, Northeast Brazil." *Tsukuba Studies in Human Geography* 15 (1991): 269–300.

———. "Reviving Latifundios and Agribusiness in the Coastal Cashew-Meon Belt of Northeast Brazil [in Japanese]." *Tsukuba Studies in Human Geography* 23 (1999): 81–104.

Saito, Isao, and Hiraoki Maruyama. "Some Types of Livestock Ranching in Sao Joao do Cariri on the Upper Paraiba Valley, Northeast Brazil." *Latin American Studies, Japan* 10 (1988): 101–20.

Saito, Isao, and Noritake Yagasaki. "Drought, Irrigation, and Change in the Sertao of North-East Brazil." In *The Fragile Tropics of Latin America: Sustainable Management of Changing Environments*, edited by Toshie Nishizawa and J. Vitto, 301–23. Tokyo, Japan: UN University Press, 1995.

———. "Zonal Patterns of Agricultural Land Use in the State of Paraiba, Northeast Brazil." *Geographical Review of Japan* 60B, no. 1 (1987): 66–82.

De Saldanha, Morilia, et al. "A evolucao da agricultura na regiao Nordester na decada de 70." *Revista Brasileira de Geografia* 49, no. 1 (1987): 47–106.

Sanders, John, and Frederick Bein. "Agricultural Development in the Brazilian Frontier: Southern Mato Grosso." *Economic Development and Cultural Change* 24, no. 3 (1976): 539–610.

Santos, Milton. "La culture de cacao dans l'Etat de Bahia." *Les Cahiers d'Outre Mer* 16 (1963): 360–78.

Da S. Gomes, A., et al. "Large-Scale Capitalist and Family-Based Agriculture in Bahia State of Brazil [in French]." *Annales de Geographie* 120, no. 1 (2011): 3–25.

Siren, Anders H., and Eduardo S. Brandizio. "Detecting Subtle Land Use Change in Tropical Forests." *Applied Geography* 29, no. 2 (2009): 201–11.

Smith, Harry, and J. Raemaekers. "Land Use Pattern and Transport in Curitiba." *Land Use Policy* 15, no. 3 (1998): 233–51.

Smith, Nigel J. H. "Agricultural Productivity along Brazil's Transamazon Highway." *Agro-Ecosystems* 4 (1978): 415–32.

Socorro Brito, Maria. "A face destruidora da apropriacao e uso do territorio amazonica." *Revista Brasileira de Geografia* 59, no. 1 (2005): 93–104.

Socorro Brito, Maria, and Mitijo Y. Une. "A Evolucao da Agricultura na Regiao Norte na Decada de 70." *Revista Brasileira de Geografia* 49, no. 1 (1987): 11–46.

Sorrenson, Cynthia. "Contributions of Use Study to Land Use/Cover Change Frameworks: Understanding Change in Agricultural Frontiers." *Human Ecology* 32, no. 4 (2004): 395–420.

Strain, Warren. "Carnauba Wax: Brazil's Monopoly Product." *Journal of Geography* 41 (1942): 121–29.

Taravella, Romain, and Xavier Arnauld de Sartre. "The Symbolic and Political Appropriation of Scales: A Critical Analysis of the Amazonian Ranchers' Narrative." *Geoforum* 43, no. 3 (2012): 645–56.

Uttley, Marguerite. "Brazil's Place in the Rubber World." *Journal of Geography* 22 (1923): 336–41.

Valverde, Orlando. "Land Use in the Northeast State of Rio Grande do Norte, Brazil." *Revista Geografica* 55 (1961): 13–20.

Van Den Berg, M. *Land Use Systems Research on Strongly Weathered Soils in South and South-East Brazil.* Utrecht: Netherlands Geographical Studies 271, 2000.

Voeks, Robert A. "The Brazilian Fiber Harvest and Management of the Piasana Fiber Palm (*Attalea funifera*) Mart." *Advances in Economic Botany* 6 (1988): 262–75.

Waibel, Leo. "Vegetation and Land Use in the Planalto Central of Brazil." *Geographical Review* 38, no. 4 (1948): 529–54.

Ward, Robert. "Climate and Coffee in Brazil." *Journal of Geography* 10 (1911): 16–20.

Watkins, C. "Dedezeira: African Oil Palm Agroecologies in Bahia, Brazil, and Implications for Development." *Journal of Latin American Geography* 10, no. 1 (2011): 9–34.

Watson, Kelly, and Moira Achinelli. "Context and Contingency: The Coffee Crisis for Conventional Small-Scale Coffee Farmers in Brazil." *Geographical Journal* 174, no. 3 (2008): 223–34.

Wesche, Rolf. "Planned Rainforest Family Farming on Brazil's Transamazonic Highway." *Revista Geografica* 81 (1974): 105–14.

WinklerPrins, Antoinette M. "Why Context Matters: Local Soil Knowledge and Management along an Indigenous Peasantry on the Lower Amazon Floodplain, Brazil." *Ethnoecologica* 5, no. 7 (2001): 6–20.

WinklerPrins, Antoinette M., and D. G. McGrath. "Smallholder Agriculture along the Lower Amazon Floodplain, Brazil." *PLEC News and Views* 16 (2000): 34–42.

Wolford, Wendy. "Environmental Justice and the Construction of Scale in Brazilian Agriculture." *Society and Natural Resources* 21, no. 7 (2008): 641–55.

Yagasaki, Noritake, et al. "Irrigation Farming in Teixeira, Northeast Brazil." *Journal of the Yokohama National University* 1, no. 35 (1989): 71–98.

Commerce and Trade

Coxe, George R. "Brazil's Foreign Trade." *Journal of Geography* 22 (1923): 318–23.

Ghizzo, Marcio, and Marcio Mendes Rocha. "Acidade para o consumo: Mobilidade e centralidade em Marginga-PR. Brazil: O caso do Hipermercado big." *Revista Geografica* 138 (2005): 29–46.

Haman de Figueiredo, Adma. "Credito Rural e Mudanca Tecnologica no Oeste do Parana." *Revista Geografica* 117 (1993): 119–69.

Jones, Clarence F. "The Evolution of Brazilian Commerce." *Economic Geography* 2, no. 4 (1926): 550–74.

Lustosa Costa, Maria. "La segregation des marts: la diffusion d'une idée des lumieres d'Europe occidentale jusqua Fortaleza (Bresil)." *Geographie et Cultures* 24 (1997): 51–70.

Shaw, Earl B. "The Banana Trade of Brazil." *Economic Geography* 23, no. 1 (1947): 15–21.

Da Silva, Jorge X., et al. "Um Banco de Dados Ambientais Para a Amazonia." *Revista Brasileira de Geografia* 53, no. 3 (1991): 91–124.

Takahashi, N., and N. M. Yoshikae. "Structure de l'espace financier au Bresil." *Latin American Studies*, Japan 7 (1983): 77–102.

———. "Structure de l'espace financier au Bresil [in Japanese]." *Tsukuba Studies in Human Geography* 8 (1984): 203–34.

Webb, Kempton E. *Geography of Food Supply in Central Minas Gerais.* Washington, DC: National Academy of Science, 1959.

———. "Problems of Food Supply in Brazil." *Journal of Latin American Studies* 3, no. 2 (1961): 239–48.

Manufacturing and Labor Studies

Anthony, Edward J. "The Structure of Industrial Change: Regional Incentives to the Textile Industry of Northeast Brazil." In *Spatial Analysis, Industry, and the Industrial Environment: Progress in Research and Application. Regional Economies and Industrial Systems*, edited by F. E. I. Hamilton and G. J. R. Linge, 3:485–508. Chichester, England: Wiley, 1981.

Antunes, Ricardo. "Global Economic Restructuring and the World of Labor in Brazil: The Challenges to Trade Union and Social Movements." *Geoforum* 32, no. 4 (2001): 449–58.

Borello, Jose A. "Evaluation and Regional Development: Sudene, Brazil and Government-Induced Industrialization." *Geoscope* 14, no. 2 (1984): 65–81.

Campos Ribeiro, Miguel A., and Roberto Schmidt de Almeida. "Analise de organizacao espacial in industria nordestina atraves de uma tipologia de centros industriais." *Revista Brasileira de Geografia* 53, no. 2 (1991): 5–32.

———. "Os Pequenos e Medios Estabeliecimentos Industriais Nordestinos: Padroes de Distribuicao e Fatores Condicionantes." *Revista Brasileira de Geografia* 53, no. 2 (1991): 5–50.

Cavalcanti, Leonardo, et al. "Multinationals, the New International Economic Order and the Spatial Industrial Structure of Brazil." In *Spatial Analysis, Industry, and the Industrial Environment: Progress in Research and Application. International Industrial Systems*, edited by F. E. I. Hamilton and G. J. R. Linge, 3:423–38. Chichester, England: Wiley, 1981.

Dickenson, John P. *Brazil: Studies in Industrial Geography*. Boulder, CO: Westview, 1978.

———. "Imbalance in Brazil's Industrialization." In *Spatial Aspects of Development*, edited by Brian S. Hoyle, 291–306. New York: Wiley, 1974.

———. "Industrial Estates in Brazil." *Geography*, 55, (1970): 326–29.

———. *Industrial Estates in Brazil: Problems of Definition*. Liverpool, England: University of Liverpool, Department of Geography, Liverpool Papers in Human Geography 7, 1982.

———. "The Iron and Steel Industry in Minas Gerais, Brazil, 1695–1965." In *Liverpool Essays in Geography*, edited by Robert W. Steel and R. Lawton, 407–22. London: Longman, 1967.

Enders, Wayne T. "Regional Disparities in Industrial Growth in Brazil." *Economic Geography* 56, no. 3 (1980): 300–310.

Geiger, Pedro P. "Industrial Boom in Southeastern Brazil." *Revista Geografica* 56 (1962): 23–42.

Geiger, Pedro P., et al. "Regional Differences in Brazil's Industrial System." *Luso-Brazilian Review* 20, no. 1 (1983): 13–43.

Graham, Douglas H. "Interstate Migration and the Industrial Labor Force in Central-South Brazil." *Journal of Developing Areas* 12, no. 1 (1977): 31–48.

Grenier, Phillippe. "The Alcohol Plan in the Development of Northeast Brazil." *GeoJournal* 11, no. 1 (1985): 61–68.

Grewe, K. "Future Aspects of New Vehicle Distribution from VW do Brasil." *GeoJournal* 3, no. 1 (1979): 27–34.

Haddad, Paulo R., and Jacques Schwartzman. "The Location of Labor-Oriented Activities: A Graphic Analysis." *Annals of Regional Science* 9, no. 2 (1975): 40–45.

———. "A Space Cost Curve of Industrial Location." *Economic Geography* 50, no. 2 (1974): 141–43.

Hamnan de Figueiredo, Adma, et al. "Difusao das Industries de beneficimento, fiacao e tecelagem de algodao e fiacao e tecelagem de fibras artificais e sinteticas na Regiao Sudeste." *Revista Geografica* 88 (1978): 89–102.

James, Preston E. "Industrial Development in the Sao Paulo State, Brazil." *Economic Geography* 11, no. 3 (1935): 258–66.

McGrath, Siobhan. "Fuelling Global Production with Slave Labour? Migrant Sugar Cane Workers in the Brazilian Ethanol GPN." *Geoforum* 44 (2013): 32–43.

Nobre Carneiro, Rosalvo. "A sinteracoe: sociedade e natureza nos espacos das redes nordestinos e as configuracoes de seus meios geograficos." *Revista Geografica Academica* 2, no. 3 (2008): 50–56.

O'hUallachain, Breandan, and David Wasserman. "Vertical Integration in a Lean Supply Chain: Brazilian Automobile Compenent Parts." *Economic Geography* 75, no. 1 (1999): 21–42.

Pintaudi, Silvana, and Ana Allesandri-Carlos. "L'espace et l'industrie dans l'etat de Sao Paulo (Bresil)." *Les Cahiers d'Outre Mer* 49, no. 1 (1996): 31–52.

Ramaldo, Jose R., and Marco Santana. "VW's Modular System and Workers' Organization in Resende, Brazil." *International Journal of Urban and Regional Research* 26, no. 4 (2002): 756–66.

Schmidt de Almeida, Roberto, and Miguel A. Campos Ribeiro. "A Questao ambiental do Nordeste na Decoda de 1980." *Revista Geografica* 116 (1992): 41–51.
Storper, Michael. "Who Benefits from Industrial Development? Social Power in the Labor Market, Income Distribution, and Spatial Policy in Brazil." *Regional Studies* 18, no. 2 (1984): 143–64.
Townroe, Peter M. "A Discriminate Analysis of the 1980 Sao Paulo Industrial Location Survey." *Environment and Planning A* 17, no. 1 (1985): 115–32.
Vining, Daniel R. "Industrial Decentralization in Brazil: A Query." *Regional Studies* 19, no. 2 (1985): 163–64.
Wehrhahn, Rainer, und Jurgen Bahr. "Industrielle Polarisierung und Dekonzentration in Sao Paulo." *Zeitschrift fur Wirtschaftsgeographie* 45, no. 1 (2001): 31–53.

Energy, Mining, Fishing, and Lumbering

Abreu, S. F. "The Mineral Wealth of Brazil." *Geographical Review* 36, no. 2 (1946): 222–46.
Anderson, Anthony B., and Mario A. G. Jardin. "Costs and Benefits of Floodplain Forest Management by Rural Inhabitants in the Amazon Estuary: A Case Study of Acai Palm Production." In *Fragile Lands of Latin America: Strategies for Sustainable Development*, edited by John O. Browder, 114–29. Boulder, CO: Westview, 1989.
Barbosa de Almeida, Sergio. "O potencial hidrelechtrico Brasileiro." *Revista Brasileira de Geografia* 53, no. 3 (1991): 183–204.
Barham, Bradford L., and Oliver T. Coomes. *Prosperity's Promise: The Amazon Rubber Boom and Distorted Economic Development*. Boulder, CO: Westview, 1996.
Bodenman, John E. "Brazilian Development and Deforestation of Amazonia: An Environmental Valuation Approach." *Middle States Geographer* 33 (2000): 20–30.
Brown, K., and S. Rosendo. "The Institutional Architecture of Extractive Reserves in Rondonia, Brazil." *Geographical Journal* 166, no. 1 (2000): 35–48.
Cesar de Lima Ramires, Juilo. "As grandes corporacoes e a dinamica socio-espacial-a Acao da Petrobas em Macae." *Revista Brasileira de Geografia* 53, no. 4 (1991): 115–52.
Cooper, J. C. "Brazil: A Major Bauxite Producer of the 1980s." *Geography* 66, no. 3 (1981): 222–25.
Fearnside, Philip M. "Deforestation in Brazilian Amazonia: The Rates and Causes of Forest Destruction." *The Ecologist* 19 (1989): 214–18.
———. "Deforestation in the Brazilian Amazon: How Fast Is It Occurring?" *Ambio* 7 (1982): 82–88.
———. "Jari Revisited: Changes and the Outlook for Sustainability in Amazonia's Largest Silvacultural Estate." *Interciencia* 10 (1985): 121–29.
———. "Land Use Trends in the Brazilian Amazon Region as Factors in Accelerating Deforestation." *Environmental Conservation* 10, no. 2 (1983): 141–47.
———. "The Rate and Extent of Deforestation in the Brazilian Amazonia." *Environmental Conservation* 17 (1990): 213–26.
———. "Spatial Concentration of Deforestation in the Brazilian Amazon." *Ambio* 15 (1986): 74–81.
Garrida Filha, Irene. "Estudos geograficos da area mineradora garimpeira de ouro do norte de Mato Grosso e da area mineradora garimpeira e empressarial de ouro do Amapa." *Revista Brasileira de Geografia* 59, no. 1 (2005): 105–34.
Garrida Filha, Irene, et al. "A mineracao da bauxita no vale do Trombetus: Estudo de meio ambiente e use do solo." *Revista Brasileira de Geografia* 52, no. 3 (1990): 41–82.
James, Preston E. "Water Power Resources in Brazil." *Economic Geography* 18, no. 1 (1942): 13–16.
Jordan, C. F. "Jari: A Pulp Plantation in the Brazilian Amazon." *GeoJournal* 19, no. 4 (1989): 429–36.
Lerat, S. "Le Bresil et le petrole." *Les Cahiers d'Outre Mer* 41 (1988): 317–30.
Moran, Emilio. "Deforestation and Land Use in the Brazilian Amazon." *Human Ecology* 21, no. 1 (1993): 1–22.

Parayil, G., and F. Tong. "Pasture-Led to Logging-Led Deforestation in the Brazilian Amazon: The Dynamics of Socio-environmental Change." *Global Environmental Change* 8, no. 1 (1998): 63–79.

Pontius, Robert G., et al. "Accuracy Assessment for a Simulation Model of Amazonian Deforestation." *Annals of the Association of American Geographers* 97, no. 4 (2007): 677–95.

Schneider, Laura, et al. "Method for Spatially Explicit Calculations of Potential Biomass Yields and Assessment of Land Availability for Biomass Energy Production in Northeastern Brazil. "*Agriculture, Ecosystems, and Environments* 84 (2001): 207–26.

Simmons, Cynthia S., et al. "Tree Planting by Small Producers in the Tropics: A Comparative Study of Brazil and Panama." *Agroforestry Systems* 56 (2002): 89–105.

Slinger, Vanessa A. "Peri-urban Agroforestry in the Brazilian Amazon." *Geographical Review* 90, no. 2 (2000): 177–90.

Smith, Nigel J. H., et al. "Agroforestry Development and Potential in the Brazilian Amazon." *Land Degradation and Rehabilitation* 6 (1995): 251–63.

———. "Agroforestry Trajectories among Smallholders in the Brazilian Amazon: Innovation and Resiliency in Pioneer and Older Settled Areas." *Ecological Economics* 18 (1996): 15–27.

Sternberg, Hilgard O. "The Distribution of Water Power Resources in Brazil with References to the Participation Ration Concept." *Annals of the Association of American Geographers* 38, no. 2 (1948): 133–44.

Sternberg, Rolf. "Hydroelectric Power in the Context of Brazilian Urban and Industrial Planning." In *Latin America: Case Studies*, edited by Richard Boehm and Sent Visser, 187–98. Dubuque, IA: Kendall-Hunt, 1984.

———. "Large Scale Hydroelectric Projects and Brazilian Politics." *Revista Geografica* 101 (1985): 29–44.

———. "Perceptions, Plans, Projects, and Policies: The Hydroelectric Sector in Brazil." *Proceedings, Conference of Latin Americanist Geographers* 9 (1983): 172–73.

———. "Perspectivas geograficas nos sistemas hidroelectricos." *Revista Brasileira de Geografia* 52, no. 2 (1990): 157–88.

Teulieres, R. "Les diamants du Minas (Bresil)." *Les Cahiers d'Oure Mer* 9 (1956): 389–411.

Tourreau, Francois M., and Annal Greissing. "A Quest for Sustainability: Brazil Nut Gatherers of Sao Francisco do Iratapura and the Natura Corporation." *Geographical Journal* 176, no. 4 (2010): 334–49.

Vadjunec, Jacqueline. "Extracting a Livelihood: Institutional and Social Dimention of Deforestation in the Chico Mendes Extrative Reserve, Acre, Brazil." *Journal of Latin American Geography* 10, no. 1 (2011): 151–74.

Walker, Robert. "The Impact of Brazilian Biofuel Production on Amazonia." *Annals of the Association of American Geographers* 101, no. 4 (2011): 929–38.

Wilcox, Kellie, et al. "The Influence of Socioeconomic Status and Fuelwood Access on Domestic Fuelwood Use in the Brazilian Atlantic Forest." *Journal of Latin American Geography* 10, no. 1 (2011): 195–216.

Transportation and Communications

Arima, Eugenio, et al. "Loggers and Forest Fragmentation: Behavioral Models of Road Building in the Amazon Basin." *Annals of the Association of American Geographers* 95, no. 3 (2005): 525–41.

Cappa, Josmar, and Jose H. Souza. "A importancia do aeroporto international de vira copos para as estrategias emprasariais no Mercado international." *Revista Geografica Academica* 4, no. 1 (2010): 71–82.

Cardeiro, Helena K., and Denise Aparecida Bovo. "A modernidade do espaco Brasileiro atraves de rede nacional de telex." *Revista Brasileira de Geografia* 42, no. 2 (1990): 107–56.

Correa Galvao, Meria. "Caracteriticas da geografia dos transportes no Brasil." *Revista Geografica* 65 (1966): 69–92.

Fenley, Claudio, et al. "Air Transport and Sustainability: Lessons from Amazonas." *Applied Geography* 27, no. 2 (2007): 63–77.

Galvao, Marilia. "Characteristics of the Geography of Transportation in Brazil." *Revista Geografica* 65 (1966): 69–93.

Gauthier, Howard L. "Transportation and the Growth of the Sao Paulo Economy." *Journal of Regional Science* 8 (1968): 77–94.

———. "Transportation and the Growth of the Sao Paulo Economy." In *Transport and Development*, edited by Brian S. Hoyle, 23–42. New York: Harper & Row, 1973.

Guimares, Andre L., and Christopher Uhl. "Rural Transport in Eastern Amazonia: Limitations, Options, and Opportunities." *Journal of Rural Studies* 13, no. 4 (1997): 429–40.

Momsen, Richard P. "Routes over the Serra do Mar: The Evolution of Transportation in the Highlands of Rio de Janeiro and Sao Paulo." *Revista Geografica* 58 (1963): 5–167.

Perz, Stephen G., et al. "Road Networks and Forest Fragmentation in the Amazon: Explanations for Local Differences with Implications for Conservation and Development." *Journal of Latin American Geography* 7, no. 2 (2008): 85–104.

Thery, H. "Routes transamazoniennes et marginasation de l'espace: Le cas de Rondonia." *Les Cahiers d'Outre Mer* 34 (1981): 5–22.

Timmons-Roberts, J. "Expansion of Television in Eastern Amazonia." *Geographical Review* 85, no. 1 (1995): 41–49.

De Vasconcellos, E. A. "Transport Metabolism, Social Diversity, and Equity: The Case of Sao Paulo, Brazil." *Journal of Transport Geography* 13, no. 4 (2005): 329–39.

Theses and Dissertations

Adams, J. G. U. "The Banana Industry in the Paulista Litoral." Master's thesis, University of Western Ontario, London, Ontario, 1967.

Allderdice, William H. "The Expansion of Agriculture along the Belem-Brasilia Road in Northern Goias, Brazil." PhD diss., Columbia University, New York, 1972.

Arguello, Heliodoro. "Evaluation of Some Alternatives for Economic Development Activities in Acre, Brazil." PhD diss., University of Florida, Gainesville, 1996.

Baxto, Michael W. "Garimpieros of Poxoreo: Small Scale Diamond Miners and Their Environment in Brazil." PhD diss., University of California, Berkeley, 1975.

Britten, Charles E. "The Coffee Industry of Sao Paulo." Master's thesis, University of Tennessee, Knoxville, 1947.

Cole, Ernest W. "A Study of the Development of the Interior Resources of Brazil." Master's thesis, Marshall University, Huntington, WV, 1953.

Dawsey, Cyrus B. "Mode of Transportation and the Income-Distance to Work Relationship in Piracicaba, Sao Paulo, Brazil." PhD diss., University of Florida, Gainesville, 1975.

DeWitt, John W. "Food Production and Regional Development in Bahia, Brazil." PhD diss., University of Florida, Gainesville, 1973.

Dias, Nelson W. "Spatial and Temporal Dimensions of Land Use Change in the Castanhal Region, State of Para, Brazil." Master's thesis, Indiana State University, Terre Haute, 1996.

Dickensori, John P. "Zona Metalurigia: A Study of the Geography of Industrial Development in Minas Gerais, Brazil." PhD diss., University of London, London, 1970.

Donnelly-Morrison, Duane N. "Defining Agricultural Land Use in Rondonia, Brazil by Examination of SPOT Multispectral Data." Master's thesis, Virginia Polytechnic and State University, Blacksburg, 1994.

Faissol, Speridiao. "Problema do Desenvolbinumto Agricola do Sudeste do Planalto Central (Brasil)." PhD diss., Syracuse University, Syracuse, NY, 1956.

Glen, J. "The Technological Structure of Sao Paulo's Agricultural Economy." Master's thesis, University of Western Ontario, London, Ontario, 1971.

Gomez, C. V. "Twenty Years after Chico Mendes: Extractive Reserves' Expansion, Cattle Adoption, and Evolving Self-Definition among Rubber Tappers in the Brazilian Amazon." PhD diss., University of Florida, Gainesville, 2009.

Goulding, W. Michael. "The Ecology and Management of the Rio Madeira Fisheries." PhD diss., University of California, Los Angeles, 1978.

Hansen, Dean L. "Acquiring High Technology: The Case of the Brazilian Informatics Industry." PhD diss., University of Washington, Seattle, 1990.

Haskins, Edward C. "An Agricultural Geography of the Reconcavo of Bahia." PhD diss., University of Minnesota, Minneapolis, 1956.

Hicks, David R. "Agricultural Land Use and Related Innovations and Government Assistance in Rio Grande do Sul." PhD diss., Michigan State University, East Lansing, 1980.

Johns, Norman. "Brazil's Chocolate Forest: Environment and Economic Rates of Conservation in Bahia's Cocoa Agro-System." PhD diss., University of Texas, Austin, 1996.

Johnson, Dennis V. "The Cashew of Northeast Brazil: A Geographical Study of a Tropical Tree Crop." PhD diss., University of California, Los Angeles, 1972.

Kallicott, Signy. "Cotton Production and Potentialities in Northern Brazil." Master's thesis, University of Maryland, College Park, 1947.

Lavernere-Wanderly, Laura L. "Employment and Housing Conditions in the Industrial Area of Santa Luzia (State of Minas Gerais, Brazil.)." PhD diss., University of Pittsburgh, Pittsburgh, PA, 1981.

Long, Robert. "The Middle Paraiba Valley of Brazil: A Study of Land Utilization." PhD diss., Northwestern University, Evanston, IL, 1949.

Mandell, Paul I. "The Development of the Southern Goias Brasilia Region: Agricultural Development in a Landrich Economy." PhD diss., Columbia University, New York, 1969.

Muniz, Osvaldo. "Innovation and Diffusion of Electrical Equipment for Agricultural Development in Minas Gerais, Brazil." PhD diss., University of Tennessee, Knoxville, 1991.

Poccard-Chapuis, Rene. "Les reseaux de la conquete filiere bovine et structuration de l'espace sur les fronts pionniers d'Amazonie oriental Bresilienne." PhD diss., University of Paris, 2004.

Schmidt, M. "Farming and Patterns of Agrobiodiveristy on the Amazon Floodplain in the Vicinity of Mamiraua, Amazonas, Brazil." Master's thesis, University of Florida, Gainesville, 2003.

Sheldon, Nancy. "Characteristics and Potentials of Commercial Oceanic Fisheries in Brazil." Master's thesis, Columbia University, New York, 1968.

Silva, G. M. "Peanut Diversity Management by the Kaiabi (Tupi Guarani) Indigenous People, Brazilian Amazon." PhD diss., University of Florida, Gainesville, 2009.

Steininger, Marc K. "Monitoring the Fate of Cleared Tropical Lands: A Test of Remote Sensing Methods around Manaus, Brazil." Master's thesis, University of Maryland, College Park, 1994.

Stillwell, H. Daniel. "A Geography of Itatiaca National Park, Brazil." PhD diss., Michigan State University, East Lansing, 1961.

Sullinger, Clara B. "Cotton Production in Sao Paulo State, Brazil." Master's thesis, Oklahoma State University, Stillwater, 1950.

Taylor, Harry W. "Areal Patterns of Industry, Sao Paulo State." PhD diss., University of Maryland, College Park, 1962.

Warren, Lynne. "Enterpre Palms in Northern Brazil: Market Structure and Socioeconomic Implications for Sustainable Management." Master's thesis, University of Florida, Gainesville, 1992.

HISTORICAL GEOGRAPHY

General Works

Almeida Vasconcelos, Pedro. "Culture, region et exclavage a Bahia (1549–1898)." *Bulletin de l'Association de Geographes Francais* 73, no. 1 (1996): 21–28.

Bell, Stephen. "Early Industrialization in the South Atlantic: Political Influences on the Charqueadas of Rio Grande do Sul before 1860." *Journal of Historical Geography* 19, no. 4 (1993): 399–411.

Birkland, Ceres. "The Seventeenth-Century Jesuit Missions in Rio Grande do Sul." In *Plantation Traits in the New World*, edited by Roland E. Chardon, 1–120. Baton Rouge: Louisiana State University Studies in Geography, 1983.

Boone, Christopher. "Streetcars and Politics in Rio de Janeiro: Private Enterprise versus Municipal Government in the Provision of Mass Transit, 1903–1920." *Journal of Latin American Studies* 27, no. 2 (1995): 343–65.

———. "Streetcars and Popular Protest in Rio de Janeiro: The Case of the Rio de Janeiro Tramway, Light and Power Company, 1903–1920." *Yearbook, Conference of Latin Americanist Geographers* 20 (1994): 49–58.

Brannstrom, Christian. "Brazilian County-Level Juridical Documents as Sources for Historical Geography: A Case Study from Western Sao Paulo State." *Yearbook, Conference of Latin Americanist Geographers* 23 (1997): 41–50.

———. "Forests for Cotton: Institutions and Organizations in Brazil's Mid-Twentieth-Century Cotton Boom." *Journal of Historical Geography* 36, no. 2 (2010): 169–82.

———. "Polluted Soil, Polluted Souls: The Rockefeller Hookworm Eradication Campaign in Sao Paulo, Brazil, 1907–1926." *Historical Geography* 25 (1997): 25–45.

———. "The Timber Trade in Southeastern Brazil, 1920–1960." *Bulletin of Latin American Research* 24, no. 3 (2005): 288–310.

———. "Was Brazilian Industrialization Fuelled by Wood? Evaluating Sao Paulo's Energy Hinterlands 1900–1960." *Environmental History* 11, no. 4 (2005): 395–430.

Clothier, Marcel. "The Influence of Brazil's Earliest Captaincies on the Present Boundaries of Its Eastern and Northeastern States." In *Studies in Historical Geography: Diffusion of Plantation Traits in the New World*, edited by Roland E. Chardon, 101–14. Baton Rouge: Louisiana State University, Geosciences and Man, 1981.

Dawsey, Cyrus B. *The Confederados: Old South Immigration in Brazil*. Tuscaloosa: University of Alabama Press, 1995.

Delson, Robert M. "Colonization and Modernization in Eighteenth-Century Brazil." In *Social Fabric and Spatial Structure in Colonial Latin America*, edited by David J. Robinson, 281–314. Syracuse, NY: Syracuse University, Department of Geography, 1979.

Dickenson, John P. *The Past in a Foreign Country: A Case Study of Minas Gerais, Brazil*. Liverpool, England: University of Liverpool, Department of Geography, Liverpool Papers in Human Geography, New Series 3, 1992.

Dickenson, John P., and Robert M. Delson. *Enterprise under Colonialism: A Study of Pioneer Industrialization in Brazil, 1700–1930*. Liverpool, England: University of Liverpool, Institute of Latin American Studies, Working Paper 12, 1991.

Frank, Zephyr, and Whitney Berry. "The Slave Market in Rio de Janeiro Circa 1869: Context, Movement, and Social Experience." *Journal of Latin American Geography* 9, no. 3 (2010): 85–110.

Galloway, J. H. "The Last Years of Slavery on the Sugar Plantations of Northeastern Brazil." *Hispanic American Historical Review* 51, no. 4 (1971): 586–605.

———. "Northeast Brazil, 1700–1750: The Agricultural Crisis Re-Examined." *Journal of Historical Geography* 1, no. 1 (1975): 21–38.

———. "The Sugar Industry of Pernambuco during the Nineteenth Century." *Annals of the Association of American Geographers* 58, no. 2 (1968): 285–303.

Hecht, Susana B. "The Last Unfinished Page of Genesis: Euclides da Cunha and the Amazon." *Historical Geography* 32 (2004): 43–69.

Jepson, Wendy. "Private Agricultural Colonization on a Brazilian Frontier, 1970–1980." *Journal of Historical Geography* 32, no. 4 (2006): 839–63.

Lehr, John, et al. "A Tale of Two Frontiers: Ukrainian Settlement in Canada and Brazil, 1891–1914." In *The Estevan Papers*, edited by Bernard Thraves et al., 130–47. Regina, Saskatchewan: University of Regina, Department of Geography, Regina Geographical Studies 6, 1997.

Lobb, C. Gary. "The Sesmaria in Rio Grande do Sul: A Successful Frontier Institute, 1737–1823." *Yearbook, Association of Pacific Coast Geographers* 38 (1976): 49–63.

Marchant, Alexander. "Colonial Brazil as a Way Station for the Portuguese India Fleet." *Geographical Review* 31, no. 3 (1941): 454–65.

————. "The Discovery of Brazil: A Note on Interpretation." *Geographical Review* 35, no. 2 (1945): 296–300.

Martins, Luciana. "Illusions of Power: Vision, Technology, and the Geographical Exploration of the Amazon, 1924–1925." *Journal of Latin American Cultural Studies* 16, no. 3 (2007): 285–307.

Martins, Luciana, and Mauricio A. Abreau. "Paradoxes of Modernity: Imperial Rio de Janeiro, 1808–1821." *Geoforum* 32, no. 4 (2001): 533–50.

Monasterio, Leonardo. "Brazilian Spatial Dynamics in the Long Term (1872–2000): Path Dependency or Reversal of Fortune?" *Journal of Geographical Systems* 12, no. 1 (2010): 51–67.

Monbeig, Pierre. "The Colonial Nucleus of Barao de Antonia, Sao Paulo." *Geographical Review* 30, no. 2 (1940): 260–71.

Pereira de Mello, Mauro. "A questao de limites entre os estados do Acre, do Amazonas, e de Rondonia: Aspectos historicos e formacao da territorio." *Revista Brasileira de Geografia* 52, no. 4 (1990): 5–72.

Seachrist, Robert P. "Agricultural Adaptation in a New Environment: Diffusion of Coffee Replacement Crops in Parana, Brazil." In *Studies in Historical Geography: Diffusion of Plantation Traits in the New World*, edited by Roland E. Chardon, 115–45. Baton Rouge: Louisiana State University, Publications in Geography, 1981.

De Souza, Marques. "The Last Exploration of Lieutenant Marges de Souza: Diary of a Journey on the Ananaz River, Brazil." *Geographical Review* 8 (1919): 243–58.

Thery, H. "Ambiguous Impressions of a Bygone Country, the Brazil in Conrad Malte-Brun's Geographie universelle [in French]." *Annales de Geographie* 121, no. 1 (2012): 66–85.

Theses and Dissertations

Boone, Christopher. "The Rio de Janeiro Tramway, Light, and Power Company and the Modernization of Rio de Janeiro during the Old Republic." PhD diss., University of Toronto, Ontario, 1994.

Lobb, C. Gary. "The Historical Geography of the Cattle Regions along Brazil's Southern Frontier." PhD diss., University of California, Berkeley, 1970.

Momsen, Richard P. "Route over the Serra do Mar: An Historical Geography of Transportation in the Rio de Janeiro-Sao Paulo Area, Brazil." PhD diss., University of Minnesota, Minneapolis, 1960.

Sternberg, Rolf. "Farms and Farmers in an Estaciero World, 1856–1914." PhD diss., Syracuse University, New York, 1971.

PHYSICAL GEOGRAPHY

General Works

Assuncao-Borsato, Victor da, and Edward E. DeSouza Filho. "Mudacas climaticas na bacia hidrografica do Alto Rio Parana, Brazil." *Revista Geografica* 142 (2007): 143–64.

Bataila, Angel B. "La importancia geografico-fisica del Brasil." *Revista Geografica* 48 (1958): 101–18.

Behling, H. "Holocene Environmental Dynamics in Coastal, Eastern, and Central Amazonia and the Role of the Atlantic Sea-Level Change." *Geographica Helvetica* 66, no. 3 (2011): 208–16.

Betart, F., et al. "Biodiversity, Geodiversity, and Conservation Challenges in the Humid Mountains of Northeast Brazil [in French]." *Bulletin de l'Association de Geographes Francais* 88, no. 1 (2011): 17–26.

Eyles, C. H., et al. "Glaciation and Tectonics in an Active Intracratonic Basin: The Late Palaeozoic Itarare Group, Parana Basin, Brazil." *Sedimentology* 40, no. 1 (1993): 1–25.

Fearnside, Philip M. "Environmental Destruction in the Brazilian Amazon." In *The Future of Amazonia*, edited by D. Goodman and A. Hall, 179–225. London: Macmillan, 1990.

Ferrz, J. "Irrupcoes de aguas frias no Atlantico sul Longo da Costa Meridional Brasileira, ate cafo frio." *Revista Geografica*, 9–10 (1949–50): 81–95.

Hartt, C. F. "Contributions to the Geology and Physical Geography of the Lower Amazonas: The Evere-Monte Allegre District and the Table-Topped Hills." *Bulletin of the Buffalo Society of Natural History* 26 (1874): 210–35.

James, Preston E. "Observations on the Physical Geography of Northeastern Brazil." *Annals of the Association of American Geographers* 42, no. 2 (1952): 153–76.

———. "The Sao Francisco Basin: A Brazilian Sertao." *Geographical Review* 38, no. 4 (1948): 658–61.

Mortatti, J., et al. "Biogeochemistry of the Madeira River Basin." *GeoJournal* 19, no. 4 (1989): 391–97.

Da Silva, G. M., et al. "Foredune Vegetation Patterns and Alongshore Environmental Gradients: Mozambique Beach, Santa Catarina Island, Brazil." *Earth Surface Processes and Landforms* 33, no. 10 (2008): 1537–73.

De Souza Soler, Luciana, et al. "Quantifying Deforestation and Secondary Forest Determinants for Different Spatial Extents on an Amazonian Colonization Frontier (Rondonia)." *Applied Geography* 29, no. 2 (2009): 182–93.

Sternberg, Hilgard O. "Contemporary Frontiers in Brazilian Amazonia: Some Environmental Consequences [in Japanese]." *Tsukuba Studies in Human Geography* 8 (1984): 273–93.

Biogeography

Abraco, Marcia B., et al. "Ethnobotanical Ground-Truthing: Indigenous Knowledge, Floristic Inventories, and Satellite Imagery in the Upper Rio Negro, Brazil." *Journal of Biogeography* 35, no. 12 (2008): 2237–48.

Albernz, Ana L., et al. "True Species Compositional Change and Conservation Implications in the White-Water Flooded Forests of the Brazilian Amazon." *Journal of Biogeography* 39, no. 5 (2012): 869–83.

Almeida-Neto, Mario, et al. "Harvest Man (Arachnida: Opiliones) Species Distribution along Three Neotropical Elevation Gradients: An Alternative Rescue Effect to Explain Rapoport's Rule?" *Journal of Biogeography* 33, no. 2 (2006): 361–75.

Almeida Correia Junior, Paulo, et al. "Caracterizacao atual do uso da terra e da cobertura vegetal na regiao da terra indigena Sangradouro/Volto Grande-Mato Grosso, Brasil." *Investigaciones Geograficas* 53 (2004): 27–38.

Alves, D. "Space-Time Dynamics of Deforestation in Brazilian Amazonia." *International Journal of Remote Sensing* 23, no. 14 (2002): 2903–8.

Amancio Alves, Jose J., and Sebastiano Santos do Nascimento. "Levanamento fitogeografico das plantas medicinais nativas do Cariri Paraibono." *Revista Geografica Academica* 4, no. 2 (2010): 73–85.

Atlas Forestal do Brasil. Rio de Janeiro, Brazil: Ministerio de Agricultura, 1966.

Bell, Rayna C., et al. "Evolutionary History of Scinax Treefrogs on Landbridge Islands in South-Eastern Brazil." *Journal of Biogeography* 39, no. 9 (2012): 1733–42.

Brinkman, W. L. F. "Studies in Hydrobiogeochemistry of a Tropical Lowland Forest System." GeoJournal 11, no. 1 (1985): 89–101.

Brown, J. S., and C. Albrecht. "The Effect of Tropical Deforestation on Stingless Bees of the Genus Melipana (Insecta: Hymenspter: Apdiau: Meliponini) in Central Rondonia, Brazil." *Journal of Biogeography* 28, no. 5 (2001): 623–34.

Carnaval, Ana C., and Craig Mortiz. "Historical Climate Modelling Predicts Patterns of Current Biodiversity in the Brazilian Atlantic Forest." *Journal of Biogeography* 36, no. 7 (2008): 1187–201.

Castilho da Costa, Nadja M., and Claudia Segand. "Plano de Manejo Ecologico como forma de Gestao de Unidades de conservacao-badoquena: um estudo de case." *Revista Brasileira de Geografia* 54, no. 2 (1992): 5–26.

Chapman, Frank M. "Problems of the Roraima-Duida Region as Presented by the Bird Life." *Geographical Review* 21, no. 3 (1931): 363–72.

Cintra, Danielle P., et al. "Classifying Successional Forest Stages Using IKONOS in the Atlantic Forest of Rio de Janeiro." *Revista Geografica Academica* 5, no. 1 (2011): 21–33.

Coelho Netto, A. L. "Overland/Low Production in a Tropical Rainforest Catchment: The Role of Litter Cover." *Catena* 14, no. 2 (1987): 213–31.

Cole, M. M. "The Brazilian Savanna." *Revista Geografia* 52 (1960): 5–40.

———. "Cerrado, Caatinga, and Pantanal: The Distribution and Origin of Savanna Vegetation of Brazil." *Geographical Journal* 126 (1960): 168–79.

Compton, J. T., et al. "Intensive Forest Clearing in Rondonia, Brazil as Detected by Satellite Remote Sensing." *Remote Sensing of the Environment* 15, no. 3 (1984): 245–51.

Costa, Leonora P. "The Historical Bridge between the Amazon and the Atlantic Forest of Brazil: A Study of Molecular Phylogeography with Small Mammals." *Journal of Biogeography* 30, no. 1 (2003): 71–86.

Dansereau, Pierre. *The Distribution and Structure of Brazilian Forests.* Montreal, Quebec: Universite de Montreal, Service de Biogeographie, 1948.

De Amarante Romariz, Dora. "La Vegetation du Bresil dans le cadre de la vegetation du Monde." *Revista Geografica* 81 (1974): 165–74.

De Araujo Diniz, Almerio. "Roteiro Amazanico (Estado do Amazonas e territorio do Rio Branco)." *Revista Geografica* 42 (1955): 1–87.

Denevan, William M. "Campo Cerrado Vegetation of Central Brazil." *Geographical Review* 55, no. 1 (1965): 112–15.

Eloy, L., and F. M. Le Tourneau. "Is Urbanization Responsible for Deforestation in Amazonia? Territorial and Agricultural Innovations in the North-West Amazonia (Brazil) [in French]." *Annales de Geographie* 118, no. 3 (2009): 204–27.

Fearnside, Philip M. "Amazonian Deforestation and Global Warmaing: Carbon Stocks in Vegetation Replacing Brazil's Amazon Forest." *Forest Ecology and Management* 80, nos. 1–3 (1996): 21–34.

———. "Deforestation in the Brazilian Amazon: How Fast Is It Occurring?" *Interciencia* 7 (1982): 82–88.

———. "Environmental Change and Deforestation in the Brazilian Amazon." In *Change in the Amazon Basin: Man's Impact on the Forests and Rivers*, edited by John Hemming, 70–79. Manchester, England: Manchester University Press, 1985.

———. "Spatial Concentration of Deforestation in the Brazilian Amazon." *Ambio* 15, no. 2 (1986): 74–81.

Fernandes, Alexandre, et al. "Phylogeography of the Chestnut-Tailed Antbird (Myrmeciza hemimelaena) Clarifies the Role of Rivers in Amazonian Biogeography." *Journal of Biogeography* 39, no. 8 (2012): 1524–35.

Frohn, R. C., et al. "Using Satellite Remote Sensing Analysis to Evaluate a Socio-economic and Ecological Model of Deforestation in Rondonia, Brazil." *International Journal of Remote Sensing* 17, no. 16 (1996): 3233–55.

Furlan, Sueli A., et al. "Corredores ecologicas de Mata Atlantica: Visao integradara do planejamiento territorial no continuo de Paranpiacaba-Estado de Sao Paulo, Brasil." *Revista Geografica de America Central* 43 (2009): 49–79.

Furley, Peter A. "The Nature and Diversity of Neotropical Savanna Vegetation with Particular Reference to the Brazilian Cerrados." *Global Ecology and Biogeography* 8, nos. 3–4 (1999): 223–42.

Gardner, Toby A. "Tree-Grass Coexistence in the Brazilian Cerrado: Demographic Consequences of Environmental Instability." *Journal of Biogeography* 33, no. 3 (2006): 448–63.

Goes-Filho, Luiz, and Ricardo Lisboa Braga. "A Vegetacao do Brasil-desanatamento e queimadas." *Revista Brasileira de Geografia* 53, no. 2 (1991): 135–41.

Goldsmith, F. B. "Multivariate Analyses of Tropical Grassland Communities in Mato Grosso, Brazil." *Journal of Biogeography* 1, no. 2 (1974): 111–22.

Guerra, Francisco. "A problematica floresta Amazonia." *Revista Brasileira de Geografia* 53, no. 3 (1991): 125–32.

Haggett, Peter. "Regional and Local Components in the Distribution of Forested Areas in Southeast Brazil: A Multivariate Approach." *Geographical Journal* 130, no. 3 (1964): 365–78.

Hayashi, Ichiroku. "Changing Aspects of Drought-Deciduous Vegetation in the Semi-arid Region of North-East Brazil." In *The Fragile Tropics of Latin America: Sustainable Management and Changing Environments*, edited by Toshie Nishizawa and J. Vitto, 268–79. Tokyo, Japan: UN University Press, 1995.

Irmler, V. "Inundation Forest Types in the Vicinity of Manaus." *Biogeographica* 8 (1977): 17–30.

Jambes, Jean-Pierre. "Le recul de la foret dans le norteste bresilien: L'exemple des Caatingas de Cariris Velhos." *Les Cahiers d'Outre Mer* 43 (1990): 219–38.

Jepson, Wendy. "A Disappearing Biome? Reconsidering Land-Cover Change in the Brazilian Savanna." *Geographica l Journal* 171, no. 2 (2005): 99–111.

Johnson, Dennis V. "Geography and Ecology of Native Cashew in Northeastern Brazil." *Revista Brasileira de Biologia* 33 (1973): 485–94.

Kosinski, Leszak, ed. *Ecological Disorder and Amazonia*. Rio de Janeiro, Brazil: International Social Science Council and Editora Universitoria Candido Mendes, 1991.

Laurance, W. F., et al. "Predictors of Deforestation in the Brazilian Amazon." *Journal of Biogeography* 29, nos. 5–6 (2002): 737–48.

Lee, David, et al. "Classificacao da cobertura vegetal utilizando dados do sur aerotransportado e do TM/Landsat: Regiao da floresta nacional do Tapajoz." *Investigaciones Geograficas*, Especial (1996): 19–25.

Lima, Andre, et al. "Land Use and Land Cover Changes Determine the Spatial Relationship between Fire and Deforestation in the Brazilian Amazon." *Applied Geography* 34 (2012): 239–46.

Luizao, F. J. "Litter Production and Mineral Element Imput to Forest Floor in a Central Amazonian Forest." *GeoJournal* 19, no. 4 (1989): 407–18.

Maeda, Eduardo, et al. "Fire Risk Assessment in the Brazilian Amazon Using MODIS Imagery and Change Vector Analysis." *Applied Geography* 31, no. 1 (2011): 76–84.

Maillard, Philippe, and Priscilla S. Costa-Pereira. "Estimacao da idade da regeneracao da vegetacao de cerrado a partir de imagens landsat." *Revista Geográfica Academica* 5, no. 1 (2011): 34–47.

Melo Santos, Andre M., et al. "Biogeographical Relationships among Tropical Forests in North-Eastern Brazil." *Journal of Biogeography* 34, no. 3 (2007): 437–46.

Melo Souza, Rosemeri, and Jailton J. Costa. "Genero e espacao na apropriacao fitogeografica de honcornia speciosa gomes em Sergipe, Brasil." *Revista Geografica Academica* 5, no. 2 (2011): 5–16.

Mistry, J. "Corticolous Lichens as Potential Bioindicators of Fire History: A Study in the Cerrado of the Distrito Federal, Central Brazil." *Journal of Biogeography* 25, no. 3 (1998): 409–42.

———. "Fire in the Cerrado (Savannas) of Brazil: An Ecological Review." *Progress in Physical Geography* 22, no. 4 (1998): 425–48.

———. "A Preliminary Lichen Fire History (LFH) Key for the Cerrado of the Distrito Federal, Central Brazil." *Journal of Biogeography* 25, no. 3 (1998): 443–52.

Monzon Nunes, Gustavo, et al. "Discriminacao de fitofisionomias na Amazonia central por mai de indices de vegetacao de imagens com resolucao especial moderada." *Revista Geografica Academica* 6, no. 1 (2012): 5–14.

Morato de Carvalho, Celso. "O lavrado da serra da lua em Roraima e perspectives para studios da herpeto fauna na regiao." *Revista Geografica Academica* 3, no. 1 (2009): 1–17.

Moreira, A. G. "Effects of Fire Protection on Savanna Structure in Central Brazil." *Journal of Biogeography* 27, no. 4 (2000): 1021–30.

Mors, Michael A., et al. "The Brazilian Caatinga in South American Zoogeography: Tropical Mammals in a Dry Region." *Journal of Biogeography* 12, no. 1 (1985): 57–69.

Muller, P. "Campo Cerrado-Forest or Savanna?" *GeoJournal* 3, no. 1 (1979): 27–34.

Nascimento, Marcelo, and John Procter. "Soil and Plant Changes across a Mono-Dominant Rainforest Boundary on Maraca Island, Roraima, Brazil." *Global Ecology and Biogeography Letters* 6, no. 5 (1997): 387–96.

Nedel, Jose C. "Florestas nacionais." *Revista Brasileira de Geografia* 53, no. 3 (1996): 205–28.

Nishizawa, Toshie, et al. "Characteristics and Utilization of Tree Species in the Semi-arid Woodland of North-East Brazil." In *The Fragile Tropics of Latin America: Sustainable Management of Changing Environments*, edited by Toshie Nishizawa and J. Vitto, 280–300. Tokyo, Japan: UN University Press, 1995.

Nogueira, Cristiano, et al. "Vicariance and Endemism in a Neotropical Savanna Hotspot: Distribution Patterns of Cerrado Squamate Reptiles." *Journal of Biogeography* 38, no. 10 (2011): 1907–22.

Novo, E. M., et al. "Relationship between Macrophyte Stand Variables and Radar Backscatter at Land C Band-Tucurui Reservoir, Brazil." *International Journal of Remote Sensing* 23, no. 7 (2002): 1241–60.

Oliveira Malta, Judson A., et al. "Fitogeografia e regeneracao natural em florestas urbanas de Sao Cristovao/SE-Brasil." *Investigaciones Geograficas* 77 (2012): 48–62.

Palubinskas, G., et al. "An Evaluation of Fuzzy and Texture-Based Classification Approaches for Mapping Regenerating Tropical Forest Classes from Landsat-TM Data." *International Journal of Remote Sensing* 16, no. 4 (1995): 747–59.

Perz, Stephen G., and David Skole. "Secondary Forest Expansion in the Brazilian Amazon and the Refinement of Forest Transition Theory." *Society and Natural Resources* 16, no. 4 (2003): 277–94.

Pinherio, Carlos, and Jose Ortiz. "Communities of Fruit-Feeding Butterflies along a Vegetation Gradient in Central Brazil." *Journal of Biogeography* 19, no. 5 (1992): 505–12.

Ponce, Victor, and Latia daCunha. "Vegetated Earthmounds in Tropical Savannas of Central Brazil: A Synthesis with Special Reference to the Pantonal do Mato Grosso." *Journal of Biogeography* 20, no. 2 (1993): 219–26.

Rangel, Thiago, et al. "Human Development and Biodiversity Conservation in Brazilian Cerrado." *Applied Geography* 27, no. 2 (2006): 14–27.

Rocha, L. A. "Patterns of Distribution and Processes of Speciation in Brazilian Reef Fishes." *Journal of Biogeography* 30, no. 8 (2003): 1161–72.

Rodrigues da Silva, Fernando, et al. "Humidity Levels Drive Reproductive Modes and Phylogenetic Diversity of Amphibians in the Brazilian Atlantic Forest." *Journal of Biogeography* 39, no. 9 (2012): 1720–32.

Safford, Hugh D. "Brazilian Paramos I: An Introduction to the Physical Environment and Vegetation of the Compos de Altitude." *Journal of Biogeography* 26, no. 4 (1999): 693–712.

———. "Brazilian Paramos II: Macro- and Mesoclimatic of the Campos de Altitude and Affinities with High Mountain Climates of the Tropical Andes and Costa Rica." *Journal of Biogeography* 26, no. 4 (1999): 713–38.

Silva, J. F., et al. "Spatial Heterogeneity, Land Use, and Conservation in the Cerrado Region of Brazil." *Journal of Biogeography* 33, no. 3 (2006): 536–48.

Silva Pinheiro, Eduardo, et al. "Uso de geotechnologias para o mapeamento das alteracoes na paisagem da Rodovia BR-174." *Revista Geografica Academica* 5, no. 2 (2011): 17–29.

Smith, Nigel J. H. "Further Advances of House Sparrows (*Passer domesticus*) into the Brazilian Amazon." *Condor* 82 (1980): 109–11.

Struwe, Lena, et al. "Spatial Evolutionary and Ecological Vicariance Analysis (SEEVA), a Novel Approach to Biogeography and Speciation Research, with an Example from Brazilian *Gentiana ceac*." *Journal of Biogeography* 38, no. 10 (2011): 1841–54.

Trapasso, L. M. "Deforestation of the Amazon: A Brazilian Perspective." *GeoJournal* 26, no. 3 (1992): 311–22.

Tsuchiya, Akio. "Preliminary Study on the Relationship between Vessel Growth of Thorny Shrubs and Water Balance in the Semi-arid Region, Northeastern Brazil [in Japanese]." *Chiri Kagaku/Geographical Sciences* 50, no. 2 (1995): 123–31.

Vieira Gois, Douglas, et al. "Analise de especie vegetal fitoindicadora em areas verdes publicas de Aracaju, Sergipe." *Revista Geografica Academica* 6, no. 1 (2012): 65–74.

Villela, Dora M., et al. "Effects of Selective Logging on Forest Structure and Nutrient Cycling in a Seasonably Dry Brazilian Atlantic Forest." *Journal of Biogeography* 33, no. 3 (2006): 506–16.

Voeks, Robert A. "Extraction and Tropical Rain Forest Conservation in Eastern Brazil." *Tropical Rainforest Research Current Issues* (1996): 477–87.

Walker, Robert. "The Scale of Forest Transition: Amazonia and the Atlantic Forests of Brazil." *Applied Geography* 32, no. 1 (2012): 12–20.

Werneck, Fernanda, and G. R. Colli. "The Lizard Assemblages from Seasonally Dry Tropical Forest Enclaves in the Cerrado Biome, Brazil, and Its Association with the Pleistocene Arc." *Journal of Biogeography* 33, no. 11 (2006): 1983–92.

Werneck, Fernanda, et al. "Climatic Stability in the Brazilian Cerrado: Implications for Biogeographical Connections of South American Savannas, Species Richness, and Conservation in a Biodiversity Hotspot." *Journal of Biogeography* 39, no. 9 (2012): 1695–706.

Zeilhofer, Peter, and Michael Schessl. "Relationship between Vegetation and Environmental Conditions in the Northern Pantanal of Mato Grosso, Brazil." *Journal of Biogeography* 27, no. 1 (2000): 159–68.

Climatology and Weather

Amancio Alves, Jose J., and Sebastiane Santos do Nascimento. "Ecoclimatologia do Cariri Paraibano." *Revista Geografica Academica* 2, no. 3 (2008): 28–41.

Andrade, A., and R. Rebayle. "L'impact de la secheresse dans le sertao du Sergipe." *Les Cahiers d'Outre Mer* 40 (1987): 373–87.

Andreotti, Claudio E. "Comparacao microclimatica entre dois imoveis nos municipios de Sao Paulo e Sao Roque." *Revista Geografica Academica* 3, no. 2 (2009): 44–56.

Auter, Augusto, and Peter Smart. "Late Quaternary Paleoclimate in Semiarid Northeastern Brazil from U-Series Dating of Travertine and Water-Table Specotherm." *Quaternary Research* 55, no. 2 (2001): 159–67.

Avila Rodriguez, Rafael, and Selma Simoes Castro. "A estrutura especial das chuvas nacidade de Araguari (Minas Gerais) durante a estacao cluvosa 2001–2005." *Revista Geografica Academica* 2, no. 1 (2008): 43–55.

Avila Rodriguez, Rafael, and Roziane Sobreira Santos. "Estudo de tendencia climatic na serie temporal de precipitacao pluviometrica em Araguari-Minas Gerais." *Revista Geografica Academica* 1, no. 1 (2007): 20–27.

Brooks, Reuben. "The Adversity of Brazilian Drought." *GeoJournal* 6, no. 2 (1982): 121–28.

Chung, James C. "Correlation between the Tropical Atlantic Trade Winds and Precipitation in Northeastern Brazil." *Journal of Climatology* 2, no. 1 (1982): 35–46.

Conti, Jose B. "Um episodio de chuva em territorio Paulista." *Revista Geografica* 81 (1974): 81–90.

Fearnside, Philip M. "Greenhouse Gases farom Deforestation in Brazilian Amazonia: Net Committed Emissions." *Climatic Change* 35, no. 3 (1997): 321–60.

Figueiredo Monteiro, Carlos A. "Da Necessidade de um carater genetico a clasificacao climatica: (Algunmas consideracoes metodologicas a propositodo estudo do Brasil Meridional)." *Revista Geografica* 57 (1962): 29–44.

Freise, Frederich W. "The Drought Regions of Northeastern Brazil." *Geographical Review* 28, no. 3 (1938): 363–78.

Galvani, Emerson, and Nadia G. Beserra de Lima. "Radiacao solar acima abaixo de dossel de mangue zol na barr do ribeira do Iguape, Sao Paulo." *Revista Geografica Academica* 5, no. 1 (2011): 59–67.

Galvao, Marilia. "A Classificacao Climatica de Gaussen e Bognouls e sua Alplicacao ao Centro Oeste Brasileiro." *Revista Geografica* 56 (1962): 17–22.

———. "Gausen and Bognoul's Climate Classification and Its Application to Middle Western Brazil." *Revista Geografica* 56 (1962): 5–15.

Giambelluca, T. W. "Latent and Sensible Energy Flux over Deforested Land Surfaces in the Eastern Amazon and Northern Thailand." *Singapore Journal of Tropical Geography* 21, no. 2 (2000): 107–30.

Giambelluca, T. W., et al. "Evapotranspiration and Energy Balance of Brazilian Savannas with Contrasting Tree Density." *Agricultural and Forest Meteorology* 149 (2009): 1365–76.

Giambelluca, T. W., et al. "Observations of Albedo and Radiation Balance over Post-Forest Land Surfaces in the Eastern Amazon Basin." *Journal of Climate* 10, no. 5 (1997): 919–28.

Greco, Steven, et al. "Low-Level Nocturnal Wind Maximum over the Central Amazon Basin." *Boundary-Layer Meteorology* 58, nos. 1–2 (1992): 91–116.

Hamill, John. "Seasonality of Precipitation in Brazil." *Revista Geografica* 77 (1972): 123–30.

Hamilton, Michael G., and Jose R. Torifa. "Synoptic Aspects of a Polar Outbreak Leading to Frost in Tropical Brazil, July 1972." *Monthly Weather Review* 106, no. 1 (1978): 1545–56.

Hastenrath, S. L., et al. "Towards the Monitoring and Prediction of Northeast Brazil Droughts." *Quarterly Journal of the Royal Meteorological Society* 110 (1984): 411–25.

Hastenrath, S. L., and S. Heller. "Dynamics of Climate Hazards in Northeast Brazil." *Quarterly Journal of the Royal Meteorological Society* 103 (1977): 77–92.

Jefferson, Mark. "New Rainfall Maps of Brazil." *Geographical Review* 14 (1924): 127–33.

Johnson, Dennis V. "Agroclimatological Zonation of Maize and Sorghum in Northeast Brazil." *Revista Geografica* 89 (1979): 37–44.

Kousky, Vernon. "Atmospheric Circulation Changes Associated with Rainfall Anomalies over Tropical Brazil." *Monthly Weather Review* 113 (1985): 1951–57.

———. "Diurnal Rainfall Variation in Northeast Brazil." *Monthly Weather Review* 108 (1980): 488–98.

———. "Frontal Influence in Northeast Brazil." *Monthly Weather Review* 107, no. 9 (1979): 1140–53.

Kousky, Vernon, and P. S. Chu. "Fluctuations in Annual Rainfall for Northeast Brazil." *Journal of the Meteorological Society of Japan* 57 (1978): 457–65.

Liu, W. T. H., et al. "Satellite Recorded Vegetation Response to Drought in Brazil." *International Journal of Climatology* 14, no. 3 (1994): 343–54.

Meschiatti, Monica C., et al. "Caracterizacao estatistica de tendencias em series anuais de dados hidro-climaticos no estado de Sao Paulo." *Revista Geografica Academica* 6, no. 1 (2012): 52–64.

Moura, A. D., and J. Shukla. "On the Dynamics of Drought in Northeast Brazil: Observations, Theory, and Numerical Experiments with a General Circulation Model." *Journal of the Atmospheric Sciences* 38, no. 12 (1981): 2653–75.

Namias, J. "Influences of Northern Hemisphere General Circulation on Drought in Northeast Brazil." *Tellus* 24 (1972): 336–42.

Nishizawa, Toshie, et al. "Fluctuations of Spatial Distribution of Rainfall in Northeast Brazil and Their Relationship to the Tropospheric Circulation." *Latin American Studies, Japan* 8 (1986): 11–30.

Ramos, R. P. L. "Precipitation Characteristics in the Northeast Brazil Dry Region." *Journal of Geophysical Research* 80 (1976): 1665–78.

Rao, V. B., and V. de Silva Marques. "Water Vapor Characteristics over Northeast Brazil during Two Contrasting Years." *Journal of Climate and Applied Meteorology* 23, no. 3 (1984): 440–44.

Ratisbona, C. R. "The Climate of Brazil." In *Climate of Central and South America: World Survey of Climatology*, edited by Werner Schwerdtfeger, 12:219–93. Amsterdam, Netherlands: Elsevier, 1976.

Ribeiro de Andrade, Aparecido, and Jonas Teixeira Nery. "Analise da precipitacao pluviometrica diaria, mensal e interaual da bacia hidrografica do Rio Ivai, Brasil." *Investigaciones Geograficas* 52 (2003): 7–30.

Rocha dos Santos, Diego R. "Estudo de neutralizacao dos gases de e feito estufa da universidade Fedraldo Tocantins-Reitoria e campus Universitario de Palmas: Uma forma de mitigacao ambiental." *Revista Geografica Academica* 4, no. 2 (2010): 29–40.

Rose, Newton. "Alternative Solutions to the Problem of Drought in Northeast Brazil." *Proceedings, Middle States Division, Association of American Geographers* 7 (1973): 26–28.

Roth, R. "The Effects of Change in Surface Roughness on the Precipitation Regime in the Coastal Area of Northeast Brazil." *GeoJournal* 8, no. 3 (1984): 205–9.

Tanaka, Minoru, et al. "The Empirical Orthogonal Function Analysis of the Seasonal and Interannual Rainfall in Brazil." *Latin American Studies, Japan* 10 (1988): 27–45.
Tanaka, Minoru, and Toshie Nishizawa. "The Atmospheric Circulations and the Major Drought and Flood of 1983 In Brazil." *Geographical Review of Japan*, 58B (1985): 165–71.
Une, Mitijo Y. "An Analysis of the Effects of Fronts on the Principal Coffee Areas of Brazil." *GeoJournal* 6, no. 2 (1982): 129–40.
Ward, M. N., and C. K. Folland. "Prediction of Seasonal Rainfall in the North Nordeste of Brazil Using Eigenvectors of Sea-Surface Temperature." *International Journal of Climatology* 11, no. 7 (1991): 711–44.
Ward, Robert. "The Economic Climatology of the Coffee District of Sao Paulo, Brazil." *Bulletin of the American Geographical Society* 43, no. 6 (1911): 428–45.
———. "An Outline of the Economic Climatology of Brazil: Parts 1–2." *Bulletin of the Geographical Society of Philadelphia* 7, no. 2 (1909): 7–20; 7, no. 3 (1909): 13–22.

Geomorphology, Landforms, and Volcanism

Andrades Filho, Clodis O., et al. "Identificacao de deformacoes neotectonicas atraves de imagem SRTM, e sura relacao com a genese dos aresis-sodoeste do Rio Grande do Sul." *Revista Geografica Academica* 2, no. 2 (2008): 96–110.
Auter, Augusto, et al. "Fluvial Incision Rates Derived from Magnetostratigraphy of Cave Sediments in the Cratonic Area of Eastern Brazil." *Zeitschrift fur Geomorphologie* 46, no. 3 (2002): 391–404.
Auter, Augusto, and Peter Smart. "The Influence of Bedrock-Derived Acidity in the Development of Surface and Underground Karst: Evidence from the Precambrian Carbonates of Semi-arid Northeastern Brazil." *Earth Surface Processes and Landforms* 28, no. 2 (2003): 157–68.
De Baellar, L., et al. "Controlling Factors of Gullying in the Maracuja Catchment, Southeastern Brazil." *Earth Surface Processes and Landforms* 30, no. 11 (2005): 1369–86.
Baptista Neto, J. A., et al. "Sedimentological Evidence of Human Impact on a Nearshore Environment: Jurujuba, Bouad, Rio de Janeiro State, Brazil." *Applied Geography* 19, no. 2 (1999): 153–77.
Batista Ferreira, Aline, et al. "Analise do meio fisico da sub-bacia do Rio Vacacai-Mirim-Rio Grande do Sul/Brasil." *Revista Geografica Academica* 3, no. 2 (2009): 28–34.
Becegato, Valter A., et al. "Investigaciones geofisicas aplicados no lixao desativado do municipio de saudade do Iguacu-Parana." *Revista Geografica Academica* 3, no. 1 (2009): 47–55.
Bigarella, J. J. "Slope Development in Southeast and Southern Brazil." *Zeitschrift fur Geomorphologie* 10 (1966): 150–60.
Brannstrom, Christian, and A. M. S. Oliveira. "Human Modification of Stream Valleys in the Western Plateau of Sao Paulo, Brazil: Implications for Environmental Narratives and Management." *Land Degradation and Development* 11 (2000): 535–48.
Christofolett, Antonio. "Some Recent Brazilian Studies in Fluvial Geomorphology." *Progress in Physical Geography* 4, no. 3 (1980): 414–20.
Enjalbert, H. "La terre des Palmiers du Maranhao (Bresil du Nord)." *Les Cahiers d'Outre Mer* 5 (1952): 268–78.
Favis-Ortlock, D. T., and A. J. T. Guerra. "The Implications of General Circulation Model Estimates of Rainfall for Future Erosion: A Case Study from Brazil." *Catena* 37, nos. 3–4 (1999): 329–54.
Fernandes, Nelson R., et al. "Topographic Controls of Landslides in Rio de Janeiro: Field Evidence and Modelling." *Catena* 55, no. 2 (2004): 163–81.
Forsberg, B., et al. "Development and Erosion in the Brazilian Amazon: A Geochronological Case Study." *GeoJournal* 19, no. 4 (1989): 399–406.
Fortes, Edison, et al. "Neotectonics and Channel Evolution of the Lower Ivinhema River: A Right-Bank Tributary of the Upper Parana River, Brazil." *Geomorphology* 70, nos. 3–4 (2005): 325–38.
Grohmann, Carlos H., et al. "Aplicacoes dos modelas de elevacao SRTM em geomorfologia." *Revista Geografica Academica* 2, no. 2 (2008): 73–83.

Hesp, P., et al. "Morphology of the Itapeva to Tramandai Transgressive Dunefield System and Mid- to Late-Holocene Sea Level Change." *Earth Surface Processes and Landforms* 32, no. 3 (2007): 407–14.

———. "Regional Wind Fields and Dunefield Migration, Southern Brazil." *Earth Surface Processes and Landforms* 32, no. 4 (2007): 501–73.

Igreja, Hailton, et al. "Neotectonic Influence on Fluvial Capture in the Amazon Basin, State of Amazonas, Brazil." *Science Reports, Tohoku University, 7th Series, Geography* 49, no. 2 (1999): 197–206.

James, Preston E. "The Geomorphology of Eastern Brazil." *Geographical Review* 49, no. 2 (1959): 240–46.

———. "The Surface Configuration of Southeastern Brazil." *Annals of the Association of American Geographers* 23 (1933): 165–93.

———. "The Surface Features of Southeastern Brazil." *Geographical Review* 33, no. 2 (1943): 135–39.

———. "The Tapajoz and Xingu Valleys of Brazil: A Type Study in the Evolution of Amazon Landscape." *Bulletin of the Geographical Society of Philadelphia* 28 (1930): 63–77.

Klammer, G. "Landforms, Cyclic Erosion and Deposition, and Late Cenozoic Changes in Climate in Southern Brazil." *Zeitschrift fur Geomorphologie* 25, no. 2 (1981): 146–65.

Mabersoom. J. M. "Relief of Northeastern Brazil and Its Correlated Sediments." *Zeitschrift fur Geomorphologie* 10 (1966): 419–53.

Machida, T., et al. "Land Condition in the Eastern Nordeste Region." *Tokyo Geography Papers* 20 (1976): 9–22.

Marco da Silva, Alexandre. "Rainfall Erosivity Map for Brazil." *Catena* 57, no. 3 (2004): 251–59.

De Meis, M. R., and A. F. Monteiro. "Upper Quaternary 'Rampas': Doce River Valley, Southeastern Brazilian Plateau." *Zeitschrift fur Geomorphologie* 23, no. 2 (1979): 132–51.

Moreira, I., et al. "Preliminary Investigations on the Occurance of Diagenetic Dolomite in Surface Sediments of Lagua Vermelho, Brazil." *GeoJournal* 14, no. 3 (1987): 357–60.

Nagako Shida, Claudia, and Vania R. Pivello. "Caracterizacao fisiografica e de uso de terras da regiao de Luis Antonio e Santa Rita de Passa Quatro, Sao Paulo, com o uso de sensoriamento remote e SIG." *Investigaciones Geograficas* 49 (2002): 27–42.

Neto, Jose A. B., et al. "Concentration and Bioavailability of Heavy Metals in Sediments from Niteroi Harbour (Guanabara Bay/SE Brazil)." *Journal of Coastal Research* 21, no. 4 (2005): 811–17.

De Oliveira, M. A. T. "Erosion Discomformities and Gully Morphology: A Three Dimensional Approach." *Catena* 16, nos. 4–5 (1989): 413–24.

Pinese Junior, Jose F., et al. "Monitoramento de processes erosivos em paracelas experimenatis no municipio de Uberlandia, Minas Gerais." *Revista Geografica Academica* 2, no. 3 (2008): 5–18.

De Ploey, J., and O. Cruz. "Landslides in the Serra do Mar, Brazil." *Catena* 6, no. 2 (1979): 111–22.

Ramos Martins, Fernando, et al. "Potencial eolico no Rio Grande do Sul-Distribuicao estatistica dos ventos na regiao central do estado." *Revista Geografica Academica* 6, no. 1 (2012): 38–51.

Rocha, Paulo C. "XIII SBGFA-Geomorfologia e areas inundaveis na planicie fluvial do alto Rio Parana." *Revista Geografica Academica* 5, no. 1 (2011): 98–117.

Schellmann, G., and Ulrich Radtke. "Coastal Terraces and Holocne Sea-Level Changes along the Patagonian Atlantic Coast." *Journal of Coastal Research* 19, no. 4 (2003): 983–96.

Shinji Kawakubo, Fernando. "Avaliacao das mudancas na linha de costa na foz do rio Ribeira de Iguape/desembo caura lagunar de Barrado Icapara (litoral sul de Sao Paulo-Brasil) utilizando dados do landsat MSS, TM e ETM+." *Investigaciones Geograficas* 68 (2009): 41–49.

Silva, Aguinaldo, et al. "Avaliacao da erosao na margem direita do Rio Paraguai a justante da praia do Juliao Municipio de Careres-Mato Grosso." *Revista Geografica Academica* 1, no. 1 (2007): 5–19.

Souza, Cristiano M., et al. "Zoneamento geoambiental e transformacoes da paisagem dos municipios Porto Seguro et Santa Cruz Cabralia-Bahia." *Revista Geografica Academica* 5, no. 2 (2011): 41–53.

Viera-Meideros, Rosa M. "Les assentamentos, manifestations de la lutte pour la terre au Bresil." *Les Cahiers d'Outre Mer* 49, no. 1 (1996): 95–108.

Hydrology and Glaciology

Alsdorf, Douglas E. "Water Storage of the Central Amazon Floodplain Measured with GIS and Remote Sensing Imagery." *Annals of the Association of American Geographers* 93, no. 1 (2003): 55–66.

Breunig, Fabio M., et al. "XIII SBGFA-Caracterizacao limnologica do reservatario Rodolfo Costa e Silva-Itaara/Rio Grande do Sul-Brasil." *Revista Geografica Academica* 5, no. 1 (2011): 85–97.

Carvalho Guimaraes, Jose. "Aplicacao do modelo SWAT para estudo de cenarios hipoteticos na bacia higrografica do riacho dos na reorados no cariri paraibano." *Revista Geografica Academica* 5, no. (2011): 30–40.

Carvalho Guimaraes, Jose, et al. "Simulacao hidrossedimento logica, analisando dois MDES, de duas pequenas sub-bacias no cariri paraibano utilizando o modelo SWAT." *Revista Geografica Academica* 5, no. 1 (2011): 48–58.

De Luz Netto, Fausto M., et al. "Avaliacao da qualidade da agua e do use da terra do bacia hidrografica do corrego terra branca-Uberlandia-Minas Gerais." *Revista Geografica Academica* 5, no. 2 (2011): 66–75.

Felisbino Silva, Josimar, et al. "Analise comparative entre a vazao real e avazao de referencia para outarga de agua do corrego barrerinho-Uberlandia-Minas Gerais." *Revista Geografica Academica* 4, no. 2 (2010): 86–95.

Fernandes, Nelson R., et al. "Subsurface Hydrology of Logered Colluvium Mantles in Unchannelled Valleys-South-Eastern Brazil." *Earth Surface Processes and Landforms* 19, no. 7 (1994): 609–26.

Hida, N., et al. "Annual and Daily Changes of River Water Levels at Breves and Caxinona, Amazon Estuary." *Geographical Review of Japan* 71B, no. 2 (1998): 100–105.

Lesach, Lance, and John Melack. "Flooding Hydrology and Mixture Dynamics of Lake Water Derived from Multiple Sources in an Amazon Floodplain Lake." *Water Resources Research* 31, no. 2 (1995): 329–45.

Markham, Charles. "Twenty-Six Year Cyclical Distribution of Drought and Flood in Ceara, Brazil." *Professional Geographer* 27, no. 4 (1975): 454–56.

Morato de Carvalho, Thiago. "Quantificacao dos sedimentos em suspensao e de fundo no medio Rio Araguaia." *Revista Geografica Academica* 1, no. 1 (2007): 55–64.

Mortatti, J., et al. "Hydrological and Geochemical Characteristics of the Jamari and Jiparana River Basins (Rondonia, Brazil)." *GeoJournal* 26, no. 3 (1992): 287–96.

Muasingle, Mohan, et al. "Rio Reconstruction and Flood Prevention in Brazil." *Land Use Policy* 8, no. 4 (1991): 282–87.

Novack, Tessio, et al. "Classificacao de Lagoas Salinas e nao-salinas no pantanal de nhecolandia utilizando um sistema livre de analise de imagens orientada a objecto." *Revista Geografica Academica* 4, no. 1 (2010): 32–45.

Osario de Andrade, Gilberto. "Furos, Paranas e Igarapes: Analise genetic de alguns elementos do sistema potamografico amazonico." *Revista Geografica* 48 (1958): 3–36.

Paim, Jean B., and Joao T. de Menezes. "Estimativa do balance sedimentor da bacia do Rio Tijucas (SC-Brasil) a partir da aplicacao do modelo hidrologico SWAT." *Revista Geografica Academica* 3, no. 2 (2009): 5–14.

Rios-Villamizar, Eduardo A., et al. "Caracterizacao fisico-quimica das aguas e desmantamento na bacia do Rio Purus, Amazonia Brasileira occidental." *Revista Geografica Academica* 5, no. 2 (2011): 54–65.

Sternberg, Hilgard O. *The Amazon River of Brazil*. Wiesbaden, Germany: Franz Steiner Verlag, 1975.

———. "Waters and Wetlands of Brazilian Amazonia: An Uncertain Future." In *The Fragile Tropics of Latin America: Sustainable Management of Changing Environments*, edited by Toshie Nishizawa and Juha Uitto, 113–79. Tokyo, Japan: UN University Press, 1995.

Soils

Fearnside, Philip M. "Effects of Cattle on Soil Fertility." *Tropical Ecology* 21 (1980): 125–37.
Fraser, James, et al. "Anthropogenic Soils in the Central Amazon: From Categories to a Continuum." *Area* 43, no. 3 (2011): 264–73.
Furley, Peter A. "The Nature and Sustainability of Brazilian Amazon Soils." In *The Future of Amazonia*, edited by D. Goodman and A. Hall, 309–59. London: Macmillan, 1990.
Goedert, W. J. "Management of the Cerrado Soils of Brazil: A Review." *Journal of Soil Science* 34, no. 3 (1983): 405–28.
Lepsch, I. F., et al. "Soils-Landscape Relationships in the Occidental Plateau of Sao Paulo State, Brazil. I: Geomorphic Surfaces and Soil Mapping Units." *Journal of the Soil Science Society of America* 44 (1977): 104–9.
Matsumoto, Eiji. "White Sand Soils in North-East Brazil." In *The Fragile Tropics of Latin America: Sustainable Management of Changing Environments*, edited by Toshi Nishizawara and J. Vitto, 253–67. Tokyo, Japan: UN University Press, 1995.
Nortcliff, Stephen. "A Review of Soil and Soil-Related Constraints to Development in Amazonia." *Applied Geography* 9, no. 3 (1989): 147–60.
Tengberg, Anna, et al. "The Impact of Erosion on Soil Productivity: An Experimental Design Applied in Sao Paulo State, Brazil." *Geografiska Annaler* 79A, nos. 1–2 (1997): 95–108.

Theses and Dissertations

Baptista, Sandra. "Forest Recovery and Just Sustainability in the Florinapolis City-Region." PhD diss., Rutgers University, New Brunswick, NJ, 2008.
Bastos, Therezinha X. "Delineating Agroclimatic Zones for Deforested Areas in Para State, Brazil." PhD diss., University of Hawaii, Manoa, Hawaii, 1990.
Castro, Ernesto C. "Effects of Geomorphology and Flood Regime on Landscape Structure of the Arnguaia River Floodplain, Brazil." Master's thesis, University of Wyoming, Laramie, 1995.
Dupigny-Girous, Lesley Ann. "Techniques for Rainfall Estimation and Surface Characterization over Northern Brazil." PhD diss., McGill University, Montreal, Quebec, 1996.
Gutierrez-Velez, Victor H. "Influence of Spatial Components of Uncertainty on the Prediction of Carbon Disturbed by Deforestation in the Brazilian Amazon." Clark University, Worcester, MA, 2007.
Hao, Yong Ping. "Analysis of the Spatial Scale Effects on Landscape Pattern Metrics in a Deforested Area of Rondonia, Brazil." PhD diss., University of Cincinnati, Cincinnati, OH, 2003.
Li, Hui. "Identification of Spatial and Temporal Patterns of Secondary Succession Changes in Altamira, Brazil: Integrating Remote Sensing and GIS Technology." PhD diss., Indiana State University, Terre Haute, 2004.
Lu, Deng Shang. "Estimation of Forest Stand Parameters and Application in Classifying Secondary Succession Stages and Detecting Forest Cover Changes in the Brazilian Amazon Basin." PhD diss., Indiana State University, Terre Haute, 2001.
Markham, Charles. "Climatological Aspects of Drought in Northeastern Brazil." PhD diss., University of California, Berkeley, 1967.
McCann, Joseph. "Subside from Culture: Antropogenic Soils and Vegetation in Tapajonia, Brazilian Amazona." PhD diss., University of Wisconsin, Madison, 2004.
Menzies, John. "The Effects of Scaling-Up on Remotely Derived Leaf Area Index Estimators in the Brazilian Amazon." PhD diss., Indiana State University, Terre Haute, 2005.
Oswald, David. "Estimating Resilience of the Amazonian Ecosystem Using Remote Sensing." Master's thesis, McGill University, Montreal, Quebec, 2008.

Schmidt, M. "Reconstructing Tropical Nature: Prehistoric and Modern Anthrosols (Terra Preta) in the Amazon Rainforest, Upper Xingu River, Brazil." PhD diss., University of Florida, Gainesville, 2010.

De Souza, Carlos. "Mapping and Spatiotemporal Characterization of Degraded Forests in the Brazil Amazon through Remote Sensing." PhD diss., University of California, Santa Barbara, 2005.

Thayn, Jonathon. "Locating Amazonian Dark Earths (ADE) in the Brazilian Amazon Using Satellite Imagery." PhD diss., University of Kansas, Lawrence, 2009.

Webb, Kempton E. "Thornthwaite Climate Classification of Northeast Brazil." Master's thesis, Syracuse University, Syracuse, NY, 1955.

POLITICAL GEOGRAPHY

General Works

Abakerli, Stefonia. "A Critique of Development and Conservation Policies in Environmentally Sensitive Regions in Brazil." *Geoforum* 32, no. 4 (2001): 551–65.

Aguiar, Terez C. "O papel do publico municipal e os desofios na criacao de politicas para o desenvolvimento integral e harmonico da area rural." *Revista Geografica* 114 (1991): 101–9.

Aguiar Gomes, Carlos V., et al. "Rubber Tapper Identities: Political-Economic Dynamics, Livelihood Shifts, and Environmental Implications in a Changing Amazon." *Geoforum* 43, no. 2 (2012): 260–71.

Aldrich, Stephen P. "Contested Groves: Forest Reserves and Land Conflict in the Eastern Amazon." *Journal of Latin American Geography* 11, no. 2 (2012): 73–101.

Allegretti, Mary H. "Reserves Extrativistas: Parametaros Para Uma Politicade Desenvolvimento Sustentabel na Amazonia." *Revista Brasileira de Geografia* 54, no. 1 (1992): 5–24.

Allen, Elizabeth A. "The State and Region: Some Comparative Developmental Experiences from Mexico and Brazil." *Revista Geografica* 93 (1981): 61–78.

Arnauld de Sartre, Xavier, and Romain Taravella. "National Sovereignty versus Sustainable Development Lessons from the Narrative on the Internationalization of the Brazilian Amazon." *Political Geography* 28, no. 7 (2009): 406–15.

Becker, Bertha. "Brazil's Frontier Experience and Sustainable Development: A Geopolitical Approach." In *Frontiers in Regional Development*, edited by Yehuda Gradus and Harvey Lithwick, 73–98. Lanham, MD: Rowman & Littlefield, 1996.

———. "The Crisis of the State and the Region: Regional Planning Questioned." *Environment and Planning D* 3, no. 2 (1985): 141–54.

———. "Fragmentacao do espacoe formacao de regioes na Amazonia: Um poder territorial?" *Revista Brasileira de Geografia* 52, no. 4 (1990): 117–26.

———. "Future Political Geography: A Viewpoint from Brazil." In *Political Geography of the Twentieth Century*, edited by Peter Taylor, 252–56. London: Belhaven, 1993.

———. "Geografia politica e gestao do territorio no limiar de seculo XXI. Uma representacao a partir do Brazil." *Revista Brasileira de Geografia* 53, no. 3 (1991): 169–82.

———. "Nation-State Building in a Newly-Industrializing Country: Reflections on the Brazilian Amazonia Case." In *Nationalism, Self-Determination, and Political Geography*, edited by Ronald J. Johnston et al., 40–56. London: Croom Helm, 1988.

———. "The Political Use of Territory: A Third World Perspective." *IGU Latin American Regional Conference* 2 (1982): 233–40.

———. "The State Crisis and the Region: Preliminary Thoughts from a Third World Perspective." In *Political Geography: Recent Advances and Future Directions*, edited by Peter Taylor and J. M. House, 81–97. London: Croom Helm, 1984.

———. "The State and the Land Question of the Frontier: A Geopolitical Perspective." *GeoJournal* 11, no. 1 (1985): 7–14.

———. "Tordesillas, Year 2000." *Political Geography* 20, no. 6 (2001): 709–25.

Becker, Bertha, et al. *Fronteira Amazonia: Questoes sobre a gestao do territario.* Rio de Janeiro, Brazil: Editora da UnB-Editora da UFRJ, 1990.

Bolanos, Omaira. "Redefining Identities, Redefining Landscapes: Indigenous Identity and Land Rights Struggles in the Brazilian Amazon." *Journal of Cultural Geography* 28, no. 1 (2011): 45–72.

Brannstrom, Christian. "Decentralising Brazilian Water-Resources Management." *European Journal of Development Research* 16, no. 1 (2004): 214–34.

———. "Producing Possession: Labour, Law, and Land on a Brazilian Agricultural Frontier, 1920–1945." *Political Geography* 20, no. 7 (2001): 859–83.

Brannstrom, Christian, et al. "Civil Society Participation in the Decentralisation of Brazil's Water Resources: Assessing Participation in Three Stages." *Singapore Journal of Tropical Geography* 25, no. 3 (2004): 304–21.

Brannstrom, Christian, et al. "Land Change in the Brazilian Savanna (Cerrado), 1986–2002: Comparative Analysis and Implications for Land-Use Policy." *Land Use Policy* 25, no. 4 (2008): 579–95.

Brooks, Reuben. "Alternatives to Starvation: Drought Policy and Public Policy in Northeastern Brazil." *Ekistics* 39 (1975): 30–35.

———. "Drought and Public Policy in Northeastern Brazil: Alternatives to Starvation." *Professional Geographer* 25, no. 4 (1973): 338–46.

Brown, J. C., et al. "Paving the Way to Political Change: Decentralization of Development in the Brazilian Amazon." *Political Geography* 24, no. 1 (2005): 39–52.

Brown, J. C., and M. Powell. "There's Nothing Inherent about Scale: Political Ecology, the Local Trap, and the Politics of Development in the Brazilian Amazon." *Geoforum*, 36 5 (2005): 607–24.

Castilho da Costa, Nadja M., and Claudia Rodrigues Segond. "Plano de manejo ecologico da reserve particular de Bodoquena." *Revista Geografica* 114 (1991): 91–100.

Castro Victoria, Daniel, et al. "Delimitacao de areas de preservacao permanente em topos de morros para o territorio brasileiro." *Revista Geografica Academica* 2, no. 2 (2008): 66–72.

Coelho, Djalma P. "A Localizacao da Nova Capital do Brasil." *Revista Geografica* 41 (1954): 1–31.

Costa Leite, Cristina M. "Uma analise sobre o processo de organizacao do territorio: O case do zaneavento ecologico-economico." *Revista Brasileira de Geografia* 53, no. 3 (1991): 67–90.

Cumming, B. "Dam the Rivers, Damn the People: Hydroelectric Development and Resistance in Amazonian Brazil." *GeoJournal* 35, no. 2 (1995): 151–60.

Cunningham, Susan M. "Planning Brazilian Regional Development during the 1970s." *Geography* 61 (1976): 163–67.

Davidovich, Fany. "Gestao do territorio, um tema em questao." *Revista Brasileira de Geografia* 53, no. 3 (1991): 7–32.

———. "Poder local e municipio, algumas consideracoes." *Revista Geografica* 115 (1992): 27–36.

DeCarvalho Ferreira, Carlos M. "Uma metodologia para estudo de polarizacao e selecao do polos de desenvolvimento em Minas Gerais." *Revista Geografica* 75 (1971): 127–53.

Dickenson, John P. "The Impact of Government Policy on Regional Inequalities in Brazil." In *Proceedings of the Commission on Regional Aspects of Development, IGU,* edited by Richard Thoman, 1:297–314. Toronto, Ontario: University of Toronto Press, Methodology and Case Studies, 1974.

Dunckmann, Florian. *Naturschutz und kleinbauerliche Landnutzung im Rahmen nachhaltiger Entwicklung. Untersuchungen zu regionalen und lokalen Auswirkungen von umweltpolitschen Massnahmen im Vale do Ribeira, Brasilien.* Kiel, Germany: Kieler Geographische Schriften, Band 101, 1999.

Dunckmann, Florian, and Verena Sandner. "Natuarschutz und autochthone Bevolkerung: Betrachtungen aus der Sicht der Politischen Okologie." *Geographische Zeitschrift* 91, no. 2 (2003): 75–94.

Esdras Leite, Marcos. "O uso do solo e o conflito par agua no alto Rio Riachao no norte de Minas Gerais: Uma analise auxiliada pelas geotecnologias." *Revista Geografica Academica* 4, no. 1 (2010): 46–55.

Forest, Thomas. "Brazil and Africa: Geopolitics, Trade, and Technology in the South Atlantic." *African Affairs* 81, no. 332 (1982): 3–20.

Fuks, Mario. "Environmental Conflicts and the Emergence of the Environment as a Social Problem in Rio de Janeiro." *Space and Polity* 5, no. 1 (2001): 49–68.

Garmany, Jeffrey. "The Embodied State: Governmentality in a Brazilian Favela." *Social and Cultural Geography* 10, no. 7 (2009): 721–39.

Hecht, Susanna B. "The New Amazon Geographies: Insurgent Citizenship, Amazon Nation, and the Politics of Environmentalism." *Journal of Cultural Geography* 28, no. 1 (2011): 203–23.

James, Preston E. "Forces for Union and Disunion in Brazil." *Journal of Geography* 38 (1939): 260–66.

Karriem, A. "The Rise and Transformation of the Brazilian Landless Movement into a Counter-Hegemonic Political Actor: A Gramscian Analysis." *Geoforum* 40, no. 3 (2009): 316–25.

Kearns, Kevin C. "Amazonia: The Emergence of Brazil's Northern Frontier." *Journal of Geography* 68 (1969): 535–44.

Kiracofe, Bruce E. "Development Planning and Monitoring with a Land Information System: A Proposal Case Study in the Brazilian Amazon." *Papers of the Applied Geography Conferences* 12 (1989): 152–58.

Kleinpenning, J. M. G. "Disparites internes et politique de developpement regional au Bresil." *Les Cahiers d'Outre Mer* 44 (1991): 113–28.

———. *An Evaluation of the Brazilian Policy for the Integration of the Amazon Basin (1964–1975)*. Nijmegen, Netherlands: Geografisch en Planologisch Instituut, Publication 9, 1979.

———. "Objectives and Results of the Development Policy in North-East Brazil." *Tijdschrift voor Economische en Sociale Geographie* 62, no. 5 (1971): 271–84.

Kohlhepp, Gerd. "Analysis of State and Private Regional Development Projects in the Brazilian Amazon Basin." *Applied Geography and Development* 16 (1980): 53–79.

———. "Ocupacao e valorizacao economica da Amazonia: Estrategies de desenvolvimento do govern o brasileiro e empresas privadas." *Revista Geografica* 94 (1981): 67–88.

Lardon, S., et al. "Dispositefs de gouvernance territoriale durable en agriculture: Analyse de trios situations en France et en Bresil." *Norois* 209 (2008): 17–36.

Meaney-Leckie, Anne. "Federal Response Policy toward Recurrent Drought in Northeast Brazil." *Papers of the Applied Geography Conferences* 12 (1989): 137–42.

Mistry, J., et al. "Indigenous Fire Management in the Cerrado of Brazil: The Case of the Kraho of Tocantins." *Human Ecology* 33, no. 3 (2005): 365–286.

Moran, Emilio. "Rich and Poor Ecosystems of Amazonia: An Approach to Management." In *The Fragile Tropics of Latin America: Sustainable Management of Changing Environments*, edited by Toshie Nishizawa and J. Vitto, 45–67. Tokyo, Japan: UN University Press, 1995.

Muller, Keith D., et al. "Recent Evolution of the Sem Terra (Landless) Movement: Northeast Brazil." *Papers of the Applied Geography Conferences* 22 (1999): 159–64.

———. "The Sem Terra (without Land) Movement, Northeast Brazil." *Papers of the Applied Geography Conferences* 18 (1995): 155–58.

Munoz, Juan. "La ordenacion del espacio litoral brasileno: El Plan Nacional de Gestion Costera (PNGC)." *Anales de Geografia de la Universidad Complutense* 18 (1998): 89–114.

Neli Aparecide de Mello, Herve. "The Brazilian Government and the Amazon Environment: Developments, Contradictions, and Conflicts [in French]." *L'Espace Geographique* 32, no. 1 (2003): 3–20.

Neuburger, Martina. "Global Discourses and the Local Impacts in Amazonia: Inclusion and Exclusive Processes in the Rio Negro Region." *Erdkunde* 62, no. 4 (2008): 339–56.

———. "Smallholder Vulnerability in Degraded Areas: The Political Ecology of Pioneer Frontier Processes in Brazil." *Geographische Zeitschrift*, Special Issue (2004): 58–72.

Nunes, Paulo H. "A relacoes Brasil-Japao e seus reflexos no processo de ocupacao do territorio Brasileiro." *Revista Geografica* 140 (2006): 61–78.

Perritt, Richard. "Issues of Regional Local Environment and Development Planning with Reference to Brazil." *Proceedings, Conference of Latin Americanist Geographers* 9 (1983): 112–14.

Platt, Robert S. "Brazilian Capitals and Frontiers, Part I." *Journal of Geography* 53, no. 9 (1965): 369–75.

———. "Brazilian Capitals and Frontiers, Part II." *Journal of Geography* 54, no. 1 (1955): 5–16.

Porro, Noemi, et al. "Traditional Communities in the Brazilian Amazon and the Emergence of New Political Identities: The Struggles of the Quebradeiras de coco babacu-babassu Breaker Women." *Journal of Cultural Geography* 28, no. 1 (2011): 123–46.

Richert, Aldomar. "Oestado do Rio Grande do Sul-Brasil como um territorio de internacionalizacao segmentada do espaco nacional." *Investigaciones Geograficas* 51 (2003): 125–43.

Rolnik, Raquel. "Territorial Exclusion and Violence: The Case of the State of Sao Paulo, Brazil." *Geoforum* 32, no. 4 (2001): 471–82.

Roper, Monika. *Planung und Einrichtung von Naturachutzgebieten aus sozialgeographische Sicht: Fallbeispiel aus der Pantanal-Region (Brasilein)*. Tubingen, Germany: Tubinger Geographische Studien, Heft 134, 2001.

Segebart, D. "Who Governs the Amazon? Analysing Governance in Processes of Fragmenting Development: Policy Networks and Governmentality in the Brazilian Amazon." *Die Erde* 139, no. 3 (2008): 187–204.

Simmons, Cynthia S. "The Amazon Land War in the South of Para." *Annals of the Association of American Geographers* 97, no. 3 (2007): 567–92.

———. "The Political Economy of Land Conflicts in the Eastern Brazilian Amazon." *Annals of the Association of American Geographers* 94, no. 1 (2004): 183–206.

———. "Territorializing Land Conflict: Space, Place, and Contentious Politics in the Brazilian Amazon." *GeoJournal* 64, no. 4 (2005): 307–17.

Simmons, Cynthia S., et al. "The Changing Dynamics of Land Conflict in the Brazilian Amazon: The Rural-Urban Complex and Its Environmental Implications." *Urban Ecosytems* 6, nos. 1–2 (2002): 99–122.

Simmons, Cynthia S., et al. "Spatial Processes in Scalar Context: Development and Security in the Brazilian Amazon." *Journal of Latin American Geography* 6, no. 1 (2007): 125–48.

Smith, Harry, and Emilio J. Luque-Azcona. "The Historical Development of Built Heritage Awareness and Conservation Policies: A Comparison of Two World Heritage Sites: Edinburgh and Salvador do Bahia." *GeoJournal* 77, no. 3 (2012): 399–415.

Soares Magdaleno, Fabiano. "O territorio nas Constituicoes Republicas Brasileiras." *Investigaciones Geograficas* 57 (2005): 114–32.

De Souza, Meria A. B. "Amazonia: Gestao do Territorio." *Revista Brasileira de Geografia* 53, no. 3 (1991): 133–48.

Spears, Eric K., and Harm J. deBlij. "Political Geography of Devolution in the Americas: The Case of Brazil's South." *Pennsylvania Geographer* 39, no. 1 (2001): 3–17.

Spiegal, Samuel J., et al. "Mapping Spaces of Environmental Dispute: GIS, Mining, and Surveillance in the Amazon." *Annals of the Association of American Geographers* 102, no. 2 (2012): 320–49.

Taravella, Romain, and Xavier Arnauld de Sartre. "The Symbolic and Political Appropriation of Scales: A Critical Analysis of the Amazonian Ranchers' Narrative." *Geoforum* 43, no. 3 (2012): 645–56.

Vadjunec, Jacqueline, et al. "Rubber Tapper Citizens: Emerging Places, Policies, and Shifting Rural-Urban Identities in Acre, Brazil." *Journal of Cultural Geography* 28, no. 1 (2011): 73–98.

Valencia, Marcio M. "Culture, Politics, Faith, and Poverty in Belem (Brazil)." *Geoforum* 37, no. 2 (2007): 159–61.

Wehrhahn, Rainer. *Konfliktezwischen Naturschutz und Entwicklung im Bereich des Atlantischen Regenwaldes im Bundesstadt Sao Paulo, Brasilien. Untersuchungen zur Wahrnehmung von Umweltproblem und zur Umsetzung von Schutz-Kanzepten*. Kiel, Germany: Kieler Geographische Schriften, Band 89, 1994.

Wolford, Wendy. "Agrarian Moral Economies and Neoliberalism in Brazil: Competing Worldviews and the State in the Struggle for Land." *Environment and Planning A* 37, no. 2 (2005): 241–62.

———. "Families, Fields, and Fighting for Land: The Spatial Dynamics of Contention in Rural Brazil." *Mobilization* 8, no. 2 (2003): 201–15.

———. "The Land Is Ours Now: Spatial Imaginaries and the Struggle for Land in Brazil." *Annals of the Association of American Geographers* 94, no. 2 (2004): 409–24.

Wonicz, Philippe, et al. "Du local au national la consolidation demogratique au Bresil." *Espace, Populations, Societes* 3 (2003): 501–17.

Zeilhofer, Peter, et al. "A GIS-Approach for Determining Permanent Riparian Protection Areas in Mato Grosso, Central Brazil." *Applied Geography* 31, no. 3 (2011): 990–97.

Theses and Dissertations

Andre, David J. "Security Aspects of Brazil's Post-1964 Amazonian Highway Policy." PhD diss., University of Georgia, Athens, 1979.

De Oliveira, Gustavo. "Explaining Mining Company and Community Relations in Paracatu, Brazil: Situational Context and Company Practice." Master's thesis, University of Guelph, Guelph, Ontario, 2010.

Empinotti, Vanessa. "Re-framing Participation: The Political Ecology of Water Management in the Lower Sao Francisco River Basin, Brazil." PhD diss., University of Colorado, Boulder, 2007.

Garmany, Jeffrey. "Governance without Government: Explaining Order in a Brazilian Favela." PhD diss., University of Arizona, Tucson, 2011.

Joslin, Audrey. "The Negotiation between Conservation and Development Objectives in Agrarian Reform Settlements in the South of Para, Brazil." Master's thesis, Michigan State University, East Lansing, 2008.

Lewis, Vania. "Neoliberal Reforms, Government Restructuring and Changes in Social Housing Provision in Ribeirao Preto, Brazil." PhD diss., University of Arizona, Tucson, 2010.

Moraes, Eloisa. "The Role of Local Environmental Non-governmental Organizations in the Conservation of the Brazilian Amazon." Master's thesis, George Washington University, Washington, DC, 1994.

Spears, Eric K. "The Politics of Space and Scale in the Brazilian Favela: A Case Study of San Pedro." PhD diss., West Virginia University, Morgantown, 2005.

URBAN GEOGRAPHY

General Works

Augelli, John P. "Brasilia: The Emergence of a National Capital." *Journal of Geography* 62, no. 6 (1963): 241–52.

Bandeira de Mello e Silva, Sylvio, and Jaimeval Caetano de Souza. "Analise de Hierarquia Urbana do Estado da Bahia." *Revista Brasileira de Geografia* 53, no. 1 (1991): 51–80.

Beaujeu-Garnier, J., and Milton Santos. "Le 'Centre' de Salvador." *Les Cahiers d'Outre Mer* 20 (1967): 321–44.

Beitrao Sposito, Maria E. "Novos conteudos nas periferias urbanas das cidades medias do Estado de Sao Paulo, Brasil." *Investigaciones Geograficas* 54 (2004): 114–39.

Bernardes, Lysia. "Areas of Urban Organization in the Region of Rio de Janeiro." *Revista Geografica* 55 (1961): 37–50.

———. "Sobre o Processo de metropolizacao no Brazil." *Revista Geografica* 71 (1969): 114–23.

Browder, John O., and Brian J. Godfrey. *Rainforest Cities: Urbanization, Development, and Globalization of the Brazilian Amazon.* New York: Columbia University Press, 1997.

Cavalcanti-Bernardes, Lysia. "Ensaio de delimitacao da regiao urbana do Rio de Janeiro." *Revista Geografica* 55 (1961): 5–12.

———. "Setores de organizacao urbana na regiao do Rio de Janeiro." *Revista Geografica* 55 (1961): 37–50.

Cinta, Antonio O. "Urban Development in Brazil: A Study of Policies and Unpolicies." *Luso-Brazilian Review* 17, no. 2 (1980): 213–32.

Compagnoni, Cristiane, et al. "Planejamento urbano e conflito ambiental na bacia hidrografica do Rio Penso-Municipio de Pato Branco-Estado do Parana." *Revista Geografica Academica* 3, no. 1 (2009): 28–46.

Coy, Martin. "Cuiba (Mato Grosso): Wirtschafts- und sozialraumlicher Strukturewandel einer Regional metropole im brasilianischen Mittelwesten." *Zeitschrift fur Wirtschaftsgeographie* 38, no. 4 (1992): 210–28.

Curtis, James R. "Pracas, Place, and Public Life in Urban Brazil." *Geographical Review* 90, no. 4 (2000): 475–92.

Davidovich, Fany. "Brasil Metropolitano e Brasil Urbano Nao-Metropolitano: Algumas questoes." *Revista Brasileira de Geografia* 53, no. 2 (1991): 127–34.

———. "Um foco Sobre o Processo de Urbanizacao do Estado do Rio de Janeiro." *Revista Brasileira de Geografia* 48, no. 3 (1986): 333–71.

———. "Tipos de Cidades Brasileiras." *Revista Geografica* 60 (1964): 5–16.

Davila, Julio D. "Pereira-Dos Quebradas." *Cities* 5, no. 1 (1988): 10–23.

De Azevedo, A. "Salvador et le reconcavo de Bahia." *Les Cahiers d'Outre Mer* 4 (1951): 189–203.

Deffontaines, Pierre. "The Origin and Growth of the Brazilian Network of Towns." *Geographical Review* 28, no. 3 (1938): 379–99.

Dickenson, John P. "The Future of the Past in the Latin American City: The Case of Brazil." *Bulletin of Latin American Research* 13, no. 1 (1994): 13–26.

Diniz, Alexander. "Occupation and Urbanization of Roraima State, Brazil." *Yearbook, Conference of Latin Americanist Geographers* 23 (1997): 51–62.

Faissol, Speridiao, et al. "O proceso de urbanizacao brasileiro: Uma contribuicao a formulacao de uma politica de desenvolvimento urbano/regional." *Revista Geografica* 103 (1986): 111–58.

Geiger, Pedro P. "Urban Issues and Brazilian Geography." *Proceedings, Conference of Latin Americanist Geographers* 9 (1983): 34–43.

Geraiges de Lemos, Amalia. "A mulher como forca de trabalho na cidade (exemplo: a cidade de Sao Paulo)." *Revista Geografica* 97 (1983): 86–90.

Godfrey, Brian J. "Revisiting Rio de Janeiro and Sao Paulo." *Geographical Review* 89, no. 1 (1999): 94–120.

Guerra, A. J. T. "The Catastrophic Events in Petropolis City (Rio de Janeiro State), between 1940 and 1990." *GeoJournal* 37, no. 3 (1995): 349–54.

Harrigan, John J. "Geography and Planning in Brazilian Urban and Regional Development." *Luso-Brazilian Review* 12, no. 1 (1975): 108–25.

James, Preston E. "Belo Horizonte and Ouro Preto." *Papers of the Michigan Academy of Science, Arts, and Letters* 18 (1932): 239–58.

———. "The Problems of Brazil's Capital City." *Geographical Review* 46, no. 3 (1956): 301–17.

———. "Rio de Janeiro and Sao Paulo." *Geographical Review* 23, no. 2 (1933): 271–98.

Kipnis, Baruch A. "Input-Output Tables for Medium-Sized Cities: Survey Coefficients or Short-Cut Methods? A Case Study in Brazil." *Journal of Regional Science* 24, no. 3 (1984): 443–50.

Macedo, Joseli. "Curitiba." *Cities* 21, no. 5 (2004): 537–49.

Monbeig, Pierre. "Le bas pays de Rio de Janeiro." *Les Cahiers d'Outre Mer* 5 (1952): 169–75.

———. "Une geographie de la ville de Sao Paulo." *Les Cahiers d'Outre Mer* 13 (1960): 104–9.

Morse, Richard. "Sao Paulo: Case Study of a Latin American Metropolis." In *Latin American Urban Research,* edited by F. F. Rabinovits and R. M. Trueblood, 1:151–86. Austin: University of Texas Press, 1971.

Mota de Carvalho, Ailton, and Ivan Sergei. "Hierarquia urbana do Vale do Jequitinhanha analisada atraves des fluxos telefonicos." *Revista Geografica* 130 (2001): 19–31.

Netwig Silva, Barbara C. "Analise compartiva da Posicao de Salvador e do Estado da Bahia no Cenario Nacional." *Revista Brasileira de Geografia* 53, no. 4 (1991): 49–80.

Oliveria, C. N. E. "Sao Paulo." *Cities* 11, no. 1 (1994): 10–14.

Pyle, Gerald F. "Approaches to Understanding the Urban Roots of Brazil." *Proceedings, Conference of Latin Americanist Geographers* 1 (1970): 378–96.

———. "Some Geographical Aspects Basic to Understanding Brazilian Urbanization." *Revista Geografica* 73 (1970): 5–28.

Roche, J. "Porto Alegre, metropole du Bresil meridional." *Les Cahiers d'Outre Mer* 7 (1954): 367–97.

Santos, Milton. "Modernidad, medio tecnico-cientifico y urbaniazacion en Brasil." *Anales de Geografia de la Universidad Complutense* 10 (1990): 45–60.

Scarlato, Francisco. "Ouro Preto: cidade historica da mineracao no Sertao Brasileiro." *Anales de Geografia de la Universidad Complutense* 16 (1996): 123–41.

Smith, Lisandro P., et al. "Algunas transformacoes intra-urbana nobairro nosso Senora Das Dores-Santa Maria." *Revista Geografica de America Central* 37 (1999): 81–108.

Do Socorro Alves Coelho, Maria. "O sistema urban Nordestino: Estruturacao atraves do tempo." *Revista Brasileira de Geografia* 54, no. 1 (1992): 75–94.

Souza da Silva Cilea, Jose, et al. "Saneamento basico e problemas ambientais na regiao metropolitana do Rio de Janeiro." *Revista Brasileira de Geografia* 52, no. 2 (1990): 5–106.

Wadehn, Manfred. "Urban and Regional Development in Brazil." *Applied Geography and Development* 18 (1981): 63–79.

Wallach, Bret. "Manaus." *Cities* 35, no. 1 (1985): 8–11.

Wood, Harold A. "The Role of the Middle-Sized City in Regional Development in Brazil." *Papers of the Applied Geography Conferences* 6 (1983): 198–207.

Urban Social Geography

Barata, R. B., et al. "Intra-urban Differentials in Death Rates from Homicides in the City of Sao Paulo, Brazil, 1988–1994." *Social Science and Medicine* 47, no. 1 (1998): 19–23.

Barboa, Ignez, and Aldo Paviani. "Commuting in the Brazilian Federal District." *Revista Geografica* 77 (1972): 85–94.

Botelho de Mattos, Rogerio, and Miguel A. Campos Ribeiro. "Les territories de la prostitution dans les espaces publics de la zone centrale de Rio de Janeiro." *Geographie et Cultures* 24 (1997): 9–28.

Bret, B. "Le fait urbain et les inegalites socio-spatiales au Bresil." *Mosella* 31, nos. 1–4 (2006): 9–33.

Campos Ribeiro, Miguel A., and Rogerio Botelho de Mattos. "Territorios da prostituicao nos espacos publicos da arca central do Rio de Janeiro." *Revista Brasileira de Geografia* 59, no. 1 (2005): 23–36.

Ceuato, Vania, et al. "The Geography of Homicide in Sao Paulo, Brazil." *Environment and Planning A* 39, no. 7 (2007): 1623–53.

Correia de Andrade, Manuel. "O crescimento demografico e a rede urbana do Nordeste." *Revista Geografica* 78 (1973): 103–13.

Costa Barbosa, Ignez, and Aldo Paviani. "Commuting in the Brazilian Federal District." *Revista Geografica* 77 (1972): 85–94.

Gaffney, Christopher T. "Mega-events and Socio-spatial Dynamics in Rio de Janeiro, 1919–2016." *Journal of Latin American Geography* 9, no. 1 (2010): 7–29.

Garmany, Jeffrey. "Religion and Governmentality: Understanding Governance in Urban Brazil." *Geoforum* 41, no. 6 (2010): 908–18.

Ghizzo, Marcio, and Marcio Muedes Rocha. "A cidade para o Consumo: Mobilidade e centralidade em Maringa-PR. Brasil: O caso do hipermercado Big." *Revista Geografica* 128 (2005): 23–46.

Gough, Katherine V., and M. Franch. "Spaces of the Street: Socio-spatial Mobility and Exclusion of Youth in Recife." *Children's Geographies* 3, no. 2 (2005): 149–66.

Hogan, D. J. "Population, Poverty, and Pollution in Cubatao, Sao Paulo." *Geographia Polonica* 64 (1995): 201–24.

Lima, Jose J. "Socio-spatial Segregation and Urban Form: Belem at the End of the 1990s." *Geoforum* 32, no. 4 (2001): 493–507.

Lowell, Bainbridge. "Cityward Migration in the Nineteenth Century: The Case of Recife, Brazil." *Journal of Interamerican Studies and World Affairs* 17, no. 1 (1975): 43–63.

Ludermir, A. B., and T. Harpham. "Urbanization and Mental Health in Brazil: Social and Economic Dimensions." *Health and Place* 4, no. 3 (1998): 223–32.

Martin, George. "Migration, Natural Increase, and City Growth: The Case of Rio de Janeiro." *International Migration Review* 6, no. 2 (1972): 200–215.

Moreira dos Santos, S., et al. "Saneamento bosico e problems ambientais na regiao metropolitana de Belem." *Revista Brasileira de Geografia* 54, no. 1 (1992): 25–74.

Motte-Baumvol, Benjamin, and Carlos D. Nassi. "Immobility in Rio de Janeiro, beyond Poverty." *Journal of Transport Geography* 24 (2012): 67–76.

Muller, N. L. "Demographic Growth and Urban Expansion in the Metropolitan Area of Sao Paulo." *Revista Geografica* 97 (1983): 29–30.

Netwig Silva, Barbara C., and Sylvio de Mello e Silva. "As cidades da Bahia no ano 2000." *Revista Brasileira de Geografia* 52, no. 2 (1990): 189–98.

Rodrigues, Adyr. "La urbanizacion en Brasil y el fenomero del turismo." *Anales de Geografia de la Universidad Complutense* 16 (1996): 81–99.

Rosa Souza, Romario. "A proposito de um mapeamento da epidemia de dengue na cidade de Cuiaba, Mato Grosso." *Revista Geografica Academica*, 2, no. 1 (2008): 73–87.

Santos, Simone, et al. "Ecological Analysis of the Distribution and Socio-spatial Context of Homicides in Porto Alegre, Brazil." *Health and Place* 12, no. 1 (2006): 38–47.

Skop, Emily H., et al. "Chain Migration and Residential Segregation of Internal Migrants in the Metropolitan Area of Sao Paulo, Brazil." *Urban Geography* 27, no. 5 (2006): 397–421.

Snyder, David E. "Alternative Perspectives of Brasilia." *Economic Geography* 40, no. 1 (1964): 34–35.

Stohr, Walter B. "Local Initiatives in Peripheral Areas: An Intercultural Comparison between Two Case Studies in Brazil and Austria." In *Developing Frontier Cities: Global Perspectives-Regional Contexts*, edited by Harvey Lithwick and Yehuda Gradus, 233–54. Beer Shiva, Israel: Ben-Gurion University of the Negev, Negev Center for Regional Development, 2000.

Timmons-Roberts, J. "Squatters and Urban Growth in Amazonia." *Geographical Review* 82, no. 4 (1992): 441–57.

Valladares, Licia. "Working the System: Squatter Response to Resettlement in Rio de Janeiro." *International Journal of Urban and Regional Research* 2, no. 1 (1978): 12–25.

WinklerPrins, Antoinette M., and T. S. de Souza. "Surviving the City: Urban Home Gardens and the Economy of Affection in the Brazilian Amazon." *Journal of Latin American Geography* 4, no. 1 (2005): 107–26.

Yoder, Michael L., and Glenn Fugitt. "Urbanization, Frontier Growth, and Population Redistribution in Brazil." *Luso-Brazilian Review* 16, no. 1 (1979): 67–90.

Morphology and Neighborhoods

Abreau, Mauricio A. "Le Rio de Janeiro du debut du dix-neuvieme siècle et ses differentes temporalities." *Bulletin de l'Association de Geographes Francais* 73, no. 1 (1996): 30–38.

Acioly, Claudio. "Reviewing Urban Revitalization Strategies in Rio de Janeiro: From Urban Project to Urban Management Approaches." *Geoforum* 32, no. 4 (2001): 509–20.

Aryeetey-Attoh, Samuel. "An Analysis of Household Valuations and Prefernence Structures in Rio de Janeiro, Brazil." *Growth and Change* 23, no. 2 (1992): 183–98.

Bernardes, Lysia. "The Suburban Belt of Rio de Janeiro." *Revista Geografica* 67 (1967): 69–87.

Coy, Martin, and M. Pohler. "Condominios fechados und die Fragmentierung der brasilianischen Stadt. Typen, Ahteure, Folgewritkungen." *Geographica Helvetica* 57, no. 4 (2002): 264–77.

Dawsey, Cyrus B. "Income and Residential Location on Piracicaba, Sao Paulo, Brazil." *Revista Geografica* 89 (1979): 185–90.

DeCastro, Ina E. "Housing Projects: Widening the Controversy Surrounding the Removal of Favelas." *Revista Geografica* 97 (1983): 56–69.

DeLemos, Amalia. "Tendencias en la configuration especial de la metropolis paulista." *Anales de Geografia de la Unviersidad Complutense* 10 (1990): 37–44.

Filho, Artur, and Ana Caceres Cortiz. "Ocupacao urban em areas de risco de deslizamento: ocaso das favelas em Campos do Jordao-Sao Paulo." *Revista Geografica* 145 (2009): 31–47.

Furtaldo, B. *Modeling Social Heterogeneity, Neighborhoodsd, and Local Influences on Urban Real Estate Prices: Spatial Dynamic Analyses in the Belo Horizonte Metropolitan Area, Brazil.* Utrecht: Netherlands Geographical Studies 385, 2009.

Galvao, Marilia, and Speridiao Faissol. "Areas de Pesquisa para determinacao de Areas Metropolitanas." *Revista Geografica* 70 (1969): 57–89.

Garmany, Jeffrey. "Situating Fortaleza; Urban Space and Uneven Development in Brazil." *Citie s* 28, no. 1 (2011): 45–52.

Geraiges de Lemos, Amalia. "La produccion del espacio metropolitano en San Pablo (Brasil)." *Revista Geografica* 101 (1985): 45–51.

Godfrey, Brian J., and Olivia Arguizoni. "Regulating Public Space on the Beachfronts of Rio de Janeiro." *Geographical Review* 102, no. 1 (2012): 17–34.

Koster, Martijn, and Monique Nuijten. "From Preamble to Post-project Frustrations: The Shaping of a Slum Upgrading Project in Recife, Brazil." *Antipode* 44, no. 1 (2012): 175–96.

Lanz, Stephon. "Favelas regienen: Zum Verhaltnis zwischen localstaat, Drogenkomplex und Favela in Rio de Janeiro." *Zeitschrift fur Wirtschaftsgeographie* 51, nos. 3–4 (2007): 191–205.

Lopez de Souza, Marcelo. "Metropolitan Deconcentration, Socio-political Fragmentation, and Extended Suburbanization: Brazilian Urbanisation in the 1980s and 1990s." *Geoforum* 32, no. 4 (2001): 437–47.

Machado, Jose R., and Cesar Miranda Mendes. "O processo de verticalizacao do centro de Maringa-Parana, Brasil." *Investigaciones Geograficas* 52 (2003): 53–71.

Menezes de Rodrigues, N., and Andre L. Lopes de Faria. "Utilizacao de ferramentas SIG na area urbana: Ocupacao illegal de um trecho do ribeirao Sao Bartolomeu-Vicosa/Minas Gerais." *Revista Geografica Academica* 3, no. 1 (2009): 18–27.

Moreira Alves, Thais, and Eguimar F. Chaveiro. "Metamorfose urbana: A conurbacao Goiania-Goianira e suas implicacoes socio-espaciais." *Revista Geografica Academica* 1, no. 1 (2007): 95–107.

Moreira Bento, Lilian C., and Stefania Mara de Faria. "O gerenciamento dos residues solidos urbanos do municipio de Bambui/Minas Gerais e seus possiveis reflexos no dessenvolvimento da atividade turistica." *Revista Geografica Academica* 2, no. 3 (2008): 42–49.

Nuijten, Monique, et al. "Regimes of Spatial Ordering in Brazil: Neoliberalism, Leftist Populism, and Modernist Aesthetics in Slum Upgrading in Recife." *Singapore Journal of Tropical Geography* 33, no. 2 (2012): 157–70.

O'Hare, Greg. "Urban Renaissance: New Horizons for Rio's Favela." *Geography* 86, no. 1 (2001): 51–75.

O'Hare, Greg, and M. Barke. "The Favelas of Rio de Janeiro: A Temporal and Spatial Analysis." *GeoJournal* 56, no. 3 (2002): 225–40.

Parisse, Lucien. "Les favelas dans la Ville: le cas de Rio de Janeiro." *Revista Geografica* 70 (1969): 109–30.

Riley, Elizabeth, et al. "Favela Bairro and a New Generation of Housing Programmes for the Urban Poor." *Geoforum* 32, no. 4 (2001): 521–31.

Seeman, Jorn. "Land Use Dynamics in a Latin American City: The Example of Campinas, Brazil." *Revista Geografica* 129 (2001): 83–94.

Silva, Ricardo T. "The Connectivity of Infrastructure Networks and the Urbna Space of Sao Paulo in the 1990s." *International Journal of Urban and Regional Research* 24, no. 1 (2000): 139–64.

Siqueira, Ieda, and Aida de Souza. "The Evolution of Metropolitan Spaces in Brazil." *Revista Geografica* 97 (1983): 10–28.

Soares, K. V. S., et al. "Violent Death in Young People in the City of Sao Paulo, 1991–1993." *Health and Place* 4, no. 2 (1998): 195–98.

Souza, Flavio A. M. "The Future of Informal Settlements: Lesson in the Legalization of Disputed Urban Land in Recife, Brazil." *Geoforum* 32, no. 4 (2001): 483–92.

Sternberg, Rolf. "Project Urban Infrastructure in Brazilian Amazonia." *Espacio y Desarrollo* 7, no. 9 (1997): 201–34.

Tarlis Pontes, Emilio. "Resgate do espaco public e uma nova consciencia cidada: O caso da favela cidade de dues em Fortaleza, Ceara." *Revista Geografica Academica* 3, no. 1 (2009): 84–90.

Teulieres, R. "Bidonvilles du Bresil, les favelles de Belo Horizonte." *Les Cahiers d'Outre Mer* 8 (1955): 30–55.

Whately, Maria H. "Uso da terra no municipio de Albertina-Minas Gerais: Levantamento e maseamento-situacao em 1979." *Revista Brasileira de Geografia* 48, no. 2 (1986): 219–33.

Urban Economic Geography

Almeida Vasconcelos, Pedro. "Questoes espacais sobre o 'setor informal' urbano: O caso do Brasil." *Revista Geografica* 101 (1985): 53–62.

Carvalho, Inaia M. M. "Urban Employment: A Case Study of Bahia." *Antipode* 9, no. 3 (1977): 74–85.

DeJesus, Gilmar M. "O Lugar de feira livre na Grande Didade Capitalista: Rio de Janeiro, 1964–1989." *Revista Brasileira de Geografia* 54, no. 1 (1992): 95–120.

Faissol, Speridiao. "Brazil's Urban System in 1980: Basic Dimensions and Spatial Structure in Relation to Social and Economic Development." In *World Patterns of Modern Urban Change: Essays in Honor of Chauncy D. Harris*, edited by Michael P. Conzen, 293–328. Chicago: University of Chicago, Department of Geography, Research Paper 217–218, 1986.

———. "Urban Growth and Economic Development in Brazil in the 1960s." In *Urbanization and Counterurbanization*, edited by Brian J. L. Berry, 169–18. Beverly Hills, CA: Sage, 1976.

Fernandes, Ana C., and Rovena Negreiras. "Economic Developmentalism and Change within the Brazilian Urban System." *Geoforum* 32, no. 4 (2001): 415–35.

Gardner, James A. *Urbanization in Brazil*. New York: Ford Foundation, International Urbanization Survey, 1972.

Gauthier, Howard L. "Least Cost Flows in a Capacitated Network: A Brazilian Example." In *Geographic Studies of Urban Transportation and Network Analysis*, edited by Frank Horton, 102–27. Evanston, IL: Northwestern University, Studies in Geography 16, 1968.

———. "Transportation and the Growth of the Sao Paulo Economy." *Journal of Regional Science* 8, no. 1 (1968): 77–94.

Kipnis, Baruch A. "Clusters and Complexes of Medium-Sized Urban Manufacturing Systems: Two Case Studies in Brazil." *Professional Geographer* 35, no. 1 (1983): 32–39.

———. "Plant Size and Urban Growth." *Urban Studies* 21, no. 1 (1984): 53–61.

Long, Robert. "Volta Redonda: Symbol of Maturity in Industrial Progress in Brazil." *Economic Geography* 24, no. 2 (1948): 149–54.

Roberge, R. "Changes in the Employment Structure and Urbanization Levels for Brazil's Major Regions." In *Spatial Mobility and Urban Change*, edited by Otto Verkomen and Jan Van Weesep, 77–88. Utrecht: Netherlands Geographical Studies, Band 37, 1987.

Rodriquez-Pose, Andres, and John Tomaney. "Industrial Crisis in the Center of the Periphery: Stabilisation, Economic Restructuring, and Policy Responses in the Sao Paulo Metropolitan Region." *Urban Studies* 36, no. 3 (1999): 479–98.

Rossi, Eliana, et al. "Transaction Links through Cities: Decision-Cities and Service-Cities in Outsourcing by Leading Brazilian Firms." *Geoforum* 38, no. 4 (2007): 628–42.

Rossi, Eliana, and Peter Taylor. "Gateway Cities in Economic Globalisation: How Banks Are Using Brazilian Cities." *Tijdshrift voor Economische en Sociale Geographie* 97, no. 5 (2006): 515–34.

Santos, Milton. "Retroceso metropolitan y economia segmentada: El caso de Sao Paulo." *Investigaciones Geograficas* 25 (1992): 81–112.

Strohaecker, Tonia, and Celia de Souza. "A localizacao industrial intra-urbana: Evolucao e tendencias." *Revista Brasileira de Geografia* 52, no. 4 (1990): 73–90.

Townroe, Peter M. "Spatial Policy and Metropolitan Urban Growth in Sao Paulo, Brazil." *Geoforum* 15, no. 2 (1984): 143–65.

Valencia, Marcio M. "The Closure of the Brazilian Housing Bank and Beyond." *Urban Studies* 36, no. 10 (1999): 1747–68.
Williams, Jerry R. "Recent Development in Western Amazonia: Emergence of an Industrial Complex in Manaus." *California Geographer* 13 (1972): 21–26.

Urban Environments

Batista dos Santos, Thiago, et al. "Identificacao de semidouras espantaneos de carbano do municipio de Piquete-Sao Paulo." *Revista Geografica Academica* 4, no. 2 (2010): 19–28.
Gonvalves, Fabio, et al. "Drizzle and Fog Analysis in the Sao Paulo Metropolitan Area: Changes 1933–2005 and Correlations with Other Climate Factors." *Die Erde* 139, nos. 1–2 (2008): 61–76.
Libonati Machado, Carlos R. "Trecho leste do rodoanel metropolitan de Sao Paulo: Aplicacao e comparacao de metologias de analise ambiental." *Revista Geografica Academica* 6, no. 1 (2012): 25–37.
Nishizawa, Toshie, and Jose A. Sales. "The Urban Temperature in Rio de Janeiro, Brazil." *Latin American Studies, Japan* 5 (1983): 29–38.
Powell, Rebecca, and D. A. Roberts. "Characterizing Urban Land-Cover Change in Rondonia, Brazil: 1985 to 2000." *Journal of Latin American Geography* 9, no. 3 (2010): 183–211.
Smith, B. J. "Granite Weathering in an Urban Environment: An Example from Rio de Janeiro." *Singapore Journal of Tropical Geography* 11, no. 2 (1990): 143–53.
Smyth, Conor G., and Stephen A. Royle. "Urban Landslide Hazards: Incidence and Causative Factors in Niteroi, Rio de Janeiro State, Brazil." *Applied Geography* 20, no. 2 (2000): 95–118.

Theses and Dissertations

Aves-Capelani, Rodrigo. "Gating Porto Alegre; A Study of the Changing Social and Spatial Relations in the Brazilian Metropolis." Master's thesis, Miami University, Oxford, OH, 2009.
Borgas, Manoela. "Citizenship for the Urban Poor? Inclusive through Housing Programs in Rio de Janeiro, Brazil." PhD diss., University of Colorado, Boulder, 2005.
Carmin, Robert L. "Urban Pattern and Regional Function of Anapolis, Goia, Brazil." PhD diss., University of Chicago, Chicago, 1953.
Enders, Wayne T. "The Spatial Behavior of Low Income Hospital Patients: A Case in Porto Alegre, Brazil." PhD diss., University of Texas, Austin, 1978.
Gauthier, Howard L. "Highway Development and Urban Growth in Sao Paulo, Brazil: A Network Analysis." PhD diss., Northwestern University, Evanston, IL, 1966.
Mougeot, Luc J. A. "Population Growth and Decline in Government Sponsored Urban Nuclei in Brazilian Amazonia." PhD diss., Michigan State University, East Lansing, 1981.
Powell, Rebecca. "Long-Term Monitoring of Urbanization in the Brazilian Amazon Using Remote Sensing." PhD diss., University of California, Santa Barbara, 2006.
Webb, Kempton E. "Food Supply of Belo Horizonte." PhD diss., Syracuse University, Syracuse, NY, 1958.
Williams, Jerry R. "The Functional Relationship of Manaus to the Amazon Basin." PhD diss., University of Florida, Gainesville, 1969.
Worms, Jamie. "Evaluating Social Capital as It Affects Community Development in the Favelas of Rio de Janeiro." Master's thesis, George Washington University, Washington, DC, 2009.

Chapter Five

Chile

GENERAL WORKS

Atlases and Graphic Presentations

Atlas cartografico de reino de Chile. Siglos XVII–XIX. Santiago, Chile: Instituto Geografico Militar, 1981.

Atlas de la republica de Chile. Santiago, Chile: Ministerio de Defensa Nacional, Instituto Geografico Militar, 1970.

Sagredo Baeza, Rafael. "Geography and Nation: Claudio Gay and the First Maps of Chile [in Spanish]." *Estudios Geograficos* 70, no. 266 (2009): 231–68.

Walters, Rudy F., and Patrico V. Sagrista. *Atlas: Chile y sus nuevos provincias.* Santiago, Chile: Direccion de Fronteras y limites del Estado, 1976.

Books, Monographs, and Texts

Canon, Jose R., and E. Morales. *Geografia de mar chileno.* Santiago, Chile: Colecion Geografia de Chile 9, 1985.

Crooker, Richard A. *Chile.* Broomall, PA: Chelsea, 2005.

McBride, G. M. *Chile: Land and Society.* New York: American Geographical Society, 1936.

Subercaseaux, Benjamin. *Chile: A Geographic Extravaganza.* Translated by Angel Flores. New York: Hafner, 1971.

Articles and Book Chapters

Arenas, Federico. "Chile of Regions: The Unfinished History [in Spanish]." *Estudios Geograficos* 70, no. 266 (2009): 11–40.

Aschmann, Homer. "The Turno in Northern Chile: An Institution for Defense against Drought." *Geoscience and Man* 5 (1974): 97–110.

Cunill Grau, Pedro. "Transformaciones en la geografia social y economica chilena." *Investigaciones Geograficas* 4 (1971): 213–28.

Ferrer Jimenez, Daniel. "Geographical Knowledge of Inner Patagonia and the Configuration of Torres del Paine as a Natural Heritage to Be Preserved [in Spanish]." *Estudios Geograficos* 70, no. 266 (2009): 125–54.

Fifer, J. Valerie. "Arica: A Desert Frontier in Transition." *Geographical Journal* 130, no. 4 (1964): 507–18.
———. "Chile's Pioneering Location: Pacific Rim and Southern Cone." *Geography* 79, no. 2 (1994): 129–46.
Lake, Harry. "Easter Island." *Geographical Journal* 120, no. 4 (1954): 422–32.
McBride, G. M. "Unsolved Phase of the Tacna-Arica." *Yearbook, Association of Pacific Coast Geographers* 2 (1936): 10–15.
McKay, Robert R. "A Challenge to the Doctrine of Environmentalism: Factors That Contributed to the Failure of Hispanic Pacification Policies of the Araucanian Indians." *Ecumene* 6 (1974): 25–30.
McPhail, Donald D. "Chile." In *Focus on South America*, edited by Alice Taylor, 242–64. New York: Praeger, 1973.
Paskoff, Roland P. "Les regions geographiques du Chili." *Les Cahiers d'Outre Mer* 19 (1966): 193–201.
Rudolph, William E. "Catastrophe in Chile." *Geographical Review* 50 (1960): 528–31.
———. "The Rio Loa of Northern Chile." *Geographical Review* 17 (1927): 553–85.
Skottsberg, Carl. "The Islands of Juan Fernandez." *Geographical Review* 5 (1918): 362–83.
Whitbeck, R. H. "The Chileans and Their Geographic Environment." *Annals of the Association of American Geographers* 19 (1929): 149–56.

Theses and Dissertations

Gajardo, Rodolfo. "The Rural Environs of Chillan, Central Chile." Master's thesis, University of Minnesota, Minneapolis, 1967.
De Laubenfels, David J. "The Temuco Region: A Geographical Study in South Central Chile." PhD diss., University of Illinois, Urbana, 1953.

CULTURAL AND SOCIAL GEOGRAPHY

General Works

Borsdorf, Axel, and Rodrigo Hidalgo. "Searching for Fresh Air, Tranquility, and Rural Culture in the Mountains: A New Lifestyle for Chileans?" *Die Erde* 140, no. 3 (2009): 275–92.
Butland, Gilbert J. *The Human Geography of Southern Chile*. London: George Philip, 1957.
———. "The Human Geography of Southern Chile." *Transactions, Institute of British Geographers* 24 (1957).
Echeverria, Cristian, et al. "How Landscapes Change: Integration of Spatial Patterns and Human Processes in Temperate Landscapes of Southern Chile." *Applied Geography* 32, no. 2 (2012): 822–31.
Gonzalez, Miranda S. "L'origine du Norte Grande du Chili: Frontieres ouvertes, metalites fermees." *Revue de Geographie du Alpine* 91, no. 3 (2003): 11–27.
Grenier, Philippe. *Chiloe et les chilotes: marginalite et dependence en Patagonie chilienne-etude de geographie humaine*. Aix-en-Provence, France: Edition, 1984.
Holdgate, M. W. "Man and Environment in the Southern Chilean Islands." *Geographical Journal* 127, no. 4 (1961): 401–16.

Medical Geography

Garin, Alan, and Bernarda Olea. "Distribucion geografica de las enfermedades infantiles y sus implicancias socio-ambientales en el sector amanecer de la ciudad de Temuco." *Revista Geografica* 136 (2004): 133–49.
Haynes, Robin. "The Geographical Distribution of Mortality by Cause in Chile." *Social Science and Medicine* 17, no. 6 (1983): 355–64.

Miranda, E., et al. "A Decade of HMOs in Chile: Market Behavior, Consumer Choice, and the State." *Health and Place* 1, no. 1 (1995): 51–59.

Scarpaci, Joseph L. "Restructuring Health Care Financing in Chile." *Social Science and Medicine* 21, no. 4 (1985): 415–31.

———. "Utilization of Mobile Health Care Services in Atacama Desert Villages." *Proceedings, Conference of Latin Americanist Geographers* 9 (1983): 44–56.

Ethnic, Social, and Population Geography

Bayr, Jurgen. "Rural Exodus in Oasis Regions: The Example of North Chile." *Applied Sciences and Development* 10 (1977): 146–63.

Fuller, Gary. "On the Spatial Diffusion of Fertility Decline: The Distance-to-Clinic Variables in a Chilean Community." *Economic Geography* 50 (1974): 324–32.

Golte, W. "Die deutsche Kolonisation in Sudchile." In *Europaische Ethniean im landlichen Raum der Neuen Welt*, edited by Behandelte Thenen, 87–98. Passau, Germany: Passauer Schriften zur Geographie, Band 7, 1989.

Hakenholz, Thomas. "Un people autochtane face a la 'modernite': Mapuche-Perwendhe et le barrage Ralco Alto Bio-Bio, Chile." *Les Cahiers d'Outre Mer* 57, no. 228 (2004): 347–66.

Rasheed, Khairul. "Depopulation of the Oases in Northern Chile." *Revista Geografica* 74 (1971): 101–13.

Rovira, Adriano. "Dinamica del Poblamiento en el Espacio Semiarido de Chile." *Revista Geografica* 117 (1993): 93–118.

Sanchez, Alfred. "Las estrategias de crecimiento de los paises Latinoamericanos: El caso de Chile." *Revista Geografica* 119 (1994): 107–31.

Schneider, H. J. "Drought, Demography, and Destitution Crisis in the Norte Chico." *GeoJournal* 6, no. 2 (1982): 111–20.

Silva, Euzebio F. "Comentarias sobre unos mapas de densidade de poblacion de Chile." *Revista Geografica* 42 (1955): 89–131.

Community and Settlement Studies

Bloom, Reynold J. "The Small Holder Community in Central Chile." *Geographical Survey* 6, no. 1 (1977): 21–28.

Cereceda Trancoso, Pilar, et al. "Atacama: Settling the Desert and the Semi-desert in Chile [in Spanish]." *Estudios Geograficos* 70, no. 266 (2009): 41–78.

Fifer, J. Valerie. "Chile's Pioneering Locations: Pacific Rim and Southern Cone." *Geography* 79, no. 2 (1994): 129–46.

Hanson, Earl. "Out-of-the-World Villages of Atacama." *Geographical Review* 23, no. 3 (1933): 365–77.

Jefferson, Mark. *Recent Colonization in Chile*. New York: American Geographical Society, 1924.

Madalene, Isabel M. "The Privatisation of Water and Its Impacts on Settlement and Traditional Cultural Practices in Northern Chile." *Scottish Geographical Journal* 123, no. 3 (2007): 193–208.

Olave, Didima, and Julia Fawaz. "Calidad de vida rural a inicios del siglo XXI: Analisis de caso en comunas de la provincial de Nuble, region del Bio Bio, Chile." *Revista Geografica* 143 (2008): 29–46.

Platt, Robert S. "Items in the Chilean Pattern of Occupance." *Bulletin of the Geographical Society of Philadelphia* 32 (1934): 33–41.

Pomes, A. "Les frontiers de la region de Tarapaca (Chili)." *Les Cahiers d'Outre Mer* 56, no. 222 (2003): 149–58.

Porteous, J. Douglas. "The Curious Life and Abrupt Death of the Town of Sewell." *Places* 1, no. 2 (1974): 15–19.

———. "Social Class in Atacama Company Towns." *Annals of the Association of American Geographers* 65 (1974): 409–17.

Zapato, Eduardo, et al. "Funcionalidad del sistema de centros poblados de la provincial de Talca, Region de Maule, Chile." *Revista Geografica* 100 (1984): 141–50.

Theses and Dissertations

Bloom, Reynold J. "The Influence of Agrarian Reform on the Small Holder Communities in Chile's Central Valley, 1965–1970." PhD diss., University of California, Los Angeles, 1973.
Fuller, Gary. "The Spatial Diffusion of Birth Control in Chile." PhD diss., Pennsylvania State University, University Park, 1972.
Rasheed, Khairul. "Man and the Desert: Northern Chile." PhD diss., Columbia University, New York, 1970.

ECONOMIC GEOGRAPHY

General Works

Barton, Jonathan R., et al. "Competition and Co-operation in the Semi-periphery: Closer Economic Partnership and Sectoral Transformations in Chile and New Zealand." *Geographical Journal* 173, no. 3 (2007): 224–41.
Endlicher, Wilfried, and R. Mackel. "Natural Resources, Land Use, and Degradation in the Coastal Zone of Arauco and the Nahuelbula Range, Central Chile." *GeoJournal* 11, no. 1 (1985): 43–60.
Goldsmith, W. W., and M. F. Rothschild. "The Effect of Regional Specialization on Local Economic Activity: A Study of Chile." *Papers of the Regional Science Association* 32 (1974): 183–210.
Gwynne, Robert N. "Outward Orientation and Marginal Environments: The Question of Sustainable Development in the Norte Chico, Chile." *Mountain Research and Development* 13, no. 3 (1993): 281–93.
———. "Transnational Capitalism and Local Transformation in Chile." *Tijdschrift voor Economische en Sociale Geographie* 94, no. 3 (2003): 321–33.
Porteous, J. Douglas. "The Corporation as Frontier Developer: US Enterprise in the Atacama Desert." *BC Geographical Studies* 17 (1966): 79–88.
Smith, J. Russell. "The Economic Geography of Chile." *Bulletin of the American Geographical Society* 36, no. 1 (1904): 1–21.

Agriculture and Land Use

Alaluf, David. *Problemes de la proprieded agricola en Chile*. Kiel, Germany: Kieler Geographische Schriften, Band 19, Heft 2, 1961.
Armijo, Gladys. "Agricultura urbana en metropolis iberoamericans: Estudio de casos en Santiago de Chile y Lisboa, Portugal." *Investigaciones Geograficas* 54 (2004): 36–54.
Barton, Jonathan R. "Environment, Sustainability, and Regulation in Commercial Aquaculture: The Case of Chilean Salmoid Production." *Geoforum* 28, nos. 3–4 (1997): 313–28.
———. "Salmon Aquaculture and Chile's Export-Led Economy." *Norsk Geografisk Tidsskrift* 52, no. 1 (1998): 37–47.
Bee, Anna. "Globalization, Grapes, and Gender: Women's Work in Traditional and Agro-Export Production in Northern Chile." *Geographical Journal* 166, no. 3 (2000): 255–65.
Butland, Gilbert J. *Changing Land Occupance in the Southern Chilean Provinces of Aysen and Magallanes*. London: Birkbeck College, Geographical Study 1, 1954.
Clapp, Roger A. "Regions of Refuge and the Agrarian Question: Peasant Agriculture and Plantation Forestry in Chilean Araucania." *World Development* 26, no. 4 (1998): 571–89.

――――. "Tree Farming and Forest Conservation in Chile: Do Replacement Forest Leave Any Originals Behind?" *Society and Natural Resources* 14, no. 4 (2001): 341–56.

Concha, Manuel. "El use de la Tierra en el Nucleo Central de Chile." *Revista Geografica* 61 (1964): 5–14.

Endlicher, Wilfried, and R. Mackel. "Natural Resources, Land Use, and Degradation in the Coastal Zone of Arauco and the Nahuelbuta Range, Central Chile." *GeoJournal* 11, no. 1 (1985): 43–60.

Floysand, Arnt, et al. "Global Economic Imperatives, Crisis Generation, and Local Spaces of Engagement in the Chilean Aquaculture Industry." *Norsk Geografisk Tidsskrift* 64, no. 4 (2010): 199–210.

Gwynne, Robert N. "Export-Orientation and Enterprise Development: A Comparison of New Zealand and Chilean Wine Production." *Tijdschrift voor Economische en Sociale Geographie* 97, no. 2 (2006): 138–56.

Hadarits, Monica, et al. "Adaptation in Viticulture: A Case Study of Producers in the Maule Region of Chile." *Journal of Wine Research* 21, nos. 2–3 (2010): 167–78.

Jeffries, A. "Agrarian Reform in Chile." *Geography* 56, no. 3 (1971): 221–30.

Klepeis, Peter, and Pavel Laris. "Hobby Rancing and Chile's Land-Reform Legacy." *Geographical Review* 98, no. 3 (2008): 372–94.

McBride, G. M. "The Agrarian Problem: Chile." *Geographical Review* 20, no. 4 (1930): 574–86.

Montecinos, Sonia, et al. "The Impact of Agricultural Activities on Fog Formation in an Arid Zone of Chile." *Die Erde* 139, nos. 1–2 (2008): 77–95.

Overton, John D., and Warwick Murray. "Playng the Scales: Regional Transformations and the Differentiation of Rural Space in the Chilean Wine Industry." *Journal of Rural Studies* 27, no. 1 (2011): 63–72.

Paegelow, Martin, and Enrique Toro Balbontin. "Essaret restructuration du vignoble chilien: L'exemple du Male (VIIe region): 1995–2005." *Les Cahiers d'Outre Mer* 61 (2008): 81–98.

Thompson, John. "The Chilean Dairy Industry." *California Geographer* 1 (1960): 35–40.

Winnie, William W. "Communal Land Tenure in Chile." *Annals of the Association of American Geographers* 55, no. 1 (1965): 67–86.

Wright, A. C. S. "The Soil Process and the Evolution of Agriculture in Northern Chile." *Pacific Viewpoint* 4 (1963): 65–74.

Commerce and Trade

Barton, Jonathan R., and Warwick Murray. "Grounding Geographies of Economic Globalisation: Globalised Spaces in Chile's Non-traditional Export Sector, 1980–2005." *Tijdschrift voor Economische en Sociale Geographie* 100, no. 1 (2009): 81–100.

Claro, Edmundo, and Geoff A. Wilson. "Trans-Pacific Wood Chip Exports: The Rise of Chile." *Australian Geographical Studies* 34, no. 2 (1996): 185–99.

Giuliana, Elisa. "The Selective Nature of Knowledge Networks in Clusters: Evidence from the Wine Industry." *Journal of Economic Geography* 7, no. 2 (2007): 139–68.

Gwynne, Robert N. "Globalization, Commodity Chains, and Fruit Exporting Regions in Chile." *Tijdschrift voor Economische en Sociale Geographie* 90, no. 2 (1999): 211–25.

Gwynne, Robert N., and Jose Ortiz. "Export Growth and Development in Poor Rural Regions: A Meso-scale Analysis of the Upper Limari." *Bulletin of Latin American Research* 16, no. 1 (1997): 25–42.

Jones, Clarence F. "Chilean Commerce." *Economic Geography* 3, no. 2 (1927): 139–66.

Kleine, Dorothea. "The Ideology behind the Technology-Chilean Microentrepreneurs and Public KT Policies." *Geoforum* 40, no. 2 (2009): 171–83.

――――. "Negotiating Partnerships, Understanding Power: Doing Action Research on Chilean Fairtrade Wine Value Chains." *Geographical Journal* 174, no. 2 (2008): 109–23.

Scholz, Imre. "Trade-Induced Corporate Environmental Learning: Empirical Evidence from Chile." *Geographische Zeitschrift* 84, nos. 3–4 (1996): 169–78.

Werner, F. "Wassermarkte in Chile." *Zeitschrift fur Wirtschaftsgeographie* 44, no. 1 (2000): 32–40.

Manufacturing and Labor Studies

Chermakian, Jean. "Desarrollo caracteristicas actuales de la industria del cemento en Chile." *Revista Geografica* 63 (1965): 5–32.

Gwynne, Robert N. "The Deindustrialization of Chile, 1974–1984." *Bulletin of Latin American Research* 5, no. 1 (1986): 1–23.

———. "Government Planning and the Location of the Motor Vehicle Industry in Chile." *Tijdschrift voor Economische en Sociale Geographie* 69, no. 3 (1978): 130–40.

———. *Import Substitution and the Decentralization of Industry in Less Developed Countries: The Television Industry in Chile, 1962–74.* Birmingham, England: University of Birmingham, Department of Geography, Occasional Paper 12, 1980.

———. "Industrial Development in the Periphery: The Motor Vehicle Industry in Chile." *Bulletin of the Society for Latin American Studies* 29 (1978): 47–68.

Lazcano, Jose. "L'industrie du fer au Chili: Prosperite et decline." *Les Cahiers d'Outre Mer* 43 (1990): 239–72.

Overton, John D., et al. "The Remaking of Casablanca: The Sources and Impacts of Rapid Local Transformation in Chile's Wine Industry." *Journal of Wine Research* 23, no. 1 (2012): 47–59.

Saito, H., and M. Gopinath. "Plants' Self-Selection, Agglomeration Economies and Regional Productivity in Chile." *Journal of Economic Geography* 9, no. 4 (2009): 539–58.

White, C. Langdon, and R. H. Chilcote. "Chile's New Iron and Steel Industry." *Economic Geography* 37, no. 3 (1961): 259–266.

Energy, Mining, Fishing, and Lumbering

Auty, R. M. "Chile's Economic Recovery to Best Practice." In *Sustaining Development in Mineral Economies: The Resource Curse Thesis,* edited by R. M. Auty, 110–23. London: Routledge, 1993.

Besoain, E., and G. Sepulveda. "Merkmale und Eigenschaften einiger wich tiger vulkanischer Aschenboden aus dem Seengebiet Sudchiles-Minerolie, Entstehung und Eigenschaften der Boden." *Colloquium Geographicum* 16 (1983): 15–42.

Clapp, Roger A. "Creating Competitive Advantage: Forest Policy as Industrial Policy in Chile." *Economic Geography* 71, no. 3 (1995): 273–96.

Fried, J., and H. S. Bianchi. "Small-Scale Forestry Enterprise in Three Zones of Southern Chile." *Tijdschrift voor Economische en Sociale Geographie* 61, no. 4 (1970): 223–31.

Guzman, Alfonzo. *Chilean Fishing and Aquaculture Development: Export and Market.* Roskilde, Denmark: University of Roskilde, Department of Geography and International Development Studies, 1994.

Lara, Albania L., and Thomas T. Veblen. "Forest Plantations in Chile: A Successful Model?" In *Afforestation: Policies, Planning, and Progress,* edited by A. Mather, 118–39. London: Belhaven, 1994.

Marr, Paul. "Ghosts of the Atacama: The Abandonment of Nitrate Mining in the Tarapaca Region of Chile." *Middle States Geographer* 40 (2007): 22–31.

McConnell, W. R. "The Nitrate Industry of Chile." *Journal of Geography* 16 (1918): 211–14.

McLaughlin, Marie. "Study Guide for the Nitrate Regions of Chile." *Journal of Geography* 31, no. 6 (1932): 237–44.

Mikus, Werner. "Hoch seefischerel und Unwelt in Chile." *Zeitschrift fur Wirtschaftsgeographie* 44, no. 1 (2000): 19–31.

Pederson, Leland R. *The Mining Industry of the Norte Chico, Chile.* Evanston, IL: Northwestern University, Studies in Geography 11, 1966.

Rich, John L. "The Nitrate District of Tarapaca, Chile: An Aerial Traverse." *Geographical Review* 31, no. 1 (1941): 1–22.

Sanchez, R. *Activadat forestall.* Santiago, Chile: Colecion Geografia de Chile 14, 1986.

Tesor, C., and S. C. Tesor. "Recent Chilean Copper Policy." *Geography* 58, no. 1 (1973): 9–13.

Whitbeck, R. H. "Chilean Nitrate and the Nitrogen Revolution." *Economic Geography* 7 (1931): 273–83.

Transportation and Communications

Munoz, D. *Geografía de transporte y communicaciones*. Santiago, Chile: Colecion Geografía de Chile 13, 1985.

Theses and Dissertations

Butler, Joseph H. "Manufacturing in the Concepcion Region of Chile." PhD diss., Columbia University, New York, 1959.

Clapp, Roger A. "The Forest at the End of the World: The Transition from Old-Growth to Plantation Forestry in Chile." PhD diss., University of California, Berkeley, 1993.

Elwell, Tammy. "The Effects of Chile's Coastal Management on Artisanal Fisher Livelihoods: Three Cases from Ancud, Chiloe." Master's thesis, University of California, Santa Barbara, 2010.

Hrycyk, Larissa. "An Analysis of Entry Strategies for Doing Business in Chile: Export, Foreign Direct Investment, and Strategic Alliances." Master's thesis, State University of New York, Buffalo, 2001.

Kreisman, Arnold J. "The Numerical Regionalization of Agricultural Land: A Chilean Example." PhD diss., University of Pittsburgh, Pittsburgh, PA, 1972.

Krispas, Jeanette E. "The Relationship of Globalization in the Development and Success of the Chilean Wine Industry." Master's thesis, Marshall University, Charleston, WV, 2008.

Martin, Gene E. "Land Redistribution in Central Chile." PhD diss., Syracuse University, Syracuse, NY, 1955.

Pederson, Leland R. "The Mining Industry of the Norte Chico, Chile." PhD diss., University of California, Berkeley, 1965.

Recort, Lucila M. H. "A Narrative Approach to Rural Health Service Location in Imperial, Chile." Master's thesis, Pennsylvania State University, University Park, 1969.

Smole, William J. "Owner-Cultivatorship in Middle Chile." PhD diss., University of Chicago, Chicago, 1964.

Swatz, Robert D. "The Chilean Copper Industry: Recent Developments and Prospects." Master's thesis, Columbia University, New York, 1962.

Urrea, Jorge. "Patterns of Industrial Structural Change of Biobio Region, Chile, 1974–1989." Master's thesis, University of Ottawa, Ottawa, Ontario, 1992.

Willett, Curtis L. "The Economy of Chile's Arid North: A Geographic Survey." Master's thesis, University of Oklahoma, Norman, 1965.

HISTORICAL GEOGRAPHY

General Works

Parsons, James J. *Antioquia's Corridor to the Sea: An Historical Geography of Settlement of Uraba*. Berkeley: University of California Press, Ibero-Americana 49, 1967.

Reyes-Coco, Marco A. "Los umbrales del crecimiento de Chillan en cuatro siglos." *Revista Geografica* 100 (1984): 151–61.

PHYSICAL GEOGRAPHY

General Works

Heusser, Calvin. "Vegetation and Climate of the Southern Chile Lake District during and since the Last Inter-glaciation." *Quaternary Studies* 4, no. 3 (1974): 290–315.

Heusser, Calvin, et al. "Vegetation Dynamics and Paleoclimate during Late Llanquiheu Glacia-tion in Southern Chile." In *Landschaftsenwicklung, Palaookologie, und Klimageschichta der Ariden Diagonale Sudamerikas im Jungquarter*, edited by K. Garleff, 211–28. Bamberg, Germany: Bamberger Geographische Schriften, Heft 15, 1998.

Hubbard, Alan L. "Modelling Climate, Topography, and Palaeoglacier Fluctuations in the Chilean Andes." *Earth Surface Processes and Landforms* 22, no. 1 (1997): 79–92.

Paskoff, Roland P. "Quaternary of Chile: The State of Research." *Quaternary Research* 8 (1977): 2–31.

Rasheed, Khairul. "Man-Environmental Synergism in Arid Highlands: The Case of Northern Chile." *Oriental Geographer* 15, nos. 1–2 (1971): 14–36.

Veit, Heinz. "Upper Quaternary Landscape and Climate Evolution in the Norte Chico: An Overview." *Mountain Research and Development* 13, no. 2 (1993): 138–44.

Winchester, Vanessa, and Stephen Harrison. "A Development of the Lichenometric Method Applied to the Dating of Glacially Influenced Debris Flows in Southern Chile." *Earth Surface Processes and Landforms* 19, no. 2 (1994): 137–51.

Biogeography

Aravena, J. C., et al. "Changes in Tree Species Richness, Stand Structure, and Soil Properites in a Successional Chronosequence of Forest Fragments in Northern Chiloe Island, Chile." *Revista Chilena de Historia Natural* 75 (2002): 339–60.

Aries, Elizabeth, et al. "The Canopy Beetle Fauna of Gondwana Event Trees in Chilean Tem-perate Rain Forests." *Journal of Biogeography* 35, no. 5 (2008): 914–25.

Armesto, Juan, and Javier Figueroa. "Stand Structure and Dynamics in the Temperate Rain Forests of Chiloe Archipelago, Chile." *Journal of Biogeography* 14, no. 4 (1987): 367–76.

Armesto, Juan, and R. Rozzi. "Seed Dispersal Systems in the Rain Forest of Chiloe: Evidence for the Importance of Biotic Dispersal in a Temperate Rain Forest." *Journal of Biogeogra-phy* 16, no. 3 (1989): 219–26.

Bahre, Conrad J. *Destruction of the Natural Vegetation on North-Central Chile*. Berkeley: University of California, Publications in Geography 23, 1979.

Breton, Gunnar, and Konrad Fiedler. "Faunal Compostion of Geometric Moths Changes with Altitude in an Andean Montane Rain Forest." *Journal of Biogeography* 30, no. 3 (2003): 431–40.

Bull-Herenu, K., et al. "Structure and Genetic Diversity in *Colliguaja odorifer* Mol. (Euphorbi-acaea), a Shrub Subjected to Pleisto-Holocene Natural Perturbation in a Mediterranean South American Region." *Journal of Biogeography* 32, no. 7 (2005): 1129–38.

Caiafa, C. F., et al. "Development of Individual Recognition of Female Southern Elephant Seals, Mirounga Leonina, from Punta Norte Peninsula Valdes, Applying Principal Compo-nents Analysis." *Journal of Biogeography* 32, no. 7 (2005): 1257–66.

Caviedes, Cesar N., and Augustin Iriate. "Migration and Distribution of Rodents in Central Chile since the Pleistocene: The Paleogeographic Evidence." *Journal of Biogeography* 16, no. 2 (1989): 181–88.

Clapp, Roger A. "The Unnatural History of the Monterey Pine." *Geographical Review* 85, no. 1 (1995): 1–19.

Endlicher, Wilfried. "Landscape Damage in Central Chile: Ecological Causes, Attempts at Quantification, and Suggestions for Improvement." *Applied Geography and Development* 35 (1990): 45–62.

Gonzalez, M. E., et al. "Fire History of Araucanria-Nothofagues Forests in Villarrice National Park, Chile." *Journal of Biogeography* 32, no. 7 (2005): 1187–202.

———. "Influence of Fire Severity on Stand Development of *Araucaria araucana–Nothofagus pumileo* Stands in the Andean Cordillera of South-Central Chile." *Austral Ecology* 35 (2010): 597–615.

———. "Tree Regulation Responses in a Lowland Nothofagus Dominated Forest after Bam-boo Dieback in South-Central Chile." *Plant Ecology* 161 (2002): 59–73.

Gutierrez, Alvaro, et al. "Gap-Phase Dynamics and Coexistence of a Long-Lived Pioneer and Shade-Tolerant Tree Species in the Canopy of an Old-Growth Temperate Rainforest of Chiloe Island, Chile." *Journal of Biogeography* 35, no. 9 (2008): 1674–87.

Haberle, Simon G. "Late Quaternary Vegetation Dynamics and Human Impact on Alexander Selkirk Island, Chile." *Journal of Biogeography* 30, no. 2 (2003): 239–56.

Herrmann, T. M., et al. "Roost Sites and Communal Behavior of Andean Condors in Chile." *Geographical Review* 100, no. 2 (2010): 246–62.

Heusser, Calvin, et al. "Paleoecology of the Southern Chilean Lake District–Isla Grande de Chiloe during Middle-Late Llanquihue Glaciation and Deglaciation." *Geografiska Annaler* 81A, no. 2 (1999): 231–84.

Hinojosa, Luis, et al. "Are Chilean Coastal Forests Pre-Pleistocene Relicts? Evidence from Foliar Physiognomy, Palaeoclimate, and Phytogeography." *Journal of Biogeography* 33, no. 2 (2006): 331–41.

Innes, John L. "Structure of Evergreen Temperate Rain Forest on the Taitao Peninsula, Southern Chile." *Journal of Biogeography* 19, no. 5 (1992): 555–62.

Jimenez, Alejandro, et al. "Do Climatically Similar Regions Contain Similar Alien Floras? A Comparison between the Mediterranean Areas of Central Chile and California." *Journal of Biogeography* 35, no. 4 (2008): 614–24.

Lara, Albania L., et al. "Dendroclimatology of High Elevation *Nothofagus pumilo* Forests at Their Northern Distribution Limit in the Central Andes of Chile." *Canadian Journal of Forest Research* 31 (2001): 925–36.

———. "The Potential of Tree-Rings for Streamflow and Estuary Salinity Deconstruction in the Valdivian Rainforest Eco-region, Chile." *Dendrochronologia* 22 (2005): 155–61.

Lusk, C. H., and F. Matus. "Juvenile Tree Growth Rates and Species Sorting on Fire-Scale Soil Fertility Gradients in a Chilean Temperate Rain Forest." *Journal of Biogeography* 27, no. 4 (2000): 1011–20.

Malo, J. E., and F. Suarez. "Dispersal Mechanism and Transcontinental Naturalization Proneness among Mediterranean Herbaceous Species." *Journal of Biogeography* 24, no. 3 (1997): 391–94.

Meserve, Peter, et al. "Geographical Ecology of Small Mammals in Continental Chile Chico, South America." *Journal of Biogeography* 18, no. 2 (1991): 179–88.

Meserve, Peter, and William E. Glanz. "Geographical Ecology of Small Animals in the Northern Chilean Arid Zone." *Journal of Biogeography* 5, no. 2 (1978): 135–48.

Molina-Montenagro, M. A., et al. "Cushion Plants as Microclimatic Shelters for Two Ladybird Beetle Species in Alpine Zone of Central Chile." *Arctic, Antarctic, and Alpine Research* 38, no. 2 (2006): 224–27.

Moreno, Rodrigo A., et al. "Patterns of Endemism in South-Eastern Pacific Benthic Polychaetes of the Chilean Coast." *Journal of Biogeography* 33, no. 4 (2006): 750–59.

Perez, Victor Q. "Impact o del fuego sobre la biodiversidad de bosques templados: El caso del bosque pluvial costero de Chile." *Revista Geografica* 131 (2002): 79–94.

———. "Perturbaciones a la vegetacion native par grandes fuegos de 50 anos atras, en bosques nord patagonicos: Case de estudio en Chile Meridionel." *Anales de Geografia de la Universidad Complutense* 28, no. 1 (2008): 85–104.

Pollman, Williama. "Effects of Natural Disturbance and Selective Logging on Nothofagus Forests in South-Central Chile." *Journal of Biogeography* 29, no. 7 (2002): 955–70.

Pollman, Williama, and Thomas T. Veblen. "Nothofagus Regeneration Dynamics in South-Central Chile: A Test of a General Model." *Ecological Monographs* 74, no. 4 (2004): 615–34.

Premoil, Andrea, et al. "Inzume Variation and Recent Biogeographical History of the Long-Lived Conifer *Fitzraga cupresoides*." *Journal of Biogeogrpahy* 27, no. 2 (2000): 251–60.

Quintanilla, Victor. "Problemas consecuencias ambientales sobre el Bosque de Alerce (Fitzroya Cupressoides Mol Johnst), debidoa la explotacion de la cordillera costera de Chile Austral." *Revista Geografica* 114 (1991): 55–72.

Rarito, S. M., et al. "Distributional Modelling and Parsimony Analysis of Endemicity of Senecio in the Mediterranian Type Climate Area of Central Chile." *Journal of Biogeography* 31, no. 10 (2004): 1623–36.

Richter, Michael, and H. Schroder. "Remarks on the Paleoecology of the Atacama Based on the Distribution of Recent Geomorphological and Phytogeographical Patterns." In *Landschaftsentwicklung, Palaookologie, und Klimageschichta der Ariden Diagonale Sudamerkias im Jungquarter,* edited by K. Garleff, 57–70. Bamberg, Germany: Bamberger Geographische Schriften, Heft 15, 1998.

Schulz, Jennifer, et al. "Monitoring Land Cover Change of the Dryland Forest Landscape of Central Chile (1975–2008)." *Applied Geography* 30, no. 3 (2010): 436–47.

Sede, Silvana, et al. "Phylogeography and Palaeodistribution Modelling in the Patagonian Steppe: The Case of *Mulinum spinosusm* (Apiaceae)." *Journal of Biogeography* 39, no. 6 (2012): 1041–57.

Squeo, Francisco. "Spatial Heterogeneity of High Mountain Vegeation in the Andean Desert Zone of Chile." *Mountain Research and Development* 13, no. 2 (1993): 203–9.

Szeicz, J. M. "Growth Trends and Climatic Sensitivity of Trees in the North Patagonian Rain Forest of Chile." *Canadian Journal of Forest Research* 27 (1997): 1003–14.

Thiel, M. "The Zoogeography of Algae-Associated Peracorids along the Pacific Coast of Chile." *Journal of Biogeography* 29, no. 8 (2002): 999–1008.

Veblen, Thomas T. "Degradation of Native Forest Resources in Southern Chile." In *History of Sustained Yield Forestry: A Symposium,* edited by H. K. Steen, 344–52. Santa Cruz, CA: Forest History Society, 1984.

———. "Natural Hazards and Forest Resources in the Andes of South-Central Chile." *GeoJournal* 6, no. 2 (1982): 141–50.

———. "Regeneration Patterns in Araucanria Araucana Forest in Chile." *Journal of Biogeography* 9, no. 1 (1982): 11–28.

———. "Stand Dynamics in Chilean Nothofagus Forests." In *The Ecology of Disturbances and Patch Dynamics,* edited by S. T. A. Pikett and P. S. White, 35–51. New York: Academic Press, 1985.

Veblen, Thomas T., and D. H. Ashton. "Catastrophic Influences on the Vegetation of the Valdivian Andes, Chile." *Vegetatio* 36 (1978): 149–67.

———. "The Regeneration Status of *Fizroya cupressoides* in the Cordillera Pelada, Chile." *Biological Conservation* 23 (1982): 141–61.

———. "Successional Pattern Above Timberline in South-Central Chile." *Vegetatio* 40, no. 1 (1979): 39–47.

Veblen, Thomas T., et al. "Distribution and Dominance of Species in the Understory of Mixed Evergreen-Deciduous Nothofagus Forest in South-Central Chile." *Journal of Ecology* 65 (1977): 815–30.

———. "Forest Dynamics in South-Central Chile." *Journal of Biogeography* 8, no. 3 (1981): 211–48.

———. "Nothofagus Stand Development on In-Transit Moraines, Casa Pangue Glacier, Chile." *Arctic and Alpine Research* 21, no. 4 (1989): 144–55.

———. "Plant Succession in a Timberline Depressed by Vulcanism in South-Central Chile." *Journal of Biogeography* 4, no. 3 (1977): 275–94.

———. "Structure and Dynamics of Old-Growth Nothofagus Forests in the Valdivian Andes, Chile." *Journal of Ecology* 68 (1980): 1–31.

———. "Tree Regeneration Strategies in a Lowland Nothofagus-Dominated Forest in South-Central Chile." *Journal of Biogeography* 6, no. 4 (1979): 329–40.

Viruel, Juan, et al. "Disrupted Phylogeographical Microsatellite and Chloroplast DNA Patterns Indicate a Vicarianen Rather Than Long-Distance Dispersal Origin for the Disjunct Distribution of the Chilean Endemic *Dioscarea biloba* (Dioscoreaceae) around the Atacama Desert." *Journal of Biogeography* 39, no. 6 (2012): 1073–85.

Climatology and Weather

Caviedes, Cesar N. "On the Paleoclimatology of the Chilean Littoral." *Geographical Perspectives* 29 (1972): 8–14.

———. "Perturbations during the Predominance of Anticyclonic Summer Weather in Central Chile." *Revista Geografica* 74 (1971): 71–81.

———. "Rainfall Variation, Snowline Depression and Vegetational Shifts in Chile during the Pleistocene." *Climatic Change* 16, no. 1 (1990): 99–114.

Cereceda Trancoso, Pilar, et al. "Comportamiento espacio-temporal de la nube estratocumulo, productora de niebla en la costa del desierto de Atacama (21 Degrees Latitude South, 70 Degrees Longitude West), durante un mes de invierno y otro de verano." *Investigaciones Geograficas* 56 (2005): 43–61.

Denton, G. H., et al. "Interhemispheric Linkage of Paleoclimate during the Last Glaciation." *Geografiska Annaler* 81A, no. 2 (1999): 107–54.

Fintorpusch, C. A. "Tierras Semiaridas." *Revista Geografica* 41 (1954): 33–44.

Grosjean, Martin, et al. "Mid-Holocene Climate and Culture Change in the Atacama Desert, Northern Chile." *Quaternary Research* 48, no. 2 (1997): 239–46.

Jefferson, Mark. *Chile: Rainfall, Recent Colonization.* New York: American Geographical Society, 1921.

———. *The Rainfall of Chile.* New York: American Geographical Society, Research Series 7, 1918.

Jenny, Bettina, et al. "Early to Mid-Holocene Aridity in Central Chile and the Southern Westerlies: The Laguna Aculeo Record (34 Degrees S)." *Quaternary Research* 58, no. 2 (2002): 160–70.

Kerr, Andrew, and D. E. Sugden. "The Sensitivity of the Southern Chilean Snowline to Climate Change." *Climatic Change* 28, no. 3 (1994): 255–72.

Messerli, Bruno, et al. "Climatic Change and Natural Resource Dynamics of the Atacama Altiplano during the Last 18,000 Years: A Preliminary Synthesis." *Mountain Research and Development* 13, no. 2 (1993): 117–27.

———. "The Problem of the Andean Dry Diagonal: Current Precipitation, Late Pleistocene Snowline, and Lake Level Changes in the Atacama Altiplano." In *Landschaftsentwicklung, Palaookologie, und Klimageschicha der Ariden Diagonale Sudamerikas im Jungquarter*, edited by K. Garleff, 17–34. Bamberg, Germany: Bamberger Geographische Heft 15, 1998.

Rutland, Jose, and Humberto Fuenzalida. "Synoptic Aspects of the Central Chile Rainfall Variability Associated with the Southern Oscillation." *International Journal of Climatology* 11, no. 1 (1991): 63–76.

Sanvedra, N., and A. J. Foppiano. "Monthly Mean Pressure Model for Chile." *International Journal of Climatology* 12, no. 5 (1992): 469–80.

Geomorphology, Landforms, and Volcanism

Berger, I. A., and A. V. Cooke. "The Origins and Distribution of Salts on Alluvial Fans in the Atacama Desert, Northern Chile." *Earth Surface Processes and Landforms* 22, no. 6 (1997): 581–600.

Bovis, M. J. *Rio Las Minas Basin, Southern Chile: A Geomorphic and Geotechnical Assessment.* Washington, DC: Report to Inter-American Development Bank, 1995.

Brenning, A. "Geomorphological, Hydrological, and Climatic Significance of Rock Glaciers in the Andes of Central Chile." *Permafrost and Periglacial Processes* 16, no. 3 (2005): 231–40.

Castillo, Edilia J. "Geomorfologia de la Cuenca de Ro Andalien, Chile." *Revista Geografica* 143 (2008): 97–116.

Cevo, Juan H. "Manifestaciones volcanicas en Aysen (Chile) entre 1971 7 1973." *Revista Geografica de America Central* 1 (1974): 51–74.

Correa, Carlos T. "Propuesta para el uso de dunas litorales en Chile Central." *Revista Geografica* 134 (2003): 113–28.

Denton, G. H., et al. "Geomorphology, Stratigraphy, and Radiocarbon Chronology of Llanquihue Drift in the Area of the Southern Lake District, Seno Reloncavi and Isla Grande de Chiloe, Chile." *Geografiska Annaler* 81A, no. 2 (1999): 167–230.

Fintorpusch, C. A. "Tierras Semiaridas." *Revista Geografica* 41 (1954): 33–44.

Glasser, N. F., et al. "The Geomorphology and Sedimentology of the Tempanos Moraine at Laguna San Rafael, Chile." *Journal of Quaternary Science* 21, no. 6 (2006): 629–43.

Glasser, N. F., and Stephen Harrison. "Sediment Distribution around Glacially Abraded Bedrock Landforms (Whalebacks) at Lago Tranquilo, Chile." *Geografiska Annaler* 87A, no. 3 (2005): 421–30.

Heusser, Calvin J. "Late Pleistocene Environments of the Laguna de San Rafael Area, Chile." *Geographical Review* 50, no. 4 (1960): 555–77.

MacPhail, Donald. "The Geomorphology of the Rio Tino Lahar, Central Chile." *Geographical Review* 63 (1973): 517–32.

Morgan, R., et al. "Origin of the Cerro El Dragon Coastal Dune, Iquique, Atacama Desert, Chile." *Zeitschrift fur Geomorphologie* 49, no. 2 (2005): 167–82.

Oka, A. "On the Neotectonics of the Atacama Fault Zone Region-Preliminary Notes on Late Cenozoic Faulting and Geomorphic Development of the Coast Region of Northern Chile [in Japanese]." *Bulletin, Tokyo University, Department of Geography* 2 (1971): 47–65.

Paskoff, Roland P. "Terrasses littorals et tectonique recente entre l'embaouchure du rio limari et la baie Teniete, province de Coquimbo, Chili." *Revista Geografica* 65 (1966): 57–67.

———. "Zonality and Main Geomorphic Features of the Chilean Coast." In *Zonality of Coastal Geomorphology and Ecology-Proceedings of the Sylt Symposium, August 30–Septemper 3, 1989*, edited by E. Bird and D. Kelletat, 237–67. Essen, Germany: Essener Geographische Arbeiten, Band 18, 1989.

Rudolph, William E. "Licancabur: Mountain of the Atacamenos." *Geographical Review* 45, no. 2 (1955): 151–71.

Taulis, Emilio. "De la Distribution de pluies au Chile." In *Materiaux pour l'etude des calamites*, Part 1, 3–20. Geneva, Switzerland: Societe de Geographie de Geneve, 1934.

Winchester, Vanessa, and Stephen Harrison. "Dendrochronology and Lichenometry: Colonization, Growth Rate, and Dating of Geomorphological Events on the East Side of the North Patagonian Icefield, Chile." *Geomorphology* 34, nos. 3–4 (2000): 181–94.

Hydrology and Glaciology

Anderson, B. G., et al. "Glacial Geomorphologic Maps of Llanquihue Drift in the Area of the Southern Lake District, Chile." *Geografiska Annaler* 81A, no. 2 (1999): 155–66.

Barcaza, Gonzalo, et al. "Satellite-Derived Equilibrium Lines in Northern Patagonia Icefield, Chile, and Their Implications to Glacier Variations." *Arctic, Antarctic, and Alpine Research* 41, no. 2 (2009): 174–82.

Benn, Douglas, and Chalmers M. Clapperton. "Glacial Sediment-Landform Associations and Paleoclimate during the Last Glaciation, Strait of Magellan, Chile." *Quaternary Research* 54, no. 1 (2000): 13–23.

Bentley, Michael. "The Role of Lakes in Moraine Formation, Chilean Lake District." *Earth Surface Processes and Landforms* 21, no. 6 (1996): 493–507.

Bentley, Michael, and R. D. McCulloch. "Impact of Neotectonics on the Record of Glacier and Sea Level Fluctuations, Strait of Magellan, Southern Chile." *Geografiska Annaler* 87A, no. 2 (2005): 393–402.

Caseres, L., et al. "Water Recycling in Arid Regions: Chilean Case." *Ambio* 21, no. 2 (1992): 138–44.

Caviedes, Cesar N. "Annual and Seasonal Fluctuations of Precipitation and Streamflow in the Aconcagua River Basin." *Journal of Hydrology* 120 (1990): 79–102.

Caviedes, Cesar N., and Roland P. Paskoff. "Quaternary Glaciations in the Andes of North-Central Chile." *Journal of Glaciology* 70 (1975): 155–70.

Clapperton, Chalmers M., et al. "The Last Glaciation in Central Magellan Strait, Southernmost Chile." *Quaternary Research* 44, no. 2 (1995): 133–48.

Fernandez Rivera, Alfonso, et al. "Variaciones recientes de glaciares entre 41 Degrees S y 49 Degrees S y su relacion con los cambios climaticos." *Revista Geografica* 139 (2006): 39–65.

Fogwill, C. J., and P. W. Fogwill. "A Glacial Stage Spanning the Antarctic Cold Reversal in Torres de Paine (51 Degrees S), Chile, Based on Preliminary Cosmogenaic Exposure Ages." *Geografiska Annaler* 87A, no. 2 (2005): 403–8.

Garcia, Juan L. "Late Pleistocene Ice Fluctuations and Glacial Geomorphology of the Archipelago de Chiloe, Southern Chile." *Geografiska Annaler* 94A, no. 4 (2012): 459–79.

Geyle, M. A., et al. "Radiocarbon Reservoir Effect and the Time of the Late-Glacial/Early Holocene Humid Phase in the Atacama Desert (Northern Chile)." *Quaternary Research* 52, no. 2 (1999): 143–53.

Grosjean, Martin, et al. "A Late-Holocene (< 2600) BP Glacial Advance in the South Central Andes (29 Degrees S), Northern Chile." *Holocene* 8, no. 4 (1998): 473–79.

———. "Mid- and Late-Holocene Limnology of Laguna del Negro Francisco, Northern Chile, and Its Palaeoclimatic Implications." *Holocene* 7, no. 2 (1997): 151–59.

Harrison, Stephen, et al. "Glacier Leon, Chilean Patagonia: Late-Holocene Chronology and Geomorphology." *Holocene* 18, no. 4 (2008): 643–52.

———. "Morphostratigraphy of Moraines in the Lago Tranquilo Area, Chilean Patagonia." *Bulletin of Glaciological Research* 221 (2004): 37–43.

———. "Onset of Rapid Calving and Retreat of Glaciar San Quintin, Hielo Patagonico Norte, Southern Chile." *Polar Geography* 25, no. 1 (2001): 54–61.

———. "Quantifying Rates of Paraglacial Sedimentation: An Example from Chilean Patagonia." *Zeitschrift fur Geomorphologie* 49, no. 3 (2005): 321–34.

Harrison, Stephen, and Vanessa Winchester. "Historical Fluctuations of the Guale and Reicher Glaciers, North Patagonian Icefield, Chile." *Holocene* 8, no. 4 (1998): 481–85.

———. "Nineteenth and Twentieth Century Glacier Fluctuations and Climate Implications in the Arco and Colonia Valleys, Hielo Patagonico North, Chile." *Arctic, Antarctic, and Alpine Research* 32, no. 1 (2000): 55–63.

Holmlund, Per, and Humberto Fuenzalida. "Anomalous Glacier Responses to Twentieth-Century Climatic Changes in Darwin Cordillera, Southern Chile." *Journal of Glaciology* 41, no. 1 (1995): 465–73.

Koppes, N., et al. "Synchronous Acceleration of Ice Loss and Glacier Erosion, Glaciar Marinelli, Chilean Tierra del Fuego." *Journal of Glaciology* 55, no. 190 (2009): 207–20.

Kuylenstierna, J. L., et al. "Late Holocene Glacier Variations in the Cordillera Darwin, Tierra del Fuego, Chile." *Holocene* 6, no. 3 (1996): 353–58.

Moreno, P. I., et al. "Abrupt Vegetation and Climate Change during the Last Glacial Maximum and Last Termination in the Chilean Lake District: A Case Study from Canal de la Purtilla." *Geografiska Annaler* 81A, no. 2 (1999): 285–312.

Olivares, R. B. "The Quaternary Glaciations in the West of Lake Llanquihue in Southern Chile." *Revista Geografica* 67 (1967): 101–8.

Warren, C. R. "Rapid Recent Fluctuations of the Calving San Rafael Glacier, Chilean Patagonia: Climatic or Non-climatic?" *Geografiska Annaler* 75A, no. 3 (1993): 111–26.

Warren, C. R., et al. "Boyancy-Driven Lacustrine Calving, Glacier Nef, Chilean Patagonia." *Journal of Glaciology* 47, no. 1 (2001): 135–46.

———. "Characteristics of Tide-Water Calving at Glacier San Rafael, Chile." *Journal of Glaciology* 41, no. 138 (1995): 273–89.

Waylen, Peter R., and Cesar N. Caviedes. "Annual and Seasonal Fluctuations of Precipitation and Streamflow in the Aconcagua River Basin, Chile." *Journal of Hydrology* 120 (1990): 79–102.

Waylen, Peter R., et al. "El Nino-Southern Oscillation and the Surface Hydrology of Chile: A Window on the Future?" *Canadian Water Resources Journal* 18, no. 4 (1993): 425–41.

Winchester, Vanessa, and Stephen Harrison. "Recent Oscillations of the San Quentin and San Rafael Glaciers, Patagonian Chile." *Geografiska Annaler* 78A, no. 1 (1996): 35–50.

Theses and Dissertations

Aagesen, David. "A Social and Natural History of Araucaria Araucana." Master's thesis, University of Minnesota, Minneapolis, 1993.

Bahre, Conrad J. "Relationships between Man and the Wild Vegetation in the Province of Coquimbo, Chile." PhD diss., University of California, Riverside, 1974.

Holz, Carlos A. "Tree Regeneration after Bamboo Dieback in Temperate Forests." Master's thesis, University of Colorado, Boulder, 2004.

Saavedra, Maria F. "El Morado National Park, an Attempt to Explain the Composition and Distribution of a Mountain Flora in the Central Andes of Chile." Master's thesis, University of Florida, Gainesville, 1993.

Young, Gwendolynne. "Community Vulnerability to Climate Change in the Elqui River Basin, Chile." Master's thesis, University of Guelph, Guelph, Ontario, 2006.

POLITICAL GEOGRAPHY

General Works

Azocar, Gerardo, et al. "Conflicts for Control of Mapuche-Pehuenche Land and Natural Resources in the Biobio Highlands, Chile." *Journal of Latin American Geography* 4, no. 2 (2005): 57–76.

Budds, Jessica. "Contested H2O: Science, Policy, and Politics in Water Resources Management in Chile." *Geoforum* 40, no. 3 (2009): 418–30.

———. "Power, Nature, and Neoliberalism: The Political Ecology of Water in Chile." *Singapore Journal of Tropical Geography* 25, no. 3 (2004): 322–42.

Caviedes, Cesar N. *Elections in Chile: The Road to Democratization.* Boulder, CO: Lynne Rienner, 1991.

———. *The Politics of Chile: A Sociogeographical Assessment.* Boulder, CO: Westview, 1979.

James, Preston E. "The Geographic Setting of the Tacna-Arica Dispute." *Journal of Geography* 21 (1922): 339–48.

Meza, Laura E. "Mapuche Struggles for Land and the Role of Private Protected Areas in Chile." *Journal of Latin American Geography* 8, no. 1 (2009): 149–63.

Molina Camacho, Francisco. "Competing Rationalities in Water Conflict: Mining and the Indigenous Community in Chiu Chiu, El Loa Province, Northern Chile." *Singapore Journal of Tropical Geography* 33, no. 1 (2012): 93–107.

Oseland, Stina E., et al. "Labor Agency and the Importance of the National Scale: Emergent Aquaculture Unionism in Chile." *Political Geography* 31, no. 2 (2012): 94–103.

Pittman, Howard T. "From O'Higgins to Pinochet: Applied Geopolitics in Chile." In *Geopolitics of the Southern Cone and Antarctica*, edited by Philip Kelly and Jack Child, 173–86. Boulder, CO: Lynne Rienner, 1988.

Porteus, J. Douglas. "The Annexation of Eastern Island: Geopolitics and Environmental Perception." *Canadian Journal of Latin American and Caribbean Studies* 6 (1981): 67–80.

———. "The Company State: A Chilean Case Study." *Canadian Geographer* 17, no. 2 (1973): 113–26.

Roucek, Joseph S. "Chile in Geopolitics." *Contemporary Review* 206 (1965): 127–41.

Rudolph, William E. "The New Territorial Divisions of Chile with Special Reference to Chiloe." *Geographical Review* 19, no. 1 (1929): 61–77.

Schneider, H. J. "Chile: Geographers and the Road Towards Socialism." In *Proceedings, Eighth New Zealand Geography Conference*, edited by W. M. Brochie et al., 377–80. Christchurch: New Zealand Geographical Society, 1974.

Silva, Eduardo. "The Political Economy of Forest Policy in Mexico and Chile." *Singapore Journal of Tropical Geography* 25, no. 3 (2004): 261–80.

Visser, E. J., and P. de Langen. "The Importance and Quality of Governance in the Chilean Wine Industry." *GeoJournal* 65, no. 3 (2006): 177–97.

Theses and Dissertations

Aron, Aline. "Gouvernance de la biodiversite et developpement local: le Parc National Torres del Paine en Patagonie chilienne." Master's thesis, University of Montreal, Montreal, Quebec, 2008.

Bryon, Joseph H. "By Reason or By Force: Territory, State Power, and Mapuche Land Rights in Southern Chile." Master's thesis, University of California, Berkeley, 2001.

Kanaan, Nuhad J. "A Geographic Study in Transport Planning: The Case of the Bio-Bio Region in Chile." PhD diss., Syracuse University, Syracuse, NY, 1971.

Martin, Judith P. "The Ideology, Strategy, and Recent Electoral Experience of the Chilean Socialists and Communists." Master's thesis, George Washington University, Washington, DC, 1971.

Moore, Jack E. "Political Geography of the Tacna-Arica Boundary Dispute." Master's thesis, Oklahoma State University, Stillwater, 1959.

Pratt, Kathryn. "Chilean Temperate Rainforests and the Politics of Conservation." Master's thesis, University of Minnesota, Minneapolis, 2005.

URBAN GEOGRAPHY

General Works

Borde, J. "Santiago-de-Chili." *Les Cahiers d'Outre Mer* 7 (1954): 5–24.

Bromhall, David F. "Urban Development Policy in Chile." *Proceedings, Conference of Latin Americanist Geographers* 1 (1970): 367–77.

Cevo, Juan H. "Un neuvo caso de geografia aplicada en Chile: El Lago Riesco Pueda Salvar la Ciudad de Puerto Aysen." *Revista Geografica*, 81 (1974): 139–53.

Cruz-Carrera, Hugo B. "Genesis y originalidad del desarrollo urbano chileno." *Revista Geografica* 101 (1985): 153–73.

Fernandez, Ivan, and Miguel Atienza. "Increasing Returns, Comparative Advantage, and History: The Formation of the Mining City of Antofagasta." *Urban Geography* 32, no. 5 (2011): 641–61.

Guajardo, Pedro. "Las Condiciones Naturales del Sitio de Concepcion Metropolitano." *Revista Geografica* 91–92 (1980): 141–51.

Hernandez, Hilario. "Relaciones rangotamano y etapas de la evolucion del sistema urbano chileno (1865–1970)." *Revista Geografica* 100 (1984): 9–17.

Hidalgo, Rodrigo, and Axel Borsdorf. "European Urban Growth: Concepts, Trends, and a Comparative Framework for the Metropolitan Area of Santiago de Chile [in Spanish]." *Estudios Geograficos* 70, no. 266 (2009): 181–204.

James, Preston E. "Iquique and the Atacama Desert." *Scottish Geographical Magazine* 43 (1927): 203–15.

Kopfmuller, Jurgen, et al. "Sustainable Development of Megacities: An Integrative Research Approach for the Case of Santiago Metropolitan Region." *Die Erde* 140, no. 4 (2009): 417–48.

Rojas, C., et al. "Strategic Environmental Assessment in Latin America: A Methodological Proposal for Urban Planning in the Metropolitan Area of Concepcion (Chile)." *Land Use Policy* 30, no. 1 (2013): 519–27.

Rudolph, William E. "Chuquicamata Twenty Years Later." *Geographical Review* 41, no. 1 (1951): 88–113.

Sanchez Munoz, Alfredo, et al. "Valparaiso: Geography, History, and Identity of a Town That Belongs to the Patrimony of Humankind [in Spanish]." *Estudios Geograficos* 70, no. 266 (2009): 269–94.

Sanchez Munoz, Alfredo, and Miguel Garayar. "Distribucion especial del sistema urbano costero en la Provincia de Concepcion." *Revista Geografica* 100 (1984): 119–28.

Santiago, Jacques. "Urbanisation et sous-developpement: Santiago du Chili." *Les Cahiers d'Outre Mer* 30 (1977): 153–77.

Scarpaci, Joseph L. "Chile." In *Latin American Urbanization: Historical Profiles of Major Cities*, edited by Gerald M. Greenfield, 106–33. Westport, CT: Greenwood, 1994.

Urban Social Geography

Galleguillos Araya-Schubelin, Myrian. *Moglichkeiten zum Abbau von Segregation in Armen-viertein. Die Frage nach der sozialen und okonomischen Nachhaltigkeit urbane Ballungsraume am Beispiel Santiago de Chile*. Kiel, Germany: Kieler Geographische Schriften, Band 115, 2007.

Giron, Paola, and Giulietta Fadda. "Acoustic Contamination and Its Impact on the Quality of Life in Chilean Cities [in French]." *Geocarrefour* 78, no. 2 (2003): 95–110.

Gonzalezeiva, Maria. "Evolucion de las caracteristicas demograficas de la poblacion de la ciudad de Santiago." *Revista Geografica* 100 (1984): 77–87.

Jones, Andrew. "Re-theorizing the Core: A 'Globalized' Business Elite in Santiago, Chile." *Political Geography* 17, no. 3 (1998): 295–318.

Morales, Miguel, and Pedro Labra. "Condicionales naturals, metropolizacion y problemas de planificacion del Gran Santiago, Chile." *Revista Geografica* 91–92 (1980): 179–211.

Pflieger, Geraldine, and Sarah Mattieussent. "Water and Power in Santiago de Chile: Socio-spatial Segregation through Network Integration." *Geoforum* 39, no. 6 (2008): 1907–21.

Porteus, J. Douglas. "Microspace Geography: Beggars in Santiago de Chile." *BC Geographical Studies* 22 (1976): 89–96.

———. "Urban Symbiosis: A Study of Company Town Camp Followers in the Atacama Desert." *Canadian Journal of Latin American and Caribbean Studies* 3 (1978): 210–21.

Utrilla, Severino E., and Jorge O. Veliz. "La complejidad de las procesos de reestructuracion socio especial de las civdades intermedias: persistencia y cambio en la ciudad de Puerto Montt (Chile)." *Anales de Geografia de la Universidad Complutense* 24 (2004): 79–106.

Velasquez, Pedro, and Jorge Sepulveda. "Desarrollo y gestion en el Area Metropolitan de Valparaiso." *Revista Geografica* 100 (1984): 97–118.

Morphology and Neighborhoods

Azocar, Gerardo, et al. "Urbanization Patterns and Their Impacts on Social Restructuring of Urban Space in Chilean Mid-Cities: The Case of Los Angeles, Central Chile." *Land Use Policy* 24, no. 1 (2007): 199–211.

Borsdorf, Axel, et al. "A New Model of Urban Development in Latin America: The Gated Communities and Fenced Cities in the Metropolitan Areas of Santiago de Chile and Valparaiso." *Cities* 24, no. 5 (2007): 365–78.

Borsdorf, Axel, and Rodrigo Hidalgo. "New Dimensions of Social Exclusion in Latin America: From Gated Communities to Gated Cities, the Case of Santiago de Chile." *Land Use Policy* 25, no. 2 (2008): 153–60.

———. "Open Port-Closed Residential Quarters? Urban Structural Transformation in the Metropolitan Area of Valparaiso, Chile." *Erdkunde* 62, no. 2 (2008): 1–14.

Escolano, Severino. "Recent Trends in the Spatial Organisation of Land Use in Large Latin American Cities: The Case of Greater Santiago (Chile) [in Spanish]." *Estudios Geográficos* 70, no. 266 (2009): 97–124.

Lopez-Morales, Ernesto J. "Real Estate Market, State-Entrepreneurialism, and Urban Policy in the 'Gentrification by Ground Rent Dispossession' of Santiago de Chile." *Journal of Latin American Geography* 9, no. 1 (2010): 146–73.

Messina, Rolando S. "Uso del Suelo y Estructura Urbana de Valparaiso." *Revista Geografica* 91–92 (1980): 153–77.

Naranjo Ramirez, Gloria. "The Role of the Infiltrated City in the Configuration of Santiago's Suburbs [in Spanish]." *Estudios Geograficos* 70, no. 266 (2009): 205–30.

Olave, Didma. "Los espacios abiertos en el area metropolitana de Santiago." *Revista Geografica* 100 (1984): 67–76.

Romero, Hugo, and Fernando Ordenes. "Emerging Urbanization in the Southern Andes: Environmental Impacts of Urban Sprawl in Santiago de Chile on the Andean Piedmont." *Mountain Research and Development* 24, no. 3 (2004): 197–201.

Scarpaci, Joseph L., et al. "Planning Residential Segregation: The Case of Santiago, Chile." *Urban Geography* 9, no. 1 (1988): 19–36.

Utrilla, Severino E., and Jorge O. Veliz. "Cambios de la configuracion urbana y sintaxis del espacio en ciudades intermedios: el caso de la Serena (Chile)." *Estudios Geograficos* 65, no. 255 (2004): 297–320.

Varela, Carmen. "Estudio de interaccion especial a traves de flujos de buses interurbanos entre localidades de la IX Region de la Araucania y X Region de Los Lagos, Chile." *Revista Geografica* 132 (2002): 33–43.

Vieira, Elias A. "Aspectos sobvre os residuos solidos urbanos domesticas da regiao metropolitana de Santiago, Chile." *Revista Geografica* 132 (2002): 15–31.

Urban Economic Geography

Berry, Brian J. L. "Relationship between Regional Economic Development and the Urban System: The Case of Chile." *Tijdschrift voor Economische en Sociale Geographie* 60, no. 5 (1969): 283–307.

Glassner, Martin I. "Feeding a Desert City: Antofagasta, Chile." *Economic Geography* 45 (1969): 339–48.

Misetic Yurac, Vladimir. "El transporte urbano de Antofagasta, Chile: Variables geograficas a considerara en su planification." *Revista Geografica* 100 (1984): 131–40.

Ortuzard, S. "Santiago's Metro." *Cities* 1, no. 2 (1983): 113–16.

Santiago, Jacques. "Les transport en common a Santiago du Chili: problemes et perspectives." *Le Cahiers d'Outre Mer* 31, no. 122 (1978): 152–70.

Valenzuela, Belfor P. "Caracterizacion industrial de la metropolis de Santiago de Chile." *Revista Geografica* 100 (1984): 89–96.

Zunico, Hugo M. "Power Relations in Urban Decision-Making: Neo-liberalism, Techno-Politicians and Authoritarian Redevelopment in Santiago de Chile." *Urban Studies* 43, no. 10 (2006): 1825–47.

Zunico, Hugo M., and Rodrigo Hidalgo. "Spatial and Socioeconomic Effects of Social Housing Policies Implemented in Neo-liberal Chile: The Case of Valparaiso, Chile." *Urban Geography* 30, no. 5 (2009): 514–42.

Urban Environments

Ebert, Annemarie, et al. "Socio-environmental Change and Flood Risks: The Case of Santiago de Chile." *Erdkunde* 64, no. 4 (2010): 303–14.

Quintanilla, Victor. "Elrol ecologico del arbo urbano en el medio ambiente de la Metropoli de Santiago." *Revista Geografica* 100 (1984): 49–65.

Ruiz, Cristian H. "The Process of Urbanization in the Chillan's Watershed and Its Adaptative Capacity to Stormwater [in Spanish]." *Estudios Geograficos* 70, no. 266 (2009): 155–80.

Sanchez Munoz, Alfredo, and Cecilia Jimenez. "Valparaiso: Most Important Port City in Chile: Valparaiso and the Vulnerability of Its Architecural Heritage to Seismic Risks [in Spanish]." *Estudios Geograficos* 70, no. 271 (2011): 559–89.

Theses and Dissertations

Araya-Larrea, Sebastian. "Impacts of Imperious Surfaces on Runoff Behavior: Santiago de Chile, 1973–2002." Master's thesis, University of Colorado, Boulder, 2005.

Stackhouse, Jill. "Geographies of Globalization: A Case Study of Santiago de Chile." Master's thesis, Georgia State University, Atlanta, 2001.

———. "The State of Housing, the Business of the State: The Spatial Consequences of Housing and the Urban Development Policies Developed by the Entrepreneurial State in Chile (1973–1989)." PhD diss., Syracuse University, Syracuse, NY, 2007.

Chapter Six

Colombia

GENERAL WORKS

Atlases and Graphic Presentations

Wood, Walter A. "Mapping the Sierra Nevada de Santa Marta." *Geographical Review* 31, no. 4 (1941): 639–43.

Books, Monographs, and Texts

Cuatrecasas, J., and A. Torres. *Paramos*. Bogota, Colombia: Villegas Editiones, 1988.
West, Robert C. *The Pacific Lowlands of Colombia*. Baton Rouge: Louisiana State University Press, 1957.

Articles and Book Chapters

Cabot, Thomas D. "The Cabot Expedition to the Sierra Nevada de Santa Marta of Colombia." *Geographical Review* 29, no. 4 (1939): 587–621.
Crist, Raymond E. "A Geographic Traverse across the Eastern and Central Cordilleras of Colombia." *Bulletin of the Pan American Union* 46 (1942): 132–44.
———. "Influences of Some Physical and Cultural Factors on the Socio-political Evolution of Colombia." *Revista Geografica* 68 (1968): 7–18.
Martin, F. O. "Explorations in Colombia." *Geographical Review* 19, no. 4 (1929): 621–37.
Mertins, Gunter, and H. Uhlig. "The Sierra Nevada de Santa Maria in Colombia: An Overall Geographical Survey." *Revista Geographica* 68 (1968): 33–62.
Parsons, James J. "Colombia." In *Focus on South America*, edited by Alice Taylor, 112–26. New York: Praeger, 1973.
Richardson, Howard. "Barriers to Education in Colombia." *Pennsylvania Geographer* 10 (1972): 14–17.
Seofriz, William. "The Sierra Nevada de Santa Marta." *Geographical Review* 24, no. 3 (1934): 478–85.
Townsend, Janet G. "Magdelana River of Colombia." *Scottish Geographical Magazine* 97, no. 1 (1981): 37–49.
Wilcox, H. Case. "An Exploration of the Rio de Oro, Colombia." *Geographical Review* 11 (1921): 372–83.

Theses and Dissertations

Cornish, John. "Geography of the San Juan Delta, Colombia, South America." Master's thesis, Louisiana State University, Baton Rouge, 1955.
Howard, D. E. "A Geographical Study of the Sabana de Bogota." Master's thesis, Oklahoma State University, Stillwater, 1950.

CULTURAL AND SOCIAL GEOGRAPHY

General Works

Eidt, Robert C. "Rural Society and Land Use Change in the Highland Basins of Colombia." *Latin American Studies, Japan* 3 (1981): 25–45.
Gordon, Burton. *Human Geography and Ecology of the Sinu Country of Colombia.* Berkeley: University of California Press, Ibero-Americana 38, 1957.

The Built Environment

Gough, Katherine V. "The Colombian Building Materials Industry: Transformation or Stagnation?" *Production of the Built Environment* 11 (1992): 79–89.

Ethnic, Social, and Population Geography

Allen, Paul H. "Indians of Southeastern Colombia." *Geographical Review* 37, no. 4 (1947): 567–82.
Dambaugh, Luella M. "Colombia's Population Resources." *Journal of Geography* 58, no. 4 (1959): 174–79.
Eidt, Robert C. "A Note on Japanese Farmers in the Cauca Valley, Colombia." *Revista Geografica* 44 (1956): 41–51.
Flinn, William L. "The Process of Migration to a Shantytown in Bogota, Colombia." *Inter-American Economic Affairs* 22 (1968): 77–88.
Guhl, Ernesto. "Algunos aspectos de la geografia demografia de Colombia." *Revista Geografica* 41 (1954): 81–104.
Heaton, T. G., et al. "Families, Jobs, and Mobility: A Comparison of Migration Streams in Thailand and Colombia." *Singapore Journal of Tropical Geography* 4, no. 2 (1983): 131–46.
Hill, A. David. "Spatial Relations and Socio-economic Change: A Preliminary Study of Differentiation of Places in the Sabuna de Bogota, Colombia." *Professional Geographer* 19 (1967): 136–43.
Meertens, Danny, and Nora Segura-Escobar. "Uprooted Lives: Gender, Violence, and Displacement in Colombia." *Singapore Journal of Tropical Geography* 17, no. 2 (1996): 165–78.
Moser, Brian, and Donald Taylor. "Tribes of the Piraporona." *Geographical Journal* 129, no. 4 (1963): 437–39.
Murphy, Robert C. "Racial Succession in the Colombian Choco." *Geographical Review* 29 (1939): 461–71.
Offen, Karl. "The Territorial Turn: Making Black Territories in Pacific Colombia." *Journal of Latin American Geography* 2, no. 1 (2003): 43–72.
Townsend, Janet G. "Perceived Worlds of the Colonists of Tropical Rainforest, Colombia." *Transactions, Institute of British Geographers* 2, no. 4 (1977): 430–58.
West, Robert C. *The Pacific Lowlands of Colombia: A Negroid Area of the American Tropics.* Baton Rouge: Louisiana State University Press, 1957.

Williams, Lyndon S., and Ernst C. Griffin. "Rural and Small-Town Depopulation in Colombia." *Geographical Review* 68, no. 1 (1978): 13–30.

Community and Settlement Studies

Crist, Raymond E., and Ernesto Guhi. "Pioneer Settlement in Eastern Colombia." In *Smithsonian Institute Report for 1956*, 391–414. Washington, DC: Smithsonian Institute, Publication 4282, 1957.

Eidt, Robert C. "Aboriginal Chibcha Settlements in Colombia." *Annals of the Association of American Geographers* 49, no. 3 (1959): 374–492.

———. "Modern Colonization as a Facet of Land Development in Colombia, South America." *Inter-American Economic Affairs* 22 (1968): 87–96.

Hill, A. David. "Simon: Simulating Problems of Rural Community Development in Colombia." *Journal of Geography* 79, no. 1 (1980): 41–42.

Parsons, James J. *Antioquenos Colonization in Western Colombia.* Berkeley: University of California Press, Ibero-Americana 32, 1949.

———. "The Settlement of the Sinu Valley of Colombia." *Geographical Review* 42, no. 1 (1952): 67–86.

Pereira, Laura. "Becoming Coca: A Materiality Approach to a Community Chain Analysis of Hoja de Coca in Colombia." *Singapore Journal of Tropical Geography* 31, no. 3 (2010): 384–400.

Rau, Herbert L., Jr. "Sequential Occupance and Settlement Patterns in the Sabana de Bogota, Colombia." *Revista Geografica* 51 (1959): 57–66.

Ricardson, Miles. "The Spanish American (Colombia) Settlement Pattern as a Societal Expression and as a Behavioral Cause." *Geoscience and Man* 5 (1974): 35–52.

Stoddart, D. R., and J. D. Trubshaw. "Colonization in Action in Eastern Colombia." *Geography* 47, no. 2 (1970): 47–53.

Taylor, Griffith. "Settlement Zones of the Sierra Nevada de Santa Marta, Colombia." *Geographical Review* 21, no. 4 (1931): 539–58.

Tourism and Recreation

Belisle, Francois J., and Don R. Hoy. "The Perceived Impact of Tourism by Residents: A Case Study in Santa Marta, Colombia." *Annals of Tourism Research* 7 (1980): 83–101.

Theses and Dissertations

Aguirre, Hector. "Demographic Change and Violence in Colombia, 1965–1995." Master's thesis, San Diego State University, San Diego, 2003.

Elbow, Gary S. "Regional Variations in Minifundia Settlement in the Ubate Valley, Colombia." Master's thesis, University of Oregon, Eugene, 1964.

Gordon, Burton. "Human Geography and Ecology in the Sinu Country of Colombia." PhD diss., University of California, Berkeley, 1954.

Hanneson, William. "Places in the Chiquinquira-Ubate Area of the Eastern Cordillera of the Colombian Andes." PhD diss., University of Oregon, Eugene, 1969.

Harlan, Glenn H. "Mitu, Colombia: A Geographical Analysis of an Isolated Border Town." PhD diss., University of Florida, Gainesville, 1970.

Moore, Eli. "Mapping for Social Change: The Radical Pedagogy of Participatory Mapping in Valle de Cauca, Colombia." Master's thesis, Syracuse University, Syracuse, NY, 2007.

Parra, Adriana. "Women in Colombia Confronting Trauma and Displacement." PhD diss., University of California, Davis, 2010.

Parsons, James J. "Antioqueno Colonization in Western Colombia." PhD diss., University of California, Berkeley, 1948.

Rucinque, Hector F. "Change in a Peasant Society: The Case of Central Highlands of Boyaca, Colombia." PhD diss., Michigan State University, East Lansing, 1977.

ECONOMIC GEOGRAPHY

General Works

Gough, Katherine V. "Home-Based Enterprises in Low-Income Settlements: Evidence from Pereira, Colombia." *Geografisk Tidsskrift* 96 (1996): 95–102.

Griffin, Ernst C. "The Changing Role of the Rio Magdalena in Colombia's Economic Growth." *Geographical Survey* 3, no. 1 (1974): 14–24.

Muller, Jan M. *Struktur und Probleme des Verkelrssystems in Kolumbien: eine integrietes Verkehrskonzept als Varaoussetzung fur eine dezentralisiete regionalentwicklung unter neoliberalen Bedingungen.* Marburg, Germany: Marburger Geographische Schriften, Heft 137, 2001.

Renner, G. T. "Colombia's Internal Development." *Economic Geography* 3, no. 3 (1927): 259–64.

Tricort, J., et al. "Preliminary Studies for Developing the Rio Lebrija Basin in Colombia." *Revista Geografica* 68 (1968): 83–140.

Agriculture and Land Use

Brunnschweiler, D. *The Llanos Frontier of Colombia: Environment and Changing Land Use in Meta.* East Lansing: Michigan State University, Latin American Studies Center, 1972.

Crist, Raymond E. *The Cauca Valley, Colombia: Land Tenure and Land Use.* Baltimore: Waverly Press, 1952.

Eidt, Robert C. "Modern Colonization as a Facet of Land Development in Colombia, South America." *Yearbook, Association of Pacific Coast Geographers* 29 (1967): 21–42.

Elbow, Gary S. "Agrarian Systems and Land Utilization in Highland Colombia." *Oregon Geographer* 1, no. 2 (1967): 1–4.

———. "Minifundia in Highland Colombia: An Agrarian System." *Proceedings of the Pennsylvania Academy of Science* 39, no. 2 (1966): 283–88.

Etter, Andres, and L. Alberto Villa. "Andean Forests and Farming Systems in Part of the Eastern Cordillera (Colombia)." *Mountain Research and Development* 20, no. 3 (2000): 236–45.

Forero-Alvarez, J. "Typologie des formes d'agriculture dans les hautes terres andines en Colombie." *Les Cahiers d'Outre Mer* 62 (2009): 419–37.

Kirby, John M. "Colombian Land-Use Change and the Settlement of the Orient." *Pacific Viewpoint* 19, no. 1 (1978): 1–25.

Meier, V. "Cut-Flower Production in Colombia: A Major Development Success Story for Women." *Environment and Planning A* 31, no. 3 (1999): 273–90.

Pruitt, Fredonia M. "The Coffee Industry of Colombia." *Journal of Geography* 35, no. 2 (1936): 73–77.

Shleman, Roy J., and L. Barry Phelps. "Dredge-Tailing Agriculture on the Rio Nechi, Colombia." *Geographial Review* 61, no. 3 (1971): 396–414.

Smith, Albert C. "Mountain Tops and Lowlands of Colombia." *Economic Geography* 6 (1930): 298–407.

Tricart, J., et al. "Etudes preliminaries pour l'amenagement du basin du Rio Lebrija (Colombie) Par le Centre de Geographie Applique (Strasbourg)." *Revista Geografica* 68 (1968): 83–139.

West, Robert C. "Ridge or Era Agriculture in the Colombian Andes." *Actas del XXXIII Congreso International de Americanistas* 1 (1959): 279–82.

Wilhemy, Herbert. "Cattle Raising on the Caribbean Seaboard of Colombia." *Revista Geografica* 68 (1968): 63–82.

Commerce and Trade

Dambaugh, Luella M. "Colombia's Foreign Trade Mirrors Her Economy." *Journal of Geography* 56 (1957): 437–41.
Pearson, Ross N. "Trade on the Rio Magdalena." *Revista Geográfica* 55 (1961): 21–35.
Ridgeley, Mark A. "Services in a Colombian Shantytown: Speculations on the Limits of Collective Self-Help." *Yearbook, Conference of Latin Americanist Geographers* 15 (1989): 59–70.

Manufacturing and Labor Studies

Brucher, Wolfgang. "Spatial Structural Influence on the Industrialization Process in Colombia." *Applied Geography and Development* 15 (1980): 82–95.

Energy, Mining, Fishing, and Lumbering

Griess, Phyliss R. "Colombia's Petroleum Resources." *Economic Geography* 22, no. 3 (1946): 245–54.
West, Robert C. "Folk Mining in Colombia." *Economic Geography* 28 (1952): 323–30.

Transportation and Communications

James, Preston E. "The Transportation Problem of Highland Colombia." *Journal of Geography* 22 (1923): 346–54.
Stokes, Charles J. "The Freight Transport System of Colombia, 1959." *Economic Geography* 43, no. 1 (1967): 71–90.

Theses and Dissertations

Balanos, Sandra. "Characterization of Coffee in Colombia Using High Resolution Imagery." Master's thesis, McGill University, Montreal, Quebec, 2008.
Eidt, Robert C. "Land Utilization in the Highland Basins of the Cordillera Oriental of Colombia." PhD diss., University of California, Los Angeles, 1954.
Guhl, A. "Coffee and Landscape Change in the Colombian Countryside, 1970–2000." PhD diss., University of Florida, Gainesville, 2004.
Lloyd-Jones, Donald J. "The Potential Economic Development of the Upper Cauca Valley, Colombia." PhD diss., Columbia, University, New York, 1961.
Pike, Robert J. "Geographic-Economic Aspects of Transportation in Colombia." Master's thesis, University of Florida, Gainesville, 1950.
Rau, Herbert L., Jr. "The Agricultural Land Use and Settlement Patterns in the Sabana de Bogota, Colombia." PhD diss., Northwestern University, Evanston, IL, 1958.
Symanski, Richard. "Periodic Markets of Andean Colombia." PhD diss., Syracuse University, Syracuse, NY, 1971.
Umos, Manuel. "Using GIS to Predict Corn Yields in Colombia." Master's thesis, University of Redlands, Redlands, CA, 2009.

HISTORICAL GEOGRAPHY

General Works

Angel, Marta H. "Las bases prehispanicas de la configuration de la Provincia de Popayan en el periodo colonial." *Journal of Latin American Geography* 5, no. 2 (2006): 53–74.

Eidt, Robert C. *Advances in Abandoned Settlement Analysis: Applications to Prehistoric Anthrosols in Colombia, South America.* Milwaukee: University of Wisconsin–Milwaukee, Center for Latin American Studies, 1984.

Etter, Andres, et al. "Historic Patterns and Drivers of Landscape Change in Colombia since 1500: A Regionalized Spatial Approach." *Annals of the Association of American Geographers* 98, no. 1 (2008): 2–23.

Herrera Angel, Marta. "Population, Territory, and Power in Eighteenth-Century New Granada: Pueblos de Indios and Authorities in the Province of Santa Fe." *Yearbook, Conference of Latin Americanist Geographers* 21 (1995): 121–32.

Parsons, James J. *Antioqueno Colonization in Western Colombia.* Berkeley: University of California Press, Ibero-Americana 32, 1949.

———. *La colonizacion antioquena en el occidente de Colombia.* Bogota, Colombia: Banco de la Republica-El Ancora Editores, 1997.

Parsons, James J., and William Bowen. "Ancient Ridged Fields of the San Jose River Floodplain, Colombia." *Geographical Review* 56, no. 3 (1966): 317–43.

Van Ausdal, Shawn. "Pattern, Profit, and Power: An Environmental History of Cattle Ranching in Colombia, 1850–1950." *Geoforum* 40, no. 5 (2009): 707–19.

Theses and Dissertations

Krogzemis, James. "A Historical Geography of the Santa Marta Area, Colombia." PhD diss., University of California, Berkeley, 1968.

Van Ausdal, Shawn. "The Logic of Livestock: An Historical Geography of Cattle Ranching in Colombia, 1850–1950." PhD diss., University of California, Berkeley, 2009.

PHYSICAL GEOGRAPHY

General Works

Bates, Marston. "Climate and Vegetation in the Villaviencio Region of Eastern Colombia." *Geographical Review* 38, no. 4 (1948): 555–74.

Jungerius, P. D. "Quaternary Landscape Development of the Rio Magdalena Basin between Neiva and Bogota, Colombia." *Paleogeography, Paleoclimatology, and Paleoecology* 19 (1976): 149–70.

Romero-Ruiz, M. H., et al. "Landscape Transformation in Savannas of Northern South America: Land Use/Cover Changes since 1987 In the Llanos Orientales of Colombia." *Applied Geography* 32, no. 2 (2012): 766–76.

Thouret, J. C., and C. Laforge. "Hazard Appraisal and Hazard-Zone Mapping of Flooding and Debris Flowage in the Rio Cambeiman Valley and Ibague City, Tolima Department, Colombia." *GeoJournal* 34, no. 4 (1994): 407–14.

Van der Hammen, T., and B. Van Geel. "Upper Quaternary Vegetation and Climate Sequence of the Fuquene Area (Eastern Cordillera, Colombia)." *Paleogeography, Paleoclimatology, and Paleoecology* 14 (1973): 9–92.

Vann, John H. "Landform-Vegetation Relationships in the Atrato Delta." *Annals of the Association of American Geographers* 49, no. 4 (1959): 345–60.

Vis, M. "Interception, Drop Size Distribution, and Rainfall Kinetic Energy in Four Colombian Forest Ecosystems." *Earth Surface Processes* 11, no. 6 (1986): 591–604.

West, Robert C. "Mangrove Swamps of the Pacific Coast of Colombia." *Annals of the Association of American Geographers* 46, no. 1 (1956): 98–121.

Biogeography

Cavelier, J., et al. "The Savanization of Moist Forests in the Sierra Nevada de Santa Maria, Colombia." *Journal of Biogeography* 25, no. 5 (1998): 901–12.

Eden, Michael J., and A. Andrade. "Colonies, Agriculture, and Adaptation in the Colombian Amazon." *Journal of Biogeography* 15, no. 1 (1988): 79–86.

Kattan, Gustavo, et al. "Biological Diversification in a Complex Region: A Spatial Analysis of Faunistic Diversity and Biogeography of the Andes of Colombia." *Journal of Biogeography* 31, no. 11 (2004): 1829–40.

Lizcano, D. J., et al. "Geographic Distribution and Population Size of the Mountain Tapir (Tapirus pinchaque) in Colombia." *Journal of Biogeography* 29, no. 1 (2002): 7–16.

Longford, M., and W. Bell. "Land Cover Mapping in a Tropical Hillside Environment: A Case Study in the Cauca Region of Colombia." *International Journal of Remote Sensing* 18, no. 6 (1997): 1289–306.

Climatology and Weather

Guhl, A. "Representations of Colombian Weather Stations in the Holdriges Life Zone Model." *Papers of the Applied Geography Conferences* 21 (1998): 439–45.

Molina, Jose M., and Concepcion M. Escobar. "Fog Collection Variability in the Andean Mountain Range of Southern Colombia." *Die Erde* 139, nos. 1–2 (2008): 127–40.

Geomorphology, Landforms, and Volcanism

Cervantes Borja, Jorge. "Algunas consideraciones geomorfologicas de la cuenca del Rio de la Magdalena." *Investigaciones Geograficas* 2 (1969): 89–108.

Eden, Michael J., et al. "Geomorphology of the Middle Caqueta Basin of Eastern Colombia." *Zeitschrift fur Geomorphologie* 26, no. 3 (1982): 343–64.

Eidt, Robert C. "Some Comments on the Geomorphology of Highland Basins in the Cordillera Oriental of Colombia." *Revista Geografica* 68 (1968): 141–56.

Gilliard, E. T. "The Cordillera Macarena, Colombia." *Geographical Review* 32, no. 3 (1942): 463–70.

Giraldo, Mario. "Spatial Scale and Land Use Fragmentation in Monitoring Water in the Colombian Andes." *Applied Geography* 34 (2012): 395–402.

Notestein, Frank B. "The Sierra Nevada de Cocuy." *Geographical Review* 22, no. 3 (1932): 423–30.

Hydrology and Glaciology

Guhl, Ernesto. "Los paramos circundantes de la Sabana de Bogota, su ecologia y su importancia para el regimen hidrologico de la misma." *Colloquium Geographicus* 9 (1968): 195–212.

Quesada, Marvin, and Cesar N. Caviedes. "Caracteristicas Estadisticas de alguno Rios de Colombia." *Revista Geografica* 116 (1992): 53–66.

Ruis, Jorge, and German Herrera. "Tropical Glacier Retreat in the Sierra Nevada del Cocuy, Boyaca, Colombia, 1986–2007." *Papers of the Applied Geography Conferences* 31 (2008): 52–57.

Van der Hammen, T., et al. "Glacial Sequence and Environmental History in the Sierra Nevada de Cocuy." *Paleogeography, Paleoclimatology, and Paleoecology* 32, nos. 3–4 (1981): 247–340.

Wood, Walter A. "Recent Glacier Fluctuations in the Sierra Nevada de Santa Maria, Colombia." *Geographical Review* 60, no. 3 (1970): 374–92.

Soils

Folster, H., et al. "Late Quaternary Paleosols in the Western and Central Cordilleras of Colombia." *Paleogeography, Paleoclimatology, and Paleoecology* 21 (1977): 245–64.
Folster, H., and W. Hetsch. "Paleosol Sequences in the Eastern Cordillera of Colombia." *Quaternary Research* 9, no. 2 (1978): 238–48.
McGregor, Duncan F. M. "An Investigation of Soil Erosion in the Colombian Rainforest Zone." *Catena* 7, no. 4 (1980): 265–73.

Theses and Dissertations

Eidt, Robert C. "The Physical Geography of the Department of Cundinamaraca, Colombia." Master's thesis, University of California, Los Angeles, 1951.
Hoyos, N. "Spatial and Temporal Patterns of Soil Erosion Potential in a Mountainous Tropical Watershed, Central Andean Cordillera of Colombia." PhD diss., University of Florida, Gainesville, 2004.
Leal, Claudia. "Black Forests: The Pacific Lowlands of Colombia, 1850–1930." PhD diss., University of California, Berkeley, 2004.

POLITICAL GEOGRAPHY

General Works

Asher, Kiran, and Diana Ojeda. "Producing Nature and Making the State: Ordenamiento Territorial in the Pacific Lowlands of Colombia." *Geoforum* 40, no. 3 (2009): 292–302.
Muller, Jan M. *Struktur und Probleme des Verkehrssystems in Kolombien. Ein integriertes Verkehrskonzept als Varaussetzung fur eine dezentralisierte Regionalentwicklung unter neolibralen Bedingungen.* Marburg, Germany: Marburger Geographische Schriften, Heft 137, 2001.
Osleander, U. "Fleshing Out the Geographies of Social Movements: Colombia's Pacific Coast Black Communities and the Aquatic Space." *Political Geography* 23, no. 8 (2004): 957–85.

Theses and Dissertations

Galvis, Juan. "The State and the Construction of Territorial Marginality: The Case of the 1961 Land Reform in Colombia." Master's thesis, University of Washington, Seattle, 2007.
Van Eyck, Kim. "Neoliberalism and Democracy? The Gendered Restructuring of Work, Unions, and the Colombian Public Sphere." PhD diss., University of Washington, Seattle, 2002.

URBAN GEOGRAPHY

General Works

Alvarez, Victor. "Un sistema de services come base de las relaciones ciudad-Campo." *Revista Geografica*, 91–92 (1980): 89–99.
Chardon, Anne. "A Geographic Approach of the Global Vulnerability in an Urban Area: Case of Manizales, Colombian Andes." *GeoJournal* 49, no. 2 (1999): 197–212.
Gilbert, Alan. "Bogota: Politics, Planning, and the Crisis of Lost Opportunities." In *Latin American Urban Research*, edited by W. A. Cornelius and R. V. Kemper, 6:87–126. Beverly Hills, CA: Sage, 1978.

————. "A Note on the Incidence of Development in the Vicinity of a Growth Center." *Proceedings, Conference of Latin Americanist Geographers* 5 (1974): 35–50.

————. "Santa Fe de Bogota: A Latin American Special Case?" In *The Mega-city in Latin America*, edited by Alan Gilbert, 241–269. Tokyo, Japan: UN University Press, 1996.

————. "Urban and Regional Development Programs in Colombia since 1951." In *Latin American Urban Research*, edited by W. A. Cornelius and F. M. Trueblood, 5:241–76. Beverly Hills, CA: Sage, 1975.

Ramirez, T. E. "The Urbanization Process in Colombia." *Revista Geografica* 68 (1968): 19–32.

Skinner, R. "Bogota." *Cities* 21, no. 1 (2004): 73–81.

Stadel, Christoph. "Colombia." In *Essays on World Urbanization*, edited by R. Jones, 238–62. London: Philipp, 1975.

Urban Social Geography

Gilbert, Alan. "Water for All: How to Combine Public Management with Commercial Practice for the Benefit of the Poor?" *Urban Studies* 44, no. 8 (2007): 1559–80.

Gouest, Vincent. "La croissance demographique de Bogota au XXe siècle." *Les Cahiers d'Outre Mer* 43 (1990): 289–306.

Nino Guerrero, Raul. *Rural to Urban Shift of the Unemployed in Colombia*. Lund, Sweden: Ab Lunda-Kopin, 1975.

Thomas, Robert N., and Kevin Byrnes. "Intervening Opportunities and the Migration Field of a Secondary Urban Center: The Case of Tunja, Colombia." *Proceedings, Conference of Latin Americanist Geographers* 5 (1974): 83–88.

Morphology and Neighborhoods

Brucher, Wolfgang, and Gunter Mertins. "Los barrios de vivienda de las estratos bajor en el modelo ideal de las grandes ciudades Latinoamericas: El ejemplo de Bogota." *Revista Geografica* 94 (1981): 7–40.

Edwards, Michael. "Residential Mobility in a Changing Housing Market: The Case of Bucaramanga, Colombia." *Urban Studies* 20, no. 2 (1983): 131–46.

Gilbert, Alan. "A Home Is for Ever? Residential Mobility and Homeownership in Self-Help Settlements." *Environment and Planning A* 31, no. 6 (1999): 1073–92.

Gough, Katherine V. "House for Sale? The Self-Help Housing Market in Pereira, Colombia." *Housing Studies* 13, no. 2 (1998): 149–60.

————. "Linking Production, Distribution, and Consumption: Self-Help Builders and the Building Materials Industry in Urban Colombia." *Third World Planning Review* 18, no. 4 (1996): 397–414.

————. "Self-Housing in Urban Colombia: Alternatives for the Production and Distribution of Building Materials." *Habitat International* 20, no. 4 (1996): 635–51.

Greenow, Linda. "Urban Form in Spanish American Colonial Cities: Cartagena de Indios, New Granada in 1777." *Middle States Geographer* 40 (2007): 47–56.

Montoya, J. W. "Nouvelle lecture de la ville Latino-Americaine. Dynamiques urbaines et changements morphologiques a Bogota." *Cahiers de Geographie du Quebec* 50, no. 141 (2006): 553–64.

Nossin, Jan J. "Monitoring of Hazards and Urban Growth in Villaviciento, Colombia, Using Scanned Air Photos and Satellite Imagery." *GeoJournal* 49, no. 2 (1999): 151–58.

Riono, Yvonne. "Spatial Perceptions and Networks of Social Interaction in Low-Income Neighborhoods of Bogota, Colombia." *Geoscope* 18, no. 2 (1988): 21–38.

Stadel, Christoph. "The Structure of Squatter Settlements in Medellin, Colombia." *Area* 7, no. 4 (1976): 249–54.

Ward, Peter. "A Patrimony for the Children: Low-Income Homeownership and Housing (Im)Mobility in Latin American Cities." *Annals of the Association of American Geographers* 102, no. 6 (2012): 1489–510.

Urban Economic Geography

Bromley, Rosemary D. F. "From Calvary to White Elephant: A Colombian Case of Urban Renewal and Marketing Reform." *Third World Planning Review* 2 (1981): 205–32.
———. "Organization, Regulation, and Exploitation in the So-Called Urban Informal Sector: The Street Traders of Cali, Colombia." *World Development* 6 (1978): 1161–71.
Gilbert, Alan. "Employment and Poverty during Economic Restructuring: The Case of Bogota, Colombia." *Urban Studies* 34, no. 7 (1997): 1047–70.
———. "The Provision of Public Services and the Debt Crisis in Latin America: The Case of Bogota." *Economic Geography* 66, no. 4 (1990): 349–61.
Gil-Beuf, A. "Sustainable City and Collective Transport: The Case of Transmilenio in Bogota [in French]." *Annales de Geographie* 116, no. 5 (2007): 533–47.
Parsons, James J. "The Historical Preconditions of Industrialization-Medellin Reconsidered." *Proceedings, Conference of Latin Americanist Geographers* 5 (1974): 119–24.
Rodriguez, D. A. "Spatial Choices and Excess Commuting: A Case Study of Bank Tellers in Bogota, Colombia." *Journal of Transport Geography* 12, no. 1 (2004): 49–61.

Urban Environments

Ridgely, Mark A. "Evaluation of Water Supply and Sanitation Options in Third World Cities: An Example from Cali, Colombia." *GeoJournal* 18, no. 2 (1989): 199–212.

Theses and Dissertations

Friberg, Justin C. "Cityward Migration Differentials in Bucaramanga, Colombia." PhD diss., Syracuse University, Syracuse, NY, 1977.
Galvis, Juan. "Managing the Living City: Public Space and Development in Bogota." PhD diss., University of Washington, Seattle, 2011.
Guarin, Alejandro. "Old Links in a New Chain: The Unlikely Resilience of Corner Stones in Bogota, Colombia." PhD diss., University of California, Berkeley, 2009.
Maclaurin, Galen. "Mapping Perception of Safety and Danger in Medellin, Colombia: A Study in the Perceptual Geography of Urban Crime." Master's thesis, University of Colorado, Boulder, 2011.
Martinez-Bolivar, Maria. "Spatial Dynamics of the Homeless in Martires, in Bogota (Colombia): Understanding Hotspots and Illegal Economies in the Localidad 14 through Kernal Density Estimation." Master's thesis, University of California, Berkeley, 2011.

Chapter Seven

Ecuador

GENERAL WORKS

Atlases and Graphic Presentations

Radcliffe, Sarah A. "National Maps, Digitalization, and Neoliberal Cartographies: Transforming Nation-State Practices and Symbols in Postcolonial Ecuador." *Transactions, Institute of British Geographers* 34, no. 4 (2009): 426–44.

Book, Monographs, and Texts

Bromley, Raymond J. *Development and Planning in Ecuador*. London: Latin American Publication Fund, 1977.

Articles and Book Chapters

Bebbington, Anthony J. "Reencountering Development: Livelihood Transitions and Place Transformations in the Andes." *Annals of the Association of American Geographers* 90, no. 3 (2000): 495–520.

Bengtson, Nels A. "Some Essential Features of the Santa Elena Peninsula, Ecuador." *Annals of the Association of American Geographers* 15 (1925): 150–58.

Bennett, Hugh H. "Some Geographical Aspects of Western Ecuador." *Annals of the Association of American Geographers* 15 (1925): 126–47.

Dambaugh, Luella M. "Santo Domingo de los Colorados, Ecuador and Its Environs." *Journal of Geography* 67, no. 3 (1968): 172–79.

Foster, Alice. "The Guayaquil Lowland." *Journal of Geography* 37, no. 6 (1938): 213–26.

Hearn, Lea. "A Geographic Study of the Village of Cotocollo, Ecuador." *Journal of Geography* 49, no. 7 (1950): 225–31.

Hermessen, J. L. "A Journey: The Rio Zamora, Ecuador." *Geographical Review* 4 (1917): 434–49.

Larren, Carlos M. "Geographical Notes on Esmeraldas, Northeastern Ecuador." *Geographical Review* 14 (1924): 373–87.

McBride, G. M. "The Galapagos Islands." *Geographical Review* 6 (1918): 229–39.

Messina, Joseph P. "A Complex Systems Based Model of Development Trajectories in the Ecuadorian Amazon." *Papers of the Applied Geography Conferences* 25 (2002): 150–56.

Preston, David A., and Anne M. S. Graham. "Some Fields of Research in Ecuador." *Professional Geographer* 13, no. 2 (1961): 16–19.

Sinclair, Joseph H. "In the Land of Cinnamon: A Journey in Eastern Ecuador." Geographical Review 19 (1929): 201–17.

Sinclair, Joseph H., and Theron Wasson. "Exploration in Eastern Ecuador." *Geographical Review* 13 (1923): 190–210.

Stadel, Christoph. "Development and Underdevelopment in the Rural Andes: A Case Study from the Eastern Cordillera of Ecuador." In *Integrated Mountain Development*, edited by T. V. Singh and J. Kaur, 193–207. New Delhi, India: Himalayan Books, 1984.

———. "Three-Dimensional Regional Geography of a Tropical Mountain Country: The Case of Ecuador." *Bulletin of the North Dakota Geographers* 38 (1990): 47–65.

CULTURAL AND SOCIAL GEOGRAPHY

General Works

Franckowiak, Gene N. "The Landowner as Change Agent in Rural Modernization: An Ecuadorian Example." *Geographical Bulletin* 19, no. 2 (1980): 14–26.

Grenier, Christophe. "Les naturalistes et les Galapagos." *Geographie et Cultures* 13 (1995): 107–32.

Himley, Matthew. "Nature Conservation, Rural Livelihoods, and Territorial Control in Andean Ecuador." *Geoforum* 40, no. 5 (2009): 832–42.

Knapp, Gregory W. *Andean Ecology: Adoptive Dynamics in Ecuador*. Boulder, CO: Westview, Dellplain Latin American Studies 27, 1991.

———. *Geografia quichua de la Sierra del Ecuador*. Quito, Ecuador: Abya Yala, 1987.

Perreault, Thomas A. "Developing Identities: Indigenous Mobilization, Rural Livelihoods, and Resource Access in Ecuadorian Amazonia." *Ecumene* 8, no. 4 (2001): 381–413.

Preston, David A. "From Hacienda to Family Farm-Changes in Environment and Society in Pimampiro, Ecuadro." *Geographical Journal* 156, no. 1 (1990): 31–38.

Saelemyr, Siren. "People, Park, and Plant Use: Perception and Use of Andean Nature in the Southern Ecuadorian Andes." *Norsk Geografisk Tidsskrift* 58, no. 4 (2004): 194–203.

Sarmiento, Fausto O. "Anthropogenic Change in the Landscapes of Highland Ecuador." *Geographical Review* 92, no. 2 (2002): 213–34.

Stadel, Christoph. "Environmental Stress and Human Activities in the Tropical Andes (Ecuador)." In *Beitrage zur vergleichen den Geographie de Hochgebirge*, edited by Erwin Grotzbach and G. Rinschele, 235–63. Regensberg, Germany: Pustet, 1984.

———. "Horizontal and Vertical Spaces: A Three-Dimension Geography of Ecuador." In *Patterns of Regional Geography: An International Perspective,* edited by R. B. Mandal, 3:93–113. New Delhi, India: Concept, 1990.

The Built Environment

Stadel, Christoph, and Luzdel Alba Moya. "Piazas and Peiras of Ambato, Ecuador." *Yearbook, Conference of Latin Americanist Geographers* 14 (1988): 43–50.

Medical Geography

Tobin, Graham A., and Linda M. Whiteford. "Chronic Hazards: Health Impacts Associated with On-Going Ash Falls around Mt. Tunguracha in Ecuador." *Papers of the Applied Geography Conferences* 27 (2004): 84–93.

Ethnic, Social, and Population Geography

Bebbington, Anthony J., and Thomas A. Perreault. "Social Capital, Development, and Access to Resources in Highland Ecuador." *Economic Geography* 75, no. 4 (1999): 395–418.

Brea, Jorge A., et al. "Circulation and Migration in Third World Settings: A Comparison in Ecuador." *Revista Geografica* 101 (1985): 97–113.

Brown, Lawrence A., et al. "Gender, Migration, and the Organization of Work under Economic Devolution: Ecuador, 1982–90." *International Journal of Population Geography* 4, no. 3 (1998): 259–74.

———. "Policy Aspects of Development and Individual Mobility: Migration and Circulation from Ecuador's Rural Sierra." *Economic Geography* 64, no. 2 (1988): 147–70.

Brown, Lawrence A., and R. Sierra. "Frontier Migration as a Multi-state Phenomenon Reflecting the Interplay of Macro Forces and Local Conditions: The Ecuador Amazon." *Papers in Regional Science* 73, no. 3 (1994): 1–22.

Jokisch, Brad D. "From Labor Circulation to International Migration: The Case of South-Eastern Ecuador." *Yearbook, Conference of Latin Americanist Geographers* 23 (1997): 63–76.

———. "Migration and Agricultural Change: The Case of Smallholder Agriculture in Highland Ecuador." *Human Ecology* 30, no. 4 (2002): 523–50.

Lawson, Victoria A. "Questions of Migration and Belonging: Understanding of Migration under Neoliberalism in Ecuador." *International Journal of Population Geography* 5, no. 4 (1999): 261–76.

Stadel, Christoph. "The Perception of Stress by Campesinos: A Profile from the Ecuadorian Sierra." *Mountain Research and Development* 9, no. 1 (1989): 35–49.

Swanson, Katherine E. *Begging as a Path to Progress: Indigenous Women and Children and the Struggle for Ecuador's Urban Spaces*. Athens: University of Georgia Press, 2010.

Valdivia, Gabriela. "Indigenous Bodies, Indigenous Minds? Towards an Understanding of Indigeneity in the Ecuadorian Amazon." *Gender, Place, and Culture* 16, no. 5 (2009): 535–51.

———. "On Indigeneity, Change, and Representation in the Northeastern Ecuadorian Amazon." *Environment and Planning A* 37, no. 2 (2005): 285–304.

Vender, JoAnn C. "Culture, Place, and School: Improving Primary Education in Rural Ecuador." *Yearbook, Conference of Latin Americanist Geographers* 20 (1994): 107–19.

Williams, Glyn B. "Embodying National Identities: Mestizo Men and White Women in Ecuadorian Racial-National Imaginaries." *Transactions, Institute of British Geographers* 24, no. 2 (1999): 213–26.

Community and Settlement Studies

Bromley, Raymond J. "The Colonization of Humid Tropical Areas in Ecuador." *Singapore Journal of Tropical Geography* 2, no. 1 (1981): 15–26.

Brown, Lawrence A., et al. "Complementary Perspectives as a Means of Understanding Regional Change: Frontier Settlement in the Ecuador Amazon." *Environment and Planning A* 24, no. 7 (1992): 939–62.

Burt, Arthur, et al. "Santo Domingo de los Colorados: A New Pioneer Zone in Ecuador." *Economic Geography* 36 (1960): 221–36.

———. "Santo Domingo de los Colorados: A New Pioneer Zone in Ecuador." *Revista Geografica* 54 (1961): 89–100.

Eastwood, David A., and H. J. Pollard. "Amazonian Colonization in Eastern Ecuador: Land Use Conflicts in a Planning Vacuum." *Singapore Journal of Tropical Geography* 13, no. 2 (1992): 103–17.

Perreault, Thomas A. "Nature Preserves and Community Conflict: A Case Study in Highland Ecuador." *Mountain Research and Development* 16, no. 2 (1996): 167–75.

Radcliffe, Sarah A. "Embodying National Identities: Mestizo Men and White Women in Ecuadorian Racial-National Imaginaries." *Transactions, Institute of British Geographers* 24, no. 2 (1999): 213–26.

———. "Gendered Notions: Nostalgia, Development, and Territory in Ecuador." *Gender, Place, and Culture* 3, no. 1 (1996): 5–22.
Rundquist, F. M., and Lawrence A. Brown. "Migrant Fertility Differentials in Ecuador." *Geografiska Annaler* 71B, no. 2 (1989): 109–24.
Wood, Harold A. "Spontaneous Agricultural Colonization in Ecuador." *Annals of the Association of American Geographers* 62, no. 4 (1972): 599–617.
Yamamoto, Shozo, et al. "Agricultural Colonization in the Nor-Oriente of Ecuador." *Science Reports, Tohoku University, 7th Series, Geography* 1 (1980): 47–82.

Tourism and Recreation

Sjoholt, Peter. "Eco-tourism and Local Development: Conceptual and Theoretical Framework and Problems in Implementation: Empirical Evidence from Costa Rica and Ecuador." *Fennia* 178, no. 2 (2000): 243–52.
Wesche, Rolf. "Ecotourism and Indigenous Peoples in the Resource Frontier of the Ecuadorian Amazon." *Yearbook, Conference of Latin Americanist Geographers* 19 (1993): 34–45.
Wood, Harold A. "Spontaneous Agricultural Colonization in Ecuador." *Annals of the Association of American Geographers* 62, no. 4 (1972): 599–626.

Place Names

McEwen, A. "The English Place-Names of the Galapagos." *Geographical Journal* 154, no. 2 (1988): 234–42.

Theses and Dissertations

Bryne-Maisto, Mary Jean. "Amenity Migration: Place Transformation in Vilcabama, Ecuador." Master's thesis, Syracuse University, Syracuse, NY, 2011.
Chaurette, Eric. "Assessing the Status of Ecotourism in the Quijos River Valley, Ecuador." Master's thesis, University of Waterloo, Waterloo, Ontario, 2002.
Gray, Clark. "Out-Migration and Rural Livelihoods in the Southern Ecuadorian Andes." PhD diss., University of North Carolina, Chapel Hill, NC, 2008.
Hansis, Richard A. "Colonization in the Upano River Valley, Ecuador." Master's thesis, University of Florida, Gainesville, 1968.
Keese, James. "The Cultural Ecology of NGO Development in Upper Conar, Ecuador." PhD diss., University of Arizona, Tucson, 1996.
Perreault, Thomas A. "Local Communities and Environmental Conservation in Northern Ecuador." Master's thesis, University of Texas, Austin, 1994.
Sanchez, Sandra. "Community-Based (Eco)Tourism: Indigenous Livelihood Development Strategies in the Ecuadorian Amazon." Master's thesis, Syracuse University, Syracuse, NY, 2007.
Sierra-Maldonado, Rodrigo. "Frontier Settlement and Cycles of Migration in Depressed Rural Settings: The Case of Ecuadorian Amazon Basin." Master's thesis, Ohio State University, Columbus, 1991.
Steinberger, Rebecca. "The Kichwa Land Reform: Shifting Identities in an Indigenous Ecuadorian Community." Master's thesis, University of Arizona, Tucson, 2006.
Swanson, Katherine E. "Begging for Dollars in Gringopampa: Geographies of Gender, Race, Ethnicity, and Childhood in the Ecuadorian Andes." PhD diss., University of Toronto, Toronto, Ontario, 2005.
Valdivia, Gabriela. "Geographies of Indigeneity in the Ecuadorian Amazon." PhD diss., University of Minnesota, Minneapolis, 2005.
Vender, JoAnn C. "Improving Rural Education in Ecuador: A Contextual Approach." Master's thesis, University of Texas, Austin, 1993.
Weinert, Julie. "Influence of Local Gender Relationships on Globalization: Ecotourism in Ecuador." PhD diss., Ohio State University, Columbus, 2008.

Zapato-Rios, Galo. "Linking Spatial Data with Population Viability Analysis Network Design in the Northeastern Ecuadorian Amazon." Master's thesis, Ohio University, Athens, Ohio, 2001.

ECONOMIC GEOGRAPHY

General Works

Brown, Lawrence A., et al. "Development Models, Economic Adjustment, and Occupational Composition: Ecuador, 1982–1990." *International Regional Science Review* 20, no. 3 (1997): 183–210.

Elbers, C., et al. "Imputed Welfare Estimates in Regression Analysis." *Journal of Economic Geography* 5, no. 1 (2005): 101–18.

Fletcher, Merna I. "The Baba Industry of Ecuador." *Economic Geography* 25, no. 1 (1949): 47–54.

Franckowiak, Gene N. "Systematic Rural Development: A Private Experiment in Northern Ecuador." *Ohio Geographer* 7 (1979): 1–12.

Mandel, J., and Victoria A. Lawson. "Occupational Composition and Economic Adjustment: Ecuador, 1982." *International Regional Science Review* 20 (1998): 183–209.

Morris, Arthur S. "Spatial and Sectoral Bias in Regional Development: Ecuador." *Tijdschrift voor Economische en Sociale Geographie* 72, no. 5 (1981): 279–87.

Rudel, Thomas. "Resource Partitioning and Regional Development Strategies in the Ecuadorian Amazon." *GeoJournal* 19, no. 4 (1989): 437–46.

Tobin, Graham A., and Linda M. Whiteford. "Economic Ramifications of Disaster: Experience of Displaced Persons on the Slope of Mount Tungurahua, Ecuador." *Papers of the Applied Geography Conferences* 25 (2002): 316–24.

White, Stuart, and Fausto Maldonado. "The Use and Conservation of Natural Resources in the Andes of Southern Ecuador." *Mountain Research and Development* 11, no. 1 (1991): 37–55.

Agriculture and Land Use

Basile, D. G. *Tillers of the Andes: Farmers and Farming in the Quito Basin.* Chapel Hill: University of North Carolina, Department of Geography, Studies in Geography 8, 1974.

Bossio, Deborah A., and Kenneth G. Cassman. "Traditional Rainfed Barley Production in the Andean Highlands of Ecuador: Soil and Nutrient Limitations and Other Constraints." *Mountain Research and Development* 11, no. 2 (1991): 115–26.

Breuer, T. "Agribusinesss als Entwicklungsstimulans? Das Beispiel der Milchwirtschafts in den ecuadorianschen Andes." *Zeitschrift fur Wirtschafsgeographie* 36, no. 4 (1992): 193–209.

Denevan, William M., and Kent Mathewson. "Preliminary Results of the Samborandan Raised-Rild Project, Guayas Basin, Ecuador." *British Archaeological Reports, International Series* 189 (1983): 167–81.

Fadiman, Maria. "Cultivated Food Plants, Culture, and Gendered Spaces of Colonists and the Chachi in Ecuador." *Journal of Latin American Geography* 4, no. 1 (2005): 43–57.

———. "Exploring Conservation: Piquigua Heteropsis Ecuadorensis, in Ecuador." *Papers of the Applied Geography Conferences* 30 (2007): 427–36.

Gondard, Pierre. "Land Use in the Andean Region of Ecuador: From Inventory to Analysis." *Land Use Policy* 5, no. 3 (1988): 341–48.

Harden, Carol. "Interrelationships between Land Abandonment and Land Degradation: A Case from the Ecuadorian Andes." *Mountain Research and Development* 16, no. 3 (1996): 274–90.

Hiraoka, Mario, and Shozo Yamamoto. "Agricultural Development in the Amazon of Ecuador." *Geographical Review* 70, no. 4 (1980): 423–45.

Horst, Oscar H. "Commercialization of Traditional Agriculture in Highland Guatemala and Ecuador." *Revista Geografica* 106 (1987): 5–18.

Langdon, R. "Manioc, A Long Concealed Key to the Enigma of Eastern Island." *Geographical Journal* 154, no. 1 (1988): 324–36.

Lemky, Kim. "A Comparative Study of Two Agricultural Zones in the Napo Valley, Ecuador." *Geoscope* 18, no. 2 (198): 50–58.

Levin, Gregor, and Anette Reenberg. "Land Use Driven Conditions for Habitat Structure: A Case Study from the Ecuadorian Andes." *Geografisk Tidsskrift* 102 (2002): 79–92.

Long, Brian. "Conflicting Land-Use Schemes in the Ecuadorian Amazon: The Case of Simaco." *Geography* 77, no. 4 (1992): 336–48.

Lopez, Santiago, and Rodrigo Sierra. "Agricultural Change in the Pastaza River Basin: A Spatially Explicit Model of Native Amazonian Cultivation." *Applied Geography* 30, no. 3 (2010): 355–69.

Mena, Carlos, et al. "Land Use Change and Household Farms in the Ecuadorian Amazon: Design and Implementation of an Agent-Based Model." *Applied Geography* 31, no. 1 (2011): 210–22.

Messina, Joseph P., and M. A. Cochrane. "The Forests Are Bleeding: How Land Use Change Is Creating a New Fire Regime in the Ecuadorian Amazon." *Journal of Latin American Geography* 6, no. 1 (2007): 85–100.

Messina, Joseph P., et al. "Land Tenure and Deforestation Patterns in the Ecuadorian Amazon: Conflicts in Land Conservation in Frontier Settings." *Applied Geography* 26, no. 2 (2006): 113–28.

Miller, E. V. "Agricultural Ecuador." *Geographical Review* 49, no. 2 (1959): 183–207.

Nations, James D., and Flavio C. Hinojosa. "Cuyabeno Wildlife Production Research." In *Fragile Lands of Latin America: Strategies for Sustainable Development,* edited by John O. Browder, 139–49. Boulder, CO: Westview, 1989.

Parry, M., et al., eds. *The Effect of Climate Variations on Agriculture in the Central Sierra of Ecuador.* Dordrecht, Netherlands: D. Reidel, 1988.

Parsons, James J. "Bananas in Ecuador: A New Chapter in the History of Tropical Agriculture." *Economic Geography* 33 (1957): 201–16.

Perreault, Thomas A. "Why Chacras (Swidden Gardens) Persist: Agrobiodiversity, Food Security, and Cultural Identity in the Ecuadorian Amazon." *Human Organization* 64, no. 4 (2005): 327–39.

Perreault, Thomas A., et al. "Indigenous Irrigation Organizations and the Formation of Social Capital in Northern Highland Ecuador." *Yearbook, Conference of Latin Americanist Geographers* 24 (1998): 1–16.

Preston, David A. "Changes in the Economic Geography of Banana Production in Ecuador." *Transactions, Institute of British Geographers* 37 (1965): 77–90.

Preston, David A., and G. A. Taveras. "Changes in Land Tenure and Land Distribution as a Result of Rural Emigration in Highland Ecuador." *Tijdschrift voor Economische en Sociale Geographie* 71, no. 2 (1980): 98–107.

Sarmiento, Fausto O., and L. M. Frolich. "Andean Cloud Forest Tree Lines, Naturalness, Agriculture, and the Human Dimension." *Mountain Research and Development* 22 (2002): 278–87.

Smith, B. "Systematic Descriptive Matrix Analysis of the Relationship between Environment and Agriculture: An Interpretation from Alchipichi-Puelloro, Ecuador." *Singapore Journal of Tropical Geography* 15, no. 1 (1994): 75–92.

Stadel, Christoph. "Del valle al Monte: Altitudinal Patterns of Agricultural Activities in the Patale-Pelileo Area of Ecuador." *Mountain Research and Development* 6, no. 1 (1986): 53–64.

Valdivia, Gabriela. "Colonization, Comunas, and Reserves; Assessing Land Use and Land Cover Change in Indigenous Communities of Northeastern Ecuador." *Papers of the Applied Geography Conferences* 22 (1999): 398–406.

Commerce and Trade

Belisle, Franois J., and Don R. Hoy. "The Commercial Structure of Latin American Towns: A Case Study of Sangolqui, Ecuador." *Revista Geografica* 90 (1979): 43–63.

Bromley, Raymond J. "El intercambio de productos agricolas entre la costa y la sierra Ecuatoricana." *Revista Geografica* 78 (1973): 15–33.

Krause, T. *Exportorientierte Regionalentwicklung und agroindustrielle Distrikte. Das Beispel der Garnefenindustrie in Ecuador.* Vol. 11. Hamburg, Germany: Beitrage zur Geographischen Regional forschung in Lateinamerika, 2000.

Smith, Vernon. "Marketing Agricultural Commodities in Pichinchia Province, Ecuador." *Geographical Review* 65, no. 3 (1975): 353–63.

Manufacturing and Labor Studies

Brown, Lawrence A., and Roy Ryder. "Employment in Boom Towns in the Ecuador Amazon." *Yearbook, Association of Pacific Coast Geographers* 60 (1998): 75–104.

Lawson, Victoria A. "Tailoring Is a Profession, Seamstressing Is Work! Resiting Work and Reworking Gender Identities among Artisanal Garment Workers in Quito." *Environment and Planning A* 31, no. 2 (1999): 209–28.

Energy, Mining, Fishing, and Lumbering

Fadiman, Maria. "Resource Stewardship: Rain Forest Use among Three Ethnic Groups of Ecuador." *Papers of the Applied Geography Conferences* 31 (2008): 320–28.

Lopez, Santiago, et al. "Tropical Deforestation in the Ecuadorian Choco: Logging Practices and Socio-spatial Relationships." *Geographical Bulletin* 51, no. 1 (2010): 3–22.

Morris, Arthur S. "Afforestation Projects in Highland Ecuador: Patterns of Success and Failure." *Mountain Research and Development* 17, no. 1 (1997): 31–42.

Raberg, Lena, and Thomas Rudel. "Where Are the Sustainable Forestry Projects? A Geography of NGO Interactions in Ecuador." *Applied Geography* 27, nos. 3–4 (2007): 131–49.

Rudel, Thomas. "A Tropical Forest Transition? Agricultural Change, Out-Migration, and Secondary Forests in the Ecuadorian Amazon." *Annals of the Association of American Geographers* 92, no. 1 (2002): 87–102.

Sheppard, George. "The Salt Industry of Ecuador." *Geographical Review* 22, no. 3 (1932): 403–10.

Sierra, Rodrigo. "Traditional Resource-Use Systems and Tropical Deforestation in a Multiethnic Region in North-West Ecuador." *Environmental Conservation* 26, no. 2 (1999): 136–45.

Sierra, Rodrigo, et al. "Forest-Cover Changes from Labor- and Capital-Intensive Commercial Logging in the Southern Choco Rainforests." *Professional Geographer* 55, no. 4 (2003): 477–90.

Sierra, Rodrigo, and Jody Stallings. "The Dynamics and Social Organization of Tropical Deforestation in Northeast Ecuador, 1983–1995." *Human Ecology* 26, no. 1 (1998): 135–62.

Solis, Misael A. "Los Mangares del Ecuador." *Revista Geografica* 54 (1961): 69–88.

Transportation and Communications

Bennett, F. W. "Guayaquil and Quito Railway." *Bulletin of the American Geographical Society* 35, no. 4 (1903): 361–64.

Eder, Herbert M. "The Roles of Transportation in Choco: Cultural and Environmental Change." *California Geographer* 6 (1965): 17–24.

Theses and Dissertations

Basile, D. G. "The Quito Basin: A Case Study Illustrating Rural Land Use in the Ecuadorean Highlands." PhD diss., Columbia University, New York, 1964.

Freile, Luis. "The Extractive and Manufacturing Industries of Ecuador: A Study in Industrial Geography." Master's thesis, University of Oklahoma, Norman, 1957.

Howatt, Benjamin. "Pattern of Specialization in Each Sector of the Economy in Ecuador." Master's thesis, Clark University, Worcester, MA, 1962.

Kelly, Jason. "Transportation Network Optimization for the Movement of Indigenous Goods in Amazonian Ecuador." Master's thesis, Arizona State University, Tempe, 2008.

Mena, Carlos. "Agricultural and Forest Transition in the Northern Ecuadorian Amazon." PhD diss., University of North Carolina, Chapel Hill, 2007.

Messina, Joseph P. "A Complex Systems Approach to Dynamic Spatial Simulation Modeling: Land Use and Land Cover Change in the Ecuadorian Amazon." PhD diss., University of North Carolina, Chapel Hill, 2001.

Raberg, Lena. "NGOs and Environment Intervention: Geographies of Sustainable Forestry by Ecuadorian NGOs." PhD diss., Rutgers University, New Brunswick, NJ, 2006.

Sierra, Rodrigo. "Land Use Strategies of Household-Based Micro-enterprises, Large Scale Timber Industry, and Deforestation in Northwest Ecuador: The Articulation of Market Forces, National Policies, and Local Conditions." PhD diss., Ohio State University, Columbus, 1994.

Sorrenson, Cynthia. "Environmental Implications of Land Use Change in Azuay, Ecuador: A Micro-regional Analysis Using GIS and Remote Sensing Applications." Master's thesis, Ohio State University, Columbus, 1994.

HISTORICAL GEOGRAPHY

General Works

Bromley, Rosemary D. F. "Urban-Rural Demographic Constraints in Highland Ecuador: Town Recession in a Period of Catastrophe, 1778–1841." *Journal of Historical Geography* 5, no. 3 (1979): 281–96.

Bromley, Rosemary D. F., and Raymond J. Bromley. "The Debate on Sunday Markets in Nineteenth Century Ecuador." *Journal of Latin American Studies* 7, no. 1 (1975): 85–108.

Herrera Angel, Marta, et al. "Geographies of the Name: Naming Practices among the Muisca and Pajez in the Audiencias of Santa Fe and Quito, Sixteenth and Seventeenth Centuries." *Journal of Latin American Geography*, 11, Special Issue (2012): 91–115.

Knapp, Gregory W. "Riego precolonial en la Sierra Norte." *Ecuador Debate* 14 (1987): 17–45.

Knapp, Gregory W., and William M. Denevan. "The Use of Wetland in the Prehistoric Economic of the Northern Ecuadorian Highlands." In *Prehistoric Intensive Agriculture in the Tropics*, edited by I. S. Farrington, 184–207. Oxford, England: British Archaeological Reports, International Series 232, 1985.

Knapp, Gregory W., and Patricia Mothes. "Quilotoa Ash and Human Settlements in the Equatorial Andes." In *Actividad volcanica y pueblos precolombinos en el Ecuador*, edited by Patricia Mothes, 139–55. Quito, Ecuador: Ediciones Abya-Yala, 1998.

Knapp, Gregory W., and David A. Preston. "Evidence of Prehistoric Ditched Fields in Shaping Land in Northern Highland Ecuador." In *The Ecology and Archaeology of Prehistoric Agricultural Fields in the Central Andes*, edited by William M. Denevan et al., 403–24. Oxford, England: British Archaeological Reports, International Series 359, 1987.

Knapp, Gregory W., and Roy Ryder. "Aspects of the Origin, Morphology, and Function of Ridged Fields in the Quito Altiplano, Ecuador." In *Drained Field Agriculture in Central and South America*, edited by J. P. Darch, 201–20. Oxford, England: British Achaeologial Research, International Series 189, 1983.

Mathewson, Kent. "Estimating Labor Imputs in Raised Field Complexes of the Guayas Basin, Ecuador." In *The Ecology and Archaeology of Prehistoric Agricultural Fields in the Central*

Andes, edited by William M. Denevan et al., 217–24. Oxford, England: British Archaeological Reports, International Series 359, 1987.

Newson, Linda A. *Life and Death in Early Colonial Ecuador.* Norman: University of Oklahoma Press, 1995.

Parsons, James J. "Mapping and Dating the Prehistoric Raised Fields of the Guayas Basin, Ecuador." In *The Ecology and Archaeology of Prehistoric Agricultural Fields in the Central Andes*, edited by William M. Denevan et al., 207–16. Oxford, England: British Archaeological Reports, International Series 359, 1987.

———. "Ridged Fields in the Rio Guayas Valley, Ecuador." *American Antiquity* 27 (1967): 92–100.

———. "Ridged Fields in the Rio Guayas Valley, Ecuador." *American Antiquity* 34 (1969): 76–80.

Parsons, James J., and Roy J. Shleman. "Nuevo informe sobre los campos elevados prehistoricas de la Cuenca del Guayas, Ecuador." *Miscelanea Anthropologica Ecuatoriana* 2 (1982): 31–37.

Villamarin, Juan, and Judith Villamarin. "Chibcha Settlement and Population Change in Central Highland Ecuador, 1778–1825." In *Social Fabric and Spatial Structure in Colonial Latin America*, edited by David J. Robinson, 25–84. Syracuse, NY: Syracuse University, Department of Geography, 1979.

PHYSICAL GEOGRAPHY

General Works

Bush, Mark B., et al. "Late Pleistocene Temperature Depression and Vegetation Change in Ecuadorian Amazonia." *Quaternary Research* 34, no. 3 (1990): 330–45.

Sheppard, George. "Notes on the Climate and Physiography of Southwestern Ecuador." *Geographical Review* 20 (1930): 445–53.

Staver, Charles P., et al. "Refining Soil Conservation Strategies in the Mountain Environment: The Climatic Factor: An Ecuadorian Case Study." *Mountain Research and Development* 11, no. 2 (1991): 127–44.

Villaceres, Jorge. "El Archipelago de Colon (Galapagos), Centro Cientifico International." *Revista Geografica* 52 (1960): 63–76.

Biogeography

Acosta Solis, Misael A. "Proteccion Ecovegetativa de las riberas del Rio Napo, entre el Payamino y Coca." *Revista Geografica* 117 (1993): 65–74.

Andrus, Nicole, et al. "Phylogenetics of Darminiothamnus (Asteraeae: Astereae): Molecular Evidence for Multipe Origins in the Endemci Flora of the Galapagos Islands." *Journal of Biogeography* 36, no. 6 (2009): 1055–69.

Bader, Maaike, and Johan Ruijten. "A Topography-Based Model of Forest Cover at the Alpine Tree Line in the Tropical Andes." *Journal of Biogeography* 35, no. 4 (2008): 711–23.

Baert, Leon, et al. "Spider Communities of Isla Santa Cruz (Galapagos, Ecuador)." *Journal of Biogeography* 18, no. 3 (1991): 333–55.

Blake, Stephen, et al. "Seed Dispersal by Galapagos Tortoises." *Journal of Biogeography* 39, no. 11 (2012): 1961–72.

Brauning, Achim, et al. "Climatic Control of Radial Growth of *Cedrala monana* in a Humid Mountain Rainforest in Southern Ecuador." *Erdkunde* 63, no. 4 (2009): 337–45.

Coblenz, D., and Philip L. Keating. "Topographic Controls on the Distribution of Tree Islands in the High Andes of South-Western Ecuador." *Journal of Biogeography* 35, no. 11 (2008): 2026–38.

Dislich, Claudia, et al. "Simulating Forest Dynamics of a Tropical Montane Forest in South Ecuador." *Erdkunde* 63, no. 4 (2009): 347–64.

Fadiman, Maria. "Management, Cultivation, and Domestication of Weaving Plants: Heteropsis an Astrocaryum in the Ecuadorian Rain Forest." *California Geographer* 44 (2004): 1–19.

Farley, Kathleen A. "Grassland to Tree Plantations: Forest Transition in the Andes of Ecuador." *Annals of the Association of American Geographers* 97, no. 4 (2007): 755–71.

———. "Pathways to Forest Transition: Local Case Studies from the Ecuadorian Andes." *Journal of Latin American Geography* 9, no. 2 (2010): 7–26.

Farley, Kathleen A., et al. "Soil Organic Carbon and Water Retention after Conversion of Grasslands to Pine Plantations in the Ecuadorian Andes." *Ecosystems* 7, no. 7 (2004): 729–39.

Farley, Kathleen A., and E. F. Kelly. "Effects of Afforestation of a Paramo Grassland on Soil Nutrient Status." *Forest Ecology and Management* 195 (2004): 281–90.

Fries, Andreas, et al. "Thermal Structure of a Megadiverse Andean Mountain Ecosystem in Southern Ecuador and Its Regionalization." *Erdkunde* 63, no. 4 (2009): 321–35.

Hilt, Nadine, and Konrad Fiedler. "Acrtiid Moth Ensembles along a Successional Gradient in the Ecuadorian Montane Rain Forest Zone: How Different Are Subfamilies and Tribes?" *Journal of Biogeography* 33, no. 1 (2006): 108–20.

Jokisch, Brad D., and Bridget M. Laie. "One Last Stand? Forests and Change on Ecuador's Eastern Cordillera." *Geographical Review* 92, no. 2 (2002): 235–56.

Keating, Philip L. "Changes in Paramo Vegetation along an Elevation Gradient in Southern Ecuador." *Journal of the Torrey Botanical Society* 126 (1999): 159–75.

———. "Effects of Anthropogenic Disturbances on Paramo Vegetation in Podocarpus National Park, Ecuador." *Physical Geography* 19, no. 3 (1998): 221–38.

———. "Fire Ecology and Conservation in the High Tropical Andes: Observations from Northern Ecuador." *Journal of Latin American Geography* 6, no. 1 (2007): 43–62.

Keating, Philip L., et al. "Variation in High Andean Vegetation at a Site in Southwestern Ecuador." *Pennsylvania Geographer* 40, no. 2 (2002): 15–35.

Keese, James, et al. "Identifying and Assessing Tropical Montane Forests on the Eastern Flank of the Ecuadorian Andes." *Journal of Latin American Geography* 6, no. 1 (2007): 63–84.

Kelly, D. L., et al. "Floristics and Biogeography of a Rain Forest in the Venezuelan Andes." *Journal of Biogeography* 21, no. 4 (1994): 421–40.

Knapp, Gregory W. *Andean Ecology; Adaptive Dynamics in Ecuador*. Boulder, CO: Westview, Dellplain Studies 27, 1991.

Kreft, Holger, et al. "Diversity and Biogeography of Vascular Epiphutes in Western Amazonia, Yasani, Ecuador." *Journal of Biogeography* 31, no. 9 (2004): 1463–76.

Kuper, W., et al. "Large-Scale Diversity Patterns of Vascular Epiphytes in Neotropical Montane Rain Forests." *Journal of Biogeography* 31, no. 9 (2004): 1477–87.

Malanson, George, et al. "Complexity at Advancing Ecotones and Frontiers." *Environment and Planning A* 38, no. 4 (2006): 619–32.

Mougeot, Luc J. A. "Circulacion de un combustible critico en el callejon andino: La lena en la cuenca atla del Rio Paute, Ecuador." *Revista Geografica* 101 (1985): 115–31.

Poulsen, Bent O., and Niels Krabbe. "Avifaunal Diversity of Five High-Altitude Cloud Forests on the Andean West Slope of Ecuador: Testing a Rapid Assessment Model." *Journal of Biogeography* 25, no. 1 (1998): 83–94.

Quintanilla, Victor. "Fitogeografia de las Islas Galapagos: Observaciones preliminares en la Isla de San Cristobal." *Revista Geografica* 98 (1983): 58–78.

Richter, Michael. "Using Epiphytes and Soil Temperatures for Eco-climatic Interpretations in Southern Ecuador." *Erdkunde* 57, no. 3 (2003): 161–82.

Rodriguez, Jose. "Galapagos: Un ecosistema insular intervenido." *Revista Geografica de America Central* 17–18 (1983): 301–12.

Sarmiento, Fausto O. "The Lapwing in Andean Ethnoecology: Proxy for Landscape Transformation." *Geographical Review* 100, no. 2 (2010): 229–45.

Sklenar, P., and P. M. Jargensen. "Distribution Patterns of Paramo Plants in Ecuador." *Journal of Biogeography* 26, no. 4 (1999): 681–92.

Snell, Heidi, et al. "A Summary of Geographical Characteristics of the Galapagos Islands." *Journal of Biogeography* 23, no. 5 (1996): 619–24.

Climatology and Weather

Bendix, Jorg, et al. "A Case Study on Rainfall Dynamics during El Nino/La Nina 1997–99 in Ecuador and Surrounding Areas as Inferred from GOES-8 and TRMM-PR Observations." *Erdkunde* 57, no. 2 (2003): 81–93.

Caviedes, Cesar N, and Peter R. Waylen. "Anamolous Westerly Winds during El Nino/Southern Oscillation Events: The Discovery and Colonization of Easter Island." *Applied Geography* 13, no. 2 (1993): 123–34.

Momsen, Richard P. "Precipitation Patterns of West-Central Ecuador." *Revista Geografica* 69 (1968): 1–106.

Sheppard, George. "The Rainy Season of 1932 In Southwestern Ecuador." *Geographical Review* 23, no. 2 (1933): 210–16.

Trapasso, L. M. "Meteorological Data Acquisition in Ecuador, South America: Problems and Solutions." *GeoJournal* 12, no. 1 (1986): 89–94.

Geomorphology, Landforms, and Volcanism

Clapperton, Chalmers M. "Glacial Morphology, Quaternary Glacial Sequence, and Palaeoclimatic Inferences in the Ecuadorian Andes." In *International Geomorphology,* edited by V. Gardiner, 2:843–70. Chichester, England: Wiley, 1987.

Harden, Carol, and P. D. Scrugge. "Infiltration on Mountain Slopes: A Comparison of Three Environments." *Geomorphology* 55, nos. 1–4 (2003): 5–24.

Harden, Carol, and T. R. Wallin. "Quantifying Trail-Related Erosion at Two Sites in the Humid Tropics, Jatun Sacha, Ecuador and La Selva, Costa Rica." *Ambio* 25, no. 7 (1996): 517–22.

Lane, Lucille, et al. "Volcanic Hazard or Economic Destitution: Hard Choices in Banos, Ecuador." *Global Environmental Change B: Environmental Hazards* 5, nos. 1–2 (2003): 23–34.

Lover, Wilhelm, and M. D. Rafiqpoor. "Paramo de Papallacta: A Physiogeographical Map 1:50,000 of the Ama around the Antisana (Eastern Cordillera of Ecuador)." *Erdkunde* 54, no. 1 (2000): 20–33.

Momsen, Richard P. "Precipitation Patterns of West-Central Ecuador." *Revista Geografica* 69 (1968): 91–105.

Walsh, Stephen, et al. "Complexity Theory, Spatial Simulation Models, and Load Use Dynamics in the Northern Ecuadorian Amazon." *Geoforum* 39, no. 2 (2008): 867–78.

Hydrology and Glaciology

Hastenrath, S. L. *The Glaciation of the Ecuadorian Andes.* Rotterdam, Netherlands: Balkema, 1981.

Heine, Klaus, and Jan Heine. "Late Glacial Climatic Fluctuations in Ecuador: Glacier Retreat during Younger Dryas Time." *Arctic and Alpine Research* 28, no. 4 (1996): 496–501.

Williams, Mark, et al. "Synoptic Survey of Surface Water Isotopes and Nutrient Concentrations, Paramo High-Elevation Region, Antisona Ecological Reserve, Ecuador." *Arctic, Antarctic, and Alpine Research* 33, no. 4 (2001): 397–403.

Soils

Bendix, Jorg, and M. D. Rafiqpoor. "Studies on the Thermal Conditions of Soils at the Upper Tree Line in the Paramo of Papallacta (Eastern Cordillera of Ecuador)." *Erdkunde* 55, no. 3 (2001): 257–76.

Dehn, Martin. "An Evaluation of Soil Conservation Techniques in the Ecuadorian Andes." *Mountain Research and Development* 15, no. 2 (1995): 175–82.

Harden, Carol. "Land Use, Soil Erosion, and Reservoir Sedimentation in an Andean Drainage Basin in Ecuador." *Mountain Research and Development* 13, no. 2 (1993): 177–84.

———. "Mesoscale Estimates of Soil Erosion in the Rio Ambato Drainage, Ecuadorian Sierra." *Mountain Research and Development* 8, no. 4 (1988): 331–41.

————. "Soil Erosion and Sustainable Mountain Development: Experiments, Observations, and Recommendations from the Ecuadorian Andes." *Mountain Research and Development* 21 (2001): 77–83.

Hewitt, C. N., and G. B. B. Candy. "Soil and Street Dust Heavy Metal Concentrations in and around Cuenca, Ecuador." *Environmental Pollution* 63, no. 2 (1990): 129–36.

Liess, Mareike, et al. "Digital Soil Mapping in Southern Ecuador." *Erdkunde* 63, no. 4 (2009): 309–19.

Wallin, T. R., and Carol Harden. "Estimating Trail-Related Soil Erosion in the Humid Tropics: Jatun Sacha, Ecuador and La Selva, Costa Rica." *Ambio* 25, no. 8 (1996): 517–22.

Wilcke, N., et al. "Soil Properties on a Chronosequence of Landslide in Montane Rain Forest, Ecuador." *Catena* 53, no. 1 (2003): 79–95.

Theses and Dissertations

Aragundi, Sheika. "Biogeogrpahic Study of the Polylepis Forest Remnants of the Northeastern Cordilla Oriental of Ecuador and Implications for Their Conservation." PhD diss., University of Georgia, Athens, 2008.

Echavarria, Fernando. "Remote Sensing of Montane Forest Degradation in Southern Ecuador." PhD diss., University of South Carolina, Columbia, 1993.

Hartsig, James. "The Effects of Land-Use Change on the Hydrophysical Properties of Andisols in the Ecuadorian Paramo." Master's thesis, University of Tennessee, Knoxville, 2011.

Keating, Philip L. "Disturbance Regimes and Regeneration Dynamics of Upper Montane Forests and Paramos in the Southern Ecuadorian Mountains." PhD diss., University of Colorado, Boulder, 1995.

Lair, Bridget, "One Last Stand? Remote Sensing Analysis of a Tropical Montane Cloud Forest in the Highlands of South-Central Ecuador." Master's thesis, Ohio University, Athens, Ohio, 2002.

POLITICAL GEOGRAPHY

General Works

Arsel, Murat. "Between Marx and Markets? The State, the Left-Turn, and Nature in Ecuador." *Tijdschrift voor Economische en Sociale Geographie* 103, no. 2 (2012): 164–79.

Bebbington, Anthony J., et al. "Fragile Lands, Fragile Organizations: Indian Organizations and the Politics of Sustainability in Ecuador." *Transactions, Institute of British Geographers* 18, no. 2 (1993): 179–96.

Bromley, Raymond J. *Development Planning in Ecuador.* Swansea, Wales: University of Wales, Centre for Development Studies, 1977.

Chatelain, J. L., et al. "Earthquake Risk Management Pilot Project in Quito, Ecuador." *GeoJournal* 49, no. 2 (1999): 185–96.

Elbow, Gary S. "Territorial Loss and National Image: The Case of Ecuador." *Yearbook, Conference of Latin Americanist Geographers* 22 (1996): 93–106.

Keating, Philip L. "Mapping Vegetation and Anthropogenic Disturbances in Southern Ecuador with Remote Sensing Techniques: Implications for Park Management." *Yearbook, Conference of Latin Americanist Geographers* 23 (1997): 77–90.

Kesza, G. J. "Regional Conflict in Ecuador: Quito and Guayaquil." *Inter-American Economic Affairs* 35, no. 2 (1981): 3–41.

Lawson, Victoria A. "Global Governmentality and Graduated Sovereignty: National Belonging among Poor Migrants in Ecuador." *Scottish Geographical Journal* 118, no. 3 (2002): 235–55.

————. "Government Policy Biases and Ecuadorian Agricultural Change." *Annals of the Association of American Geographers* 78, no. 3 (1988): 433–52.

Messina, Joseph P., et al. "Land Tenure and Deforestation Patterns in the Ecuadorian Amazon: Conflicts in Land Conservation in Frontier Settings." *Applied Geography* 26, no. 2 (2006): 113–29.

Metzger, Pascale, et al. "Political and Scientific Uncertainties in Volcanic Risk Management: The Yellow Alert in Quito in October, 1998." *GeoJournal* 49, no. 2 (1999): 213–21.

Perreault, Thomas A. "Changing Places: Transnational Networks, Ethnic Politics, and Community Development in the Ecuadorian Amazon." *Political Geography* 22, no. 1 (2003): 61–88.

———. "Making Space: Community Organization, Agrarian Change, and the Politics of Scale in the Ecuadorian Amazon." *Latin American Perspectives* 30, no. 1 (2003): 96–121.

———. *Movilizacion politica e identidad indigena en el Alto Napo*. Quito, Ecuador: Ediciones Abya Yala, 2002.

———. "Nature Preserves and Community Conflict: A Case Study in Highland Ecuador." *Mountain Research and Development* 16, no. 2 (1996): 167–75.

———. "A People with Our Own Identity: Toward a Cultural Politics of Development in Ecuadorian Amazonia." *Environment and Planning D* 21, no. 5 (2003): 583–606.

———. "Social Capital, Development, and Indigenous Politics in Ecuadorian Amazonia." *Geographical Review* 93, no. 3 (2003): 328–49.

Pohle, P., and A. Gerique. "Traditional Ecological Knowledge and Biodiversity Management in the Andes of Southern Ecuador." *Geographica Helvetica* 61, no. 4 (2006): 275–85.

Radcliffe, Sarah A. "Development for a Postneoliberal Era? Sumak Kawsay, Living Well, and the Limits of Decolonialisation in Ecuador." *Geoforum* 43, no. 2 (2012): 240–49.

———. "Gendered Nations: Nostalgia, Development, and Territory in Ecuador." *Gender, Place, and Culture* 3, no. 1 (1996): 5–22.

———. "Imaginative Geographies, Postcolonialism, and National Identities: Contemporary Discourses of the Nation in Ecuador." *Ecumene* 3, no. 1 (1996): 23–42.

———. "Reimagining the Nation: Community, Difference, and National Identities among Indigenous and Mestizo Provincials in Ecuador." *Environment and Planning A* 31, no. 1 (1999): 37–52.

Radcliffe, Sarah A., et al. "Reterritorialised Space and Ethnic Political Participation: Indigenous Municipalities in Ecuador." *Space and Polity* 6, no. 2 (2002): 289–305.

Roucek, Joseph S. "Ecuador in Geopolitics." *Contemporary Review* 205 (1964): 74–82.

Valdivia, Gabriela. "Governing Relations between People and Things: Citizenship, Territory, and the Political Economy of Petroleum in Ecuador." *Political Geography* 27, no. 4 (2008): 456–77.

Wesche, Rolf. "Developed Country Environmentalism and Indigenous Community Controlled Ecotourism in the Ecuadorian Amazon." *Geographische Zeitschrift* 84, nos. 3–4 (1996): 157–68.

Theses and Dissertations

Goldberg, Allison. "Contest for Space and Power in Ecuador's Amazon: Indigenas, Petroleum, and Western Environmentalism." Master's thesis, University of Texas, Austin, 1995.

Himley, Matthew. "The Politics of Land and Forest: Nature Conservation in Highland Ecuador." Master's thesis, Syracuse University, Syracuse, NY, 2005.

URBAN GEOGRAPHY

General Works

Bromley, Rosemary D. F. "The Functions and Development of Colonial Towns: Urban Change in the Central Highlands of Ecuador, 1698–1940." *Transactions, Institute of British Geographers* 4, no. 1 (1979): 30–43.

Brown, Lawrence A., et al. "Frameworks of Urban System Evolution in Frontier Settings and the Ecuador Amazon." *Geography Research Forum* 14 (1994): 72–96.

———. "Urban System Development, Ecuador's Amazon Region, and Generalization." In *Frontiers in Regional Development*, edited by Yehuda Gradus and Harvey Lithwick, 99–124. Lanham, MD: Rowman & Littlefield, 1996.

———. "Urban-System Evolution in Frontier Settings." *Geographical Review* 84, no. 3 (1994): 249–65.

Buckhofer, E. "Desareollo urbano en la zona petrolera del Oriente ecuatoriano." *Revista Geografica del Instituto Geografico Militar* 27 (1988): 89–108.

Gravelin, B. "Proceso de urbanizacion en zonas pioneras." In *El Espacio Urbano en el Ecuador*, edited by M. Porais and J. Leon, 170–81. Quito, Ecuador: Instituto Paramericano de Geografia y Historia, 1987.

Lowder, Stella. "The Context of Urban Planning in Secondary Cities: Examples from Andean Ecuador." *Cities* 8, no. 1 (1991): 54–65.

Ryder, Roy, and Lawrence A. Brown. "Urban Development at the Ecuador Amazon Frontier: Boom Towns or Gloom Towns." In *Developing Frontier Cities: Global Perspectives—Regional Contexts*, edited by Harvey Lithwick and Yehuda Gradus, 313–43. Beer Sheva, Israel: Ben-Gurion University of the Negev, Negev Center for Regional Development, 2000.

———. "Urban-System Evolution on the Frontier of the Ecuadorian Amazon." *Geographical Review* 90, no. 4 (2000): 511–35.

Schenck, Freya S. *Strukturveranderungen spanisch-amerikanischer Mittelstadte untersucht am Beispiel der Stadt Cuenca, Ecuador*. Kiel, Germany: Kieler Geographische Schriften, Band 94, 1997.

Zaaijer, Mirjom. "Quito." *Cities* 8, no. 2 (1991): 87–92.

Urban Social Geography

Rey, Walter, and Beatriz St. Larg. "El Caserio de San Miguel de Collacoto." *Revista Geografica* 84 (1976): 183–217.

Smith, Betty E. "Urban Population Density Analysis and Visualization of Central City Morphology of a City in the Andes Mountains of Ecuador." *Papers of the Applied Geography Conferences* 30 (2007): 135–46.

Swanson, Katherine E. "Revanchist Urbanism Heads South: The Regulation of Indigenous Beggars and Street Vendors in Ecuador." *Antipode* 39 (2007): 708–28.

Morphology and Neighborhoods

Bromley, Rosemary D. F., and Gareth A. Jones. "Conservation in Quito: Policies and Progress in the Historic Centre." *Third World Planning Review* 17, no. 1 (1995): 41–60.

Jones, Gareth A., and Rosemary D. F. Bromley. "The Relationship between Urban Conservation Programmes and Property Renovation: Evidence from Quito, Ecuador." *Cities* 13, no. 6 (1996): 373–85.

Klak, Thomas. "How Do Shelter Conditions Differ in Ecuadorian Cities and How Do We Know That They Do: A Comparative Locality Study." *Environment and Planning A* 25, no. 8 (1993): 1115–30.

———. "Recession, the State, and Working-Class Shelter: A Comparison of Quito and Guayaquil during the 1980s." *Tijdschrift voor Economische en Sociale Geographie* 83, no. 2 (1992): 120–37.

Klak, Thomas, and Michael Holtzclaw. "The Housing, Geography, and Mobility of Latin American Urban Poor: The Prevailing Model and the Case of Quito, Ecuador." *Growth and Change* 24, no. 2 (1993): 247–73.

Messina, Joseph P. "Rural to Urban Conversion: Landscape Modeling in the Ecuadorian Amazon." *Papers of the Applied Geography Conferences* 22 (1999): 374–82.

Ryder, Roy. "Land Use Diversification in the Elite Residential Sector of Quito, Ecuador." *Professional Geographer* 56, no. 4 (2004): 488–502.

Urban Economic Geography

Belisle, Francois J. "The Commercial Structure of Latin American Towns: A Case Study of Sangolqui, Ecuador." *Revista Geografica* 90 (1979): 43–64.

Jackiewicz, Edward. "Neoliberalism and Shifts in Occupational Structure: Quito, Ecuador, 1982–90." *Tijdschrift voor Economische en Sociale Geographie* 92, no. 4 (2001): 437–48.

Teltscher, Susanne. "Gender Differences in Ecuador's Urban Informal Economy: Survival to Upward Mobility." *Yearbook, Conference of Latin Americanist Geographers* 19 (1993): 81–92.

———. "Small Trade and the World Economy: Informal Vendors in Quito, Ecuadro." *Economic Geography* 70, no. 2 (1994): 167–87.

Urban Environments

Hannell, F. G. "Some Features of the Heat Island in an Equatorial City." *Geografiska Annaler* 58A, nos. 1–2 (1976): 95–109.

Swyngedouw, E. "Power, Nature, and the City: The Conquest of Water and the Political Ecology of Urbanization in Guayaquil, Ecuador, 1880–1990." *Environment and Planning A* 29, no. 2 (1997): 311–32.

Theses and Dissertations

Holtzclaw, Michael. "Understanding the Geography, Housing, and Mobility of Rural-Urban Migrants in Latin America through Case Studies in Quito and Guayaquil, Ecuador." Master's thesis, Ohio State University, Columbus, 1993.

Oswald, Robert. "Urban Migration among the Lower Classes in Ecuador." Master's thesis, Ohio State University, Columbus, 1991.

Riono, Yvonne. "Geography of the Socializing Patterns of Residents of Barrio Mean Del Hierro, Quito, Ecuador." PhD diss., University of Ottawa, Ottawa, Ontario, 1996.

Sheck, Ronald C. "Historical Geography of Quito." PhD diss., University of Oregon, Eugene, 1969.

Chapter Eight

French Guiana

GENERAL WORKS

Articles and Book Chapters

Gritzner, Charles F. "French Guiana: Development Trends in the Post-prison Era." *Journal of Geography* 62 (1963): 161–68.

Lowenthal, David. "French Guiana: Myths and Realities." *Transactions of the New York Academy of Sciences*, Series 2, 22, no. 7 (1960): 528–40.

Papy, L. "La Guyane francaise." *Les Cahiers d'Outre Mer* 8 (1955): 209–32.

CULTURAL AND SOCIAL GEOGRAPHY

General Works

Bassargette, Denis, and G. Meo. "Les limites du modele communal francaise Guyane: Le cas de Maripasoula." *Les Cahiers d'Outre Mer* 61, nos. 241–42 (2008): 49–80.

Ethnic, Social, and Population Geography

Hurault, Jean. "Les indiens du littoral de la Guyane francaise." *Les Cahiers d'Outre Mer* 16 (1963): 145–83.

Lasserre, G. "Noirs det Indiends des pays du Maroni (Guyane Francaise)." *Les Cahiers d'Outre Mer* 5 (1952): 84–89.

Puyo, Jean-Yves. "Mise en valeur de la Guyane francaise et peuplement blanc: Les espoirs decus du baron de Laussat (1819–1923)." *Journal of Latin American Geography* 7, no. 1 (2008): 177–202.

ECONOMIC GEOGRAPHY

General Works

Gritzner, Charles F. "French Guiana: Development Trends in Post-prison Era." *Journal of Geography* 62, no. 4 (1963): 161–67.

Agriculture and Land Use

Demaze, Moise T., and Sandrine Manusset. "L'agriculture itinerante sur brulis en Guyane Francaise: La fin des durabilites ecologique et socio-culturelle?" *Les Cahiers d'Outre Mer* 61 (2008): 31–48.

HISTORICAL GEOGRAPHY

General Works

Gritzner, Charles F. "French Guiana Penal Colony: Its Role in Colonial Development." *Journal of Geography* 63, no. 7 (1964): 314–19.

PHYSICAL GEOGRAPHY

Biogeography

Jullien, Mathilde, and Jean-Marc Thiollay. "Effects of Rain Forest Disturbance and Fragmentation: Comparative Changes of the Raptor Community along Natural and Human-Made Gradients in French Guiana." *Journal of Biogeography* 23, no. 1 (1996): 7–26.

Geomorphology, Landforms, and Volcanism

Anthony, Edward J., and Franck Dolique. "Intertidal Subsidence and Collapse Features on Wave-Exposed, Drift-Aligned Sandy Beaches Subject to Amazon Mud: Cayenne, French Guiana." *Earth Surface Processes and Landforms* 31, no. 8 (2006): 1051–57.
Dolique, Franck, and Edward J. Anthony. "Short-Term Profile Changes of Sandy Pocket Beaches Affected by Amazon-Derived Mud, Cayenne, French Guiana." *Journal of Coastal Research* 21, no. 6 (2005): 1195–202.

POLITICAL GEOGRAPHY

General Works

Bassargette, Denis. "The Future French Guiana National Park: An Opportunity for Re-thinking the Space Connection between a Socio-political Organization and Its Substratum [in French]." *Annales de Geographie* 112, no. 2 (2003): 188–213.
Bassargette, Denis, and Guy DiMeo. "Les limites du model communal francais en Guyane: Le cas de Maripasoula." *Les Cahiers d'Outre Mer* 61 (2008): 49–80.
Cherubini, B. "Patrimoine regional et amenagement rural en Guyane francaise." *Les Cahiers d'Outre Mer* 41 (1988): 379–408.

Chapter Nine

Guyana

GENERAL WORKS

Articles and Book Chapters

Berrange, Jevan, and Richard Johnson. "A Guide to the Upper Essequibo River, Guyana." *Geographical Journal* 138, no. 1 (1972): 41–52.

Lea, D. A. M. "Guyana and the Rupununi Savannas." *Geography Teacher* (Australia) 8, no. 2 (1968): 43–54.

Platt, Robert S. "Reconnaissance in British Guiana: With Comments on Microgeography." *Annals of the Association of American Geographers* 29 (1939): 112–24.

Ramraj, Robert. "The Omai Disaster in Guyana." *Geographical Bulletin* 43, no. 2 (2001): 83–90.

Schumburgk, Robert H. "Report of the Third Expedition into the Interior of Guyana." *Journal of the Royal Geographical Society* 10 (1841): 159–267.

Vining, James W. "Guyana." In *Focus on South America*, edited by Alice Taylor, 80–93. New York: Praeger, 1973.

CULTURAL AND SOCIAL GEOGRAPHY

General Works

Gillham, Olive. "Human Geography of British Guiana." *Journal of Geography* 25 (1926): 146–52.

Mentore, Laura H. "Waiwai Fractality and the Arboreal Bias of PES Schemes in Guyana: What to Make of the Multiplicity of Amazonian Cosmographies?" *Journal of Cultural Geography* 28, no. 1 (2011): 21–43.

Mistry, J., and A. Beradi. "The Challenges and Opportunities of Participatory Video in Geographical Research: Exploring Collaboration with Indigenous Communities in the North Rupununi, Guyana." *Area* 44, no. 1 (2012): 110–16.

Ethnic, Social, and Population Geography

Burrough, J. B. "Ethnicity as a Determinant of Peasant Farming Characteristics: The Canals Polder, Guyana." *Journal of Tropical Geography* 37 (1973): 1–8.
Holdridge, Desmond. "An Investigation of the Prospect for White Settlement in British Guiana." *Geographical Review* 29, no. 4 (1939): 622–42.
Lakhan, V. C., and P. D. Lavalle. "Assessing the Personal Concern of Indo-Guyanese for the Natural Environment." *Indo Caribbean Review* 4 (1998): 19–31.
Lasserre, G. "Des populations tribales guyanaises: Noirs Boniet Indiens Wayanna." *Les Cahiers d'Outre Mer* 19 (1966): 187–92.
Read, Jane, and Jose Frafoso. "Space, Place, and Hunting Patterns among Indigenous Peoples of the Guyanese Rupunui Region." *Journal of Latin American Geography* 9, no. 3 (2010): 213–43.
Rodway, James. "The Forest People of British Guiana." *Bulletin of the American Geographical Society* 34, no. 4 (1902): 283–94.
Stratchan, A. J. "Government Sponsored Return Migration to Guyana." *Area* 12, no. 2 (1980): 165–70.

Community and Settlement Studies

Vining, James W. "Black Rush Polder Reconsidered." *Ecumene* 11 (1979): 52–62.
———. "Economic Problems in Guyana's Program of Directed Settlement." *East Lakes Geographer* 11 (1976): 135–47.
———. "Grandiose Schemes for Foreign Colonization in Guyana: A Survey of Their Origin, Provisions, and Abandonment." *Caribbean Quarterly* 24, nos. 1–2 (1978): 75–89.
———. "Presettlement Planning in Guyana." *Geographical Review* 67, no. 4 (1977): 469–80.
———. "Site Development and Settlement Scheme Failure in Guyana." *Journal of Tropical Geography* 42 (1976): 86–97.

Tourism and Recreation

Lakhan, V. C., and P. D. LaValle. "Recreational Activities in the Coastal Environment of Guyana." *Indo Caribbean Review* 3, no. 1 (1996): 15–34.

Place Names

Rodway, James. "The River Names of British Guiana." *Bulletin of the American Geographical Society* 36, no. 7 (1904): 396–402.

Theses and Dissertations

Hope, Walter B. "The Distribution of Population in British Guyana." Master's thesis, Catholic University, Washington, DC, 1962.
Richardson, Bonham C. "Village Landscape in a Plural Society: Two Coastal Communities of Guyana." Master's thesis, University of Wisconsin, Madison, 1968.
Ronald, Kimberley. "Evaluating Rural-Urban Differences in Environmental Concern in Guyana with Log-Linear Analysis." Master's thesis, University of Windsor, Windsor, Ontario, 1997.
Tarrado, Maria. "Road Construction and Makushi Communities of Southern Guyana: Impacts and Consequences." Master's thesis, Syracuse University, Syracuse, NY, 2007.
Wilson, Emily. "Gendered Geographies and Participatory Processes: Mapping Natural Resource Use with Wapichan Women in Southern Guyana." Master's thesis, Carleton University, Ottawa, Ontario, 2005.

ECONOMIC GEOGRAPHY

General Works

Auty, R. M. "Multinational Corporate Product Strategy, Spatial Diversification, and Nationalism: A Guyanese Compromise." *Geoforum* 12, no. 4 (1981): 347–57.

Delpech, Bernard. "Les Aluku de Guyane a un tournant: De l'economie de subsistence a la societe de consummation." *Les Cahiers d'Outre Mer* 46 (1993): 175–94.

Richardson, Bonham C. "Spatial Determination of Rural Livelihood in Coastal Guyana." *Professional Geographer* 25, no. 4 (1973): 363–68.

Agriculture and Land Use

Auty, R. M. "Some Problems in Identifying Scale Economies: The Sugar Industry of Demerara, 1930–1965." *Journal of Tropical Geography* 34 (1972): 8–16.

Bereau, D., et al. "Traitement artisanal du manioc en Guyane." *Les Cahiers d'Outre Mer* 44 (1991): 281–88.

Bynoe, P. "Land Utilization in Caribbean States and Its Relevance to Guyana." *Beitrage zur Geographischen Regionalforchung in Lateinamerika* 10 (1997): 45–54.

Lakhan, V. C. "Development and Advancement of Guyana's Rice Industry." *Indo Caribbean Review* 3 (1997): 81–92.

Lall, A., V. C. Lakhan, and D. Rawana. "Resource Allocation in Agriculture: The Guyana Experience." *Canadian Journal of Development Studies* 9, no. 2 (1988): 235–48.

McConnell, R. B. "Plantation Surfaces in Guyana." *Geographical Journal* 134, no. 4 (1968): 506–20.

Merrill, Gordon C. "Recent Land Development in Coastal British Guiana." *Canadian Geographer* 5, no. 2 (1961): 24–29.

Richardson, Bonham C. "Distance Regularities in Guyanese Rice Cultivation." *Journal of Developing Areas* 8, no. 2 (1974): 235–56.

———. "Guyana's Green Revolution: Social and Ecological Problems in an Agricultural Development Program." *Caribbean Quarterly* 18 (1972): 14–23.

———. "Men, Water, and Mudflats in Coastal Guyana." *Resources Management and Optimization* 5 (1987): 213–36.

Semple, Hugh M. "An Overview of Domestic Food Production in Guyana, 1960–1993." *Caribbean Geography* 9, no. 2 (1998): 30–43.

Semple, Hugh M., and James S. Brierley. "A Logit Analysis of Problems Affecting Domestic Food Production in Guyana." *Social and Economic Studies* 49, no. 1 (2000): 211–24.

Shaw, Anthony B. "An Agroclimatic Procedure for Rice Cultivation on the Coastal Areas of Guyana." *Malaysian Journal of Tropical Geography* 5 (1982): 67–81.

———. "Impact of New Technology on the Guyanese Rice Industry: Efficiency and Equity Considerations." *Journal of Developing Areas* 18, no. 2 (1984): 191–218.

De Souza, Padma D., and V. C. Lakham. "Log-Linear Analysis of Factors Contributing to the Post-independence Decline of Guyana's Rice Industry." *Applied Geography* 15, no. 2 (1995): 147–60.

Strachan, A. J. "Water Control in Guyana." *Geography* 65, no. 4 (1984): 297–304.

Williams, Patrick E. "Drainage and Irrigation Projects in Guyana: Environmental Considerations." In *Environment and Development in the Caribbean: Geographical Perspectives*, edited by David Barker and Duncan F. M. McGregor, 233–48. Barbados: University of West Indies Press, 1995.

Energy, Mining, Fishing, and Lumbering

Desse, Michel. "Le commercialization des produits de la mer en Guyane." *Equinoxe* 22 (1986): 79–94.

———. "Les nouvelles tendences du commerce guyanis des produits de la mer." *Equinoxe* 26 (1988): 69–80.

———. *La peche cotiere en Guyane*. Brest, France: T. E. R. Universite de Bretogne Occidentale, 1984.

———. "Les peches guyanaises a la conquete de nouveaux espaces, analyse des vingt dernieres annees." *Les Cahiers d'Outre Mer* 41 (1988): 357–78.

Theses and Dissertations

Auty, R. M. "The Demerara Sugar Industry, 1930–65: A Geographical Study of Rural Change." Master's thesis, University of Toronto, Toronto, Ontario, 1968.

D'Andrea, Nicholas. "The Dominant Physical and Cultural Influences on Cattle Ranching in the North Pupununi Savanna of Guyana." Master's thesis, University of Southern Mississippi, Hattiesburg, 1967.

Richardson, Bonham C. "The Rice Culture of Coastal Guayana: A Study in Location and Livelihood." PhD diss., University of Wisconsin, Madison, 1970.

HISTORICAL GEOGRAPHY

General Works

Everitt, John C. "A Tale of Two Colonies: A Comparative Study of the Early Development of Belize and Guyana." *Geographical Perspectives* 59 (1987): 93–102.

Hills, Theo L. "The Interior of British Guiana and the Myth of El Dorado." *Canadian Geographer* 5, no. 2 (1961): 30–43.

Richardson, Bonham C. "Plantation and Village in Coastal Guyana, 1897–1969: Conflict or Complementarity?" *Journal of Historical Geography* 3, no. 4 (1977): 349–62.

Vining, James W. "Nineteenth Century Developments in the History of Organized Settlement in Guyana." *Historical Geography Newsletter* 6, no. 2 (1976): 7–12.

Theses and Dissertations

Potter, Lesley. "Internal Migration and Resettlement of East Indians in Guyana, 1870–1920." PhD diss., McGill University, Montreal, Quebec, 1976.

Wagner, Michael. "Structural Pluralism and the Portuguese in Nineteenth Century British Guiana: A Study in Historical Geography." PhD diss., McGill University, Montreal, Quebec, 1975.

PHYSICAL GEOGRAPHY

General Works

Lakhan, V. C. "Simulating the Interactions of Changing Nearshore Water Levels, Morphology, and Vegetation Growth in Guyana's Coastal Environment." In *Toward Understanding Our Environment*, edited by J. McLeod, 13–20. San Diego, CA: Simulation Councils, 1991.

Lighton, G. "British Guiana-Coastlands and Interior." *Geography* 35, no. 3 (1950): 166–76.

Ramraj, Robert. "The Omai Disaster in Guyana." *Geographical Bulletin* 43, no. 2 (2001): 83–90.

Biogeography

Eden, Michael J. *The Savanna Ecosystem: North Rupununi, British Guiana.* Montreal, Quebec: McGill University, 1964.
————. "Savanna Vegetation in the Northern Rupununi, Guyana." *Journal of Tropical Geography* 30 (1970): 17–28.
Shaw, Earl B. "The Rupununi Savannas of British Guiana." *Journal of Geography* 39, no. 3 (1940): 89–104.
Thompson, S. "Range Expansion by Alien Weeds in the Coastal Farmlands of Guyana." *Journal of Biogeography* 15, no. 1 (1988): 109–18.

Climatology and Weather

Persaud, Chander. *Potential Evapotranspiration in Different Climatic Regions of Guyana.* Montreal, Quebec: McGill University, Department of Geography, Climate Research Series 11, 1977.
Ramraj, Robert. "Rainfall Distribution along Coastal Guyana, 1901–1980." *Geographical Bulletin* 38, no. 1 (1996): 29–40.
Shaw, Anthony B. "Analysis of the Rainfall Regimes on the Coastal Regions of Guyana." *Journal of Climatology* 7, no. 3 (1987): 291–302.

Geomorphology, Landforms, and Volcanism

Cukovic, Josip, and A. S. Trenhaile. "Ripple Wavelength in a Fine-Grained Coastal Environment, Guyana, South America." In *Proceedings, 1997 Canadian Coastal Conference, Guelph, Ontario, 21–24 May 1997*, 377–83. Guelph, Ontario: Canadian Coastal Science and Engineering Association, 1997.
Ebisemiju, Fola S. "Thresholds of Gully Erosion in a Laterite Terrain, Guyana." *Singapore Journal of Tropical Geography* 10, no. 2 (1989): 136–43.
Eden, Michael J. "Some Aspects of Weathering and Landforms in Guyana." *Zeitschrift fur Geomorphologie* 15, no. 2 (1971): 181–98.
Lakhan, V. C., and D. Pepper. "Relationships between Concavity and Convexity of a Coast and Erosion and Accretion Patterns." *Journal of Coastal Research* 13, no. 1 (1997): 226–32.
Shaw, P. "Cave Development on a Coranite Inselberg, Southern Rupununi Savannas, Guayana." *Zeitschrift fur Geomorphologie* 24, no. 1 (1980): 68–76.

Hydrology and Glaciology

Kressel, Richard H. "Slope Runoff and Denudation in the Rupununi Savanna, Guyana." *Journal of Tropical Geography* 44 (1977): 33–42.
Pelling, Mark. "Coast Runoff and Denudation in the Rupununi Savanna, Guyana." *Caribbean Geography* 7, no. 1 (1996): 3–22.

Soils

Jetten, V. G., et al. "Spatial Variability of Infiltration and Related Properties of Tropical Soils." *Earth Surface Processes and Landforms* 18, no. 6 (1993): 477–88.

Theses and Dissertations

Cukovic, Josip. "Ripple Morphology in a Sand Mud Environment: Guyana, South America." Master's thesis, University of Windsor, Ontario, 1996.

Ebisemiju, Fola S. "Slope Form and Gully Morphology in the Laterites of the Northern Rupununi Savanna, Guyana." Master's thesis, McGill University, Montreal, Quebec, 1969.

Frost, David. "The Climate of the Rupununi Savannas." Master's thesis, McGill University, Montreal, Quebec, 1966.

Hutchinson, I. "Tree Growth and Edaphic Controls in the Southern Rupununi Savanna, Guayana." Master's thesis, McGill University, Montreal, Quebec, 1970.

Kagends-Atwoki, C. B. "Weather Conditions and the Climate of the Rupununi, Guayana." Master's thesis, McGill University, Montreal, Quebec, 1969.

Kirchner, Christopher L. "Use of Remotely Sensed Radar to Assess Tropical Deforestation in Guyana." Master's thesis, University of Windsor, Ontario, 1996.

Prevedel, Lloyd M. "The Distribution of Longshore Currents along Guyana's Near-Shore Zone." Master's thesis, University of Windsor, Ontario, 1997.

Sinha, N. K. P. "Geomorphic Evolution of the Northern Rupununi Savanna, Guyana." PhD diss., McGill University, Montreal, Quebec, 1967.

Windapo, J. O. "Terrain Types and Their Air-Photo Characteristics, Northern Rupununi Savanna, Guyana." Master's thesis, McGill University, Montreal, Quebec, 1968.

POLITICAL GEOGRAPHY

General Works

Crist, Raymond E. "Jungle Geopolitics in Guyana: How a Communist Utopia That Ended in Massacre Came to Be Sited." *American Journal of Economics and Sociology* 40 (1981): 107–14.

Heinprin, Angelo. "Notes Upon the Schomburgh Line and the Guyana Boundary." *Bulletin of the Geographical Society of Philadelphia* 2, no. 1 (1896): 20–30.

Lakhan, V. C. "Perspectives on Constraints to Environmental Protection in Guyana." *Journal of Indo-Caribbean Research* 5, no. 1 (2001): 14–21.

Lakhan, V. C., A. S. Trenhaile, and P. D. LaValle. *Environmental Protection Efforts in Guyana: A Review*. Toronto, Ontario: Caribbean Research Group, Technical Publication 8, 1998.

Mistry, J., et al. "Exploring the Links between Natural Resource Use and Biophysical Status in the Waterways of the North Rupununi, Guyana." *Journal of Environmental Management* 72, no. 3 (2004): 117–31.

———. "Using a Systems Viability Approach to Evaluate Integrated Conservation and Development Projects Assessing the Impact of the North Rupununi Adaptive Management Process, Guyana." *Geographical Journal* 176, no. 3 (2010): 241–52.

Pelling, Mark. "Toward a Political Ecology of Urban Environmental Risk: The Case of Guyana." In *Political Ecology: An Integrative Approach to Geography and Environment-Development Studies*, edited by Karl S. Zimmerer and Thomas Bassett, 73–93. New York: Guilford, 2003.

Ramraj, Robert. "Guyana's Border Disputes with Venezuela and Surinam." *Geographical Bulletin* 44, no. 1 (2002): 51–59.

Williams, Patrick E. "Environmental Problems in Guyana's Hinterland: Some Policy Considerations." *Caribbean Geography* 9, no. 2 (2000): 121–35.

URBAN GEOGRAPHY

General Works

Edwards, R., et al. "Georgetown, Guyana." *Cities* 22, no. 6 (2005): 446–54.

Pelling, Mark. "Household and Community Level Hazard Management Options in a Flood Prone Urban Environment: Georgetown, Guyana." In *Land, Sea, and Human Effort in the*

Caribbean, edited by B. Ratter and W. Saler, 10:81–91. Hamburg, Germany: Beitrage zur Geographischen Regionalforschung in Lateinamerika, 1997.

———. "The Political Ecology of Floor Hazard in Urban Guyana." *Geoforum* 30, no. 3 (1999): 249–61.

Urban Social Geography

Heins, J. J. F. "Spatial Inequality in Guyana: A Case Study of Georgetown." *Tijdschrift voor Economische en Sociale Geographie* 69, nos. 1–2 (1978): 46–57.

Morphology and Neighborhoods

Stratchan, A. J. "Housing Patterns and Values in a Medium Sized Third World City: Georgetown, Guyana." *Tijdschrift voor Economische en Sociale Geographie* 72, no. 1 (1981): 40–46.

Urban Economic Geography

Lighton, G. "British Guiana-Georgetown and Its Trade." *Geography* 35, no. 4 (1950): 228–38.

Urban Environments

Pelling, Mark. "What Determines Vulnerability to Floods: A Case Study in Georgetown, Guyana." *Environment and Urbanization* 9, no. 1 (1997): 203–26.

Chapter Ten

Paraguay

GENERAL WORKS

Atlases and Graphic Presentations

Ferreira, Hernan M. *Atlas Paraguay: cartografia didactica*. Asuncion, Paraguya: Fausto Cultural Ediciones, 2000.

Books, Monographs, and Texts

Ferreira Gubetich, Hugo. *Geografia del Paraguay*. Asuncion, Paraguay: El Lector, 2002.
Hopkins, E. A., et al. *Paraguay*. New York: American Geographical Society, 1968.
Kleinpenning, J. M. G. *Man and Land in Paraguay*. Amsterdam, Netherlands: Centre for Latin American Research and Documentation, Latin American Studies 41, 1987.

Articles and Book Chapters

Crist, Raymond E., and Edward P. Leahy. "Paraguay." In *Focus on South America*, edited by Alice Taylor, 205–16. New York: Praeger, 1973.
Gade, Daniel W. "Paraguay 1975: Thinking Back on the Fieldwork Moment." *Journal of Latin American Geography* 5, no. 1 (2006): 31–46.
Gaignard, Romain. "Sous-developpement et desequilibres regionaux au Paraguay." *Revista Geografica* 69 (1968): 29–74.
Grubb, W. B. "The Paraguayan Chaco and Its Possible Future." *Geographical Journal* 54 (1919): 157–78.
Kleinpenning, J. M. G. "Rural Development Policy in Paraguay since 1960." *Travaux et Documents de Geographie Tropicale* (Bordeaux, France) 55 (1986): 71–90.
Kohlhepp, Gerd. "Colonizacieny desarrollo dependiente en el oriente Paraguayo." *Revista Geografica* 99 (1984): 5–33.
———. "Problems of Dependent Regional Development in Eastern Paraguay, with Special Reference to the Brazilian Influence in the Pioneer Zone of the Amambay Plateau." *Applied Geography and Development* 22 (1983): 7–45.
Naylor, Simon. "That Very Garden of South America: European Surveyors in Paraguay." *Singapore Journal of Tropical Geography* 21, no. 1 (2000): 48–62.
Sermet, J. "Le Paraguay." *Les Cahiers d'Outre Mer* 3 (1950): 28–63.

Theses and Dissertations

Gast, Marvin. "Areal Structure of Paraguay: Present and Past." Master's thesis, University of Chicago, 1950.

CULTURAL AND SOCIAL GEOGRAPHY

General Works

Zoomers, Elisabeth B., and J. M. G. Kleinpenning. "Livelihood and Urban-Rural Relations in Central Paraguay." *Tijdschrift voor Economische en Sociale Geographie* 87, no. 1 (1996): 161–74.

Ethnic, Social, and Population Geography

Nishikawa, Daijiro. "The Expansion of Japanese Immigration in Paraguay since the Second World War [in Japanese]." *Annals, Japan Association of Economic Geographers* 10 (1963): 11–38.

Nordenskjold, Erland. *An Ethnogeographical Analysis of the Material Culture of Two Indian Tribes in the Gran Chaco.* Goteberg, Sweden: Comparative Ethnogeographical Studies 1, 1919.

Reed, Richard K. "Developing the Mbaracayu Biosphere Reserve, Paraguay: Chiripa Indians and Sustainable Economies." *Yearbook, Conference of Latin Americanist Geographers* 16 (1990): 34–40.

Zoomers, Elisabeth B., and J. M. G. Kleinpenning. "Elites, the Rural Masses, and Land in Paraguay." *Revista Geografica* 111 (1990): 129–48.

Community and Settlement Studies

Kleinpenning, J. M. G. *The Integration and Colonization of the Paraguayan Chaco.* Nijmegen, Netherlands: Nijmeeges Geografische Cahiers 24, 1984.

Kleinpenning, J. M. G., and Elisabeth B. Zoomers. "Internal Colonisation as a Policy Instrument for Changing a Country's Rural System: The Example of Paraguay." *Tijdschrift voor Economische en Sociale Geographie* 79, no. 4 (1988): 257–65.

Kohlhepp, Gerd. "Colonizacieny desarrollo dependiente en el oriente Paraguayo." *Revista Geografica* 99 (1984): 5–33.

Theses and Dissertations

Keiderling, Wallace E. "The Japanese Immigration in Paraguay." Master's thesis, University of Florida, Gainesville, 1961.

Nayant, Nicole. "Spatio-Temporal Movement of Malaria in Paraguay." Master's thesis, University of Nebraska, Lincoln, 2011.

Stewart, Norman R. "Japanese Colonization in Eastern Paraguay: A Study in the Cultural Geography of Pioneer Agricultural Settlement." PhD diss., University of California, Los Angeles, 1963.

ECONOMIC GEOGRAPHY

General Works

Hecht, A. "The Geographic Dimension of the Changing World Economic Order: with Special Reference to Paraguay." *Die Erde* 109 (1978): 397–416.

Agriculture and Land Use

Richards, Peter D. "Soy, Cotton, and the Final Atlantic Forest Frontier." *Professional Geographer* 63, no. 3 (2011): 343–63.

Energy, Mining, Fishing, and Lumbering

Evans, P. T. "Designing Agroforestry Innovations to Increase Their Adaptability: A Case Study from Paraguay." *Journal of Rural Studies* 4, no. 1 (1988): 45–56.

Stunnenberg, Peter, and J. M. G. Kleinpenning. "The Role of Extrative Industries in the Process of Colonization: The Case of Quebracho Exploitation in the Gran Chaco." *Tijdschrift voor Economische en Sociale Geographie* 84, no. 3 (1993): 220–29.

HISTORICAL GEOGRAPHY

General Works

Cowie, H. "A Creole in Paris and a Spaniard in Paraguay: Geographies of Natural History in the Hispanic World (1750–1800)." *Journal of Latin American Geography* 10, no. 1 (2011): 175–98.

Hopkins, E. A., et al. *Paraguay, 1852 and 1968.* New York: American Geographical Society, Occasional Publication 2, 1968.

Kleinpenning, J. M. G. *The Integration and Colonisation of the Paraguayan Chaco.* Nijmegen, Netherlands: Geografisch en Planologisch Institut Katholicke, Universitet Nijmeegse, Geografische Cahiers 24, 1984.

———. *Paraguay, 1515–1870: A Thematic Geography of Its Development.* Madrid, Spain: Iberoamericana, 2003.

———. *Rural Paraguay, 1870–1932: A Geography of Progress, Plunder, and Poverty.* 2 vols. Amsterdam, Netherlands: CEDLA, 1992, 2009.

Theses and Dissertations

Owens, David J. "Historical Geography of the Indian Missions in the Jesuit Province of Paraguay (1609–1768)." PhD diss., University of Kansas, Lawrence, 1977.

PHYSICAL GEOGRAPHY

General Works

Kleinpenning, J. M. G. "Degradacion ambiental en America Latina: El caso de Paraguay." *Anales de Geografia de la Universidad Complutense* 9 (1989): 37–54.

Biogeography

Gade, Daniel. "Naturalization of Plant Aliens: The Volunteer Orange in Paraguay." *Journal of Biogeography* 3, no. 3 (1976): 269–80.
Lopez-Gonzales, Celia. "Ecological Zoogeography of the Bats of Paraguay." *Journal of Biogeography* 31, no. 1 (2004): 33–46.
Spichiger, R., et al. "Geographical Zonation in the Neotropics of Tree Species Characteristics of the Paraguay-Parana Basin." *Journal of Biogeography* 31, no. 9 (2004): 1489–501.

Geomorphology, Land Forms, and Volcanism

Carnier, Karl. "Paraguay: Versuch zu einer morphologischen Betrachtung seiner Landschaftsformen." *Geographischen Gesellshaft Jena, Mitt.* 29 (1911): 1–50.

POLITICAL GEOGRAPHY

General Works

Kleinpenning, J. M. G. "Rural Development Policy in Paraguay since 1960." *Tijdschrift voor Economische en Sociale Geographie* 75, no. 3 (1984): 154–76.
Kleinpenning, J. M. G., and Elisabeth B. Zoomers. "Elites, the Rural Masses, and Land in Paraguay: The Subordination of the Rural Masses to the Ruling Class." *Development and Change* 22, no. 2 (1991): 279–96.

URBAN GEOGRAPHY

General Works

Sargent, Charles S. "Paraguay." In *Latin American Urbanization: Historical Profiles of Major Cities,* edited by Gerald M. Greenfield, 427–45. Westport, CT: Greenwood, 1994.

Theses and Dissertations

Owens, David J. "Port of Asuncion, Paraguay." Master's thesis, Southern Illinois University, Edwardsville, 1967.

Chapter Eleven

Peru

GENERAL WORKS

Atlases and Graphic Presentations

Benavides, Margarita. *Atlas de communidades nativas ee la Selva Central*. Lima, Peru: Instituto del Bien Comun, 2006.

Bernex de Falen, N., and C. Caijo. *Atlas Provincial de Quispicanchi*. Lima, Peru: Universidad Catolica de Peru, Centro de Capacitacion Agri-Industrial Jesus Ubrero/Pontifica, 1997.

Johnson, G. R., and Raye R. Platt. *Peru from the Air*. New York: American Geographical Society, Special Publication 12, 1930.

Miller, O. M. "Note on the Map of the Huallaga Gorge between Huanuco and Muna." *Geographical Review* 19, no. 2 (1929): 293–95.

Orlove, Benjamin. "The Cultural and Social Context of Cartographic Representations in Peru." *Cartographica* 30, no. 1 (1993): 29–46.

Shippe, Robert. "The Great Wall of Peru and Other Aerial Photographic Studies by the Shippe-Johnson Peruvian Expedition." *Geographical Review* 22, no. 1 (1932): 1–29.

Books, Monographs, and Texts

Bowman, Isaiah. *The Andes of Southern Peru*. New York: American Geographical Society, 1916.

———. *Los Andes del sur de Peru: reconocimiento geografico a le largo de meridiano setenta y tres*. Arequipa, Peru: Editorial la Comena, 1938.

Articles and Book Chapters

Bernex de Falen, N. "Environmental Diversity, Socioeconomic Disparity, and Regional Development in Contemporary Peru." *GeoJournal* 11, no. 1 (1985): 15–28.

Cole, John P., and P. M. Mather. "Peru Province Level Factor Analysis." *Revista Geografica* 77 (1972): 7–32.

Denevan, William M. "The 1931 Shippe-Johnson Aerial Photography Expedition to Peru." *Geographical Review* 83, no. 3 (1993): 238–51.

———. "Fieldwork as Exploration: The Rio Heath Savannas of Southeastern Peru." *Geoscience and Man* 21 (1980): 157–63.

Gade, Daniel, and Mario Escobar. "Canyons of the Apurimas." *Explorers Journal* 50, no. 3 (1972): 135–40.

Mikus, Werner. "Problems Resulting from Misinterpretation of Natural Resources and Natural Hazards in Tropical Countries: The Example of Peru." *GeoJournal* 11, no. 1 (1986): 103–9.

Miller, O. M. "The 1927–1928 Peruvian Expedition of the American Geographical Society." *Geographical Review* 19, no. 1 (1929): 1–37.

Parro, Alberto A. "Census of Peru, 1940." *Geographical Review* 32, no. 1 (1942): 1–20.

Sanderman, Christopher. "The Northern Highway of Peru." *Geographical Journal* 110, nos. 3–4 (1945): 81–99.

Shippe, Robert. "Lost Valleys of Peru." *Geographical Review* 22 (1932): 562–81.

Snyder, David E. "The Carreter Marginal de la Selva: A Geographical Review and Appraisal." *Revista Geografica* 67 (1967): 87–100.

Verkooijen, J. "Micro-regional Integrated Rural Development in the Southern Andes of Peru." In *Successful Rural Development in Third World Countries*, edited by J. Hinderink and E. Szuk-Dabrowrecka, 108–28. Utrecht: Netherlands Geographical Studies, Band 67, 1988.

Yacher, Leon. "Peru: The 1981 Census." *Geography* 72, no. 1 (1987): 68–73.

Theses and Dissertations

Ankersheil, Otto B. "The Region of the Upper Mayo River in the Department of San Marten, Peru." Master's thesis, Northwestern University, Evanston, IL, 1960.

Canjoy-Azian, Walter A. "The Role of Remote Sensing and Geographic Information Systems for the Management of Resources in the Peruvian Amazon Basin." Master's thesis, University of Waterloo, Waterloo, Ontario, 1992.

Mueller, Gregory. "Image Processing of Landsat TM Data and a Geographic Information System Approach for the Mapping and Analysis of the Nazca Lines." Master's thesis, Murray State University, Murray, KY, 1990.

Todd, Milicent. "An Investigation of Geographic Controls in Peru." PhD diss., Radcliffe University, Cambridge, MA, 1923.

CULTURAL AND SOCIAL GEOGRAPHY

General Works

Bowman, Isaiah. "The Country of the Shepherds." *Geographical Review* 1 (1916): 419–42.

Collin Delavaud, Claude. "Plantes europeennes introduites au Perou." *Les Cahiers d'Outre Mer* 53 (2000): 111–28.

Defalen, Bernez. "Environmental Diversity, Socio-economic Disparity, and Regional Development in Contemporary Peru." *GeoJournal* 11, no. 1 (1985): 15–28.

Gade, Daniel. "Cultural Geography as a Research Agenda for Peru." *Yearbook, Conference of Latin Americanist Geographers* 14 (1988): 31–37.

———. "Environment and Disease in the Land Use and Settlement of Apurima Department, Peru." *Geoforum* 15 (1973): 37–46.

———. *Plants, Man, and the Land in the Vilcanota Valley of Peru.* Hague, Netherlands: W. Bunk BV, 1974.

Godfrey, Brian J. "Regional Depiction in Contemporary Film." *Geographical Review* 83, no. 4 (1993): 428–40.

Hays-Mitchell, Maureen. "Danger, Fulfillment, and Responsibility in a Violence-Plagued Society." *Geographical Review* 91, nos. 1–2 (2001): 311–21.

Hiraoka, Mario, and N. Hida. "Human Adaptation to the Changing Economy and Ecology on the Estuarine Floodplain of the Amazon Estuary." *Geographical Review of Japan* 71B, no. 1 (1998): 45–58.

Knapp, Gregory W. "Geografia Linguistica y Cultural del Peru." *Anthropologica Peru* 6 (1988): 285–308.

White, C. Langdon. "Huaucayo and Its Famous Indian Market in the Peruvian Andes." *Journal of Geography* 50, no. 1 (1951): 1–9.

Works, Martha A. "Development and Change in the Traditional Landscape of the Mayo Aguaruna, Eastern Peru." *Journal of Cultural Geography* 6, no. 1 (1985): 1–18.

Zimmerer, Karl S. *Changing Fortunes: Biodiversity and Peasant Livelihood in the Peruvian Andes*. Berkeley: University of California Press, 1996.

Ethnic, Social, and Population Geography

Bourricaud, F., and O. Dollfus. "La population peruvienne en 1961." *Les Cahiers d'Outre Mer* 16 (1963): 184–200.

Bromley, Rosemary D. F., and P. K. Mackie. "Child Experiences as Street Traders in Peru: Contributing to a Reappraisal for Working Children." *Children's Geographies* 7, no. 2 (2009): 141–58.

Bury, Jeffrey. "Mining Migrants: Transnational Mining and Migration Patterns in the Peruvian Andes." *Professional Geographer* 59, no. 3 (2007): 378–89.

Cole, John P. "Internal Migration in Peru." *Geography Review* 3, no. 1 (1989): 25–31.

Coomes, Oliver T. "Income Formation among Amazonian Peasant Households in Northeastern Peru: Empirical Observations and Implications for Market-Oriented Conservation." *Yearbook, Conference of Latin Americanist Geographers* 22 (1996): 54–64.

———. "State Credit Programs and the Peasantry under Populist Regimes: Lessons for the APRA Experience in the Peruvian Amazon." *World Development* 24, no. 8 (1996): 1333–46.

Dyer, Donald R. *Population and Elevation in Peru*. Evanston, IL: Northwestern University Department of Geography Studies in Geography 6, 1962.

———. "Population of the Quechua Region of Peru." *Geographical Review* 52, no. 3 (1962): 337–45.

Gade, Daniel. "Cultural-Geographical Implications of Social Disorganization: Crop Theft in Highland Peru." *Geographical Survey* 1, no. 3 (1971): 1–15.

Jenkins, Katy. "Practically Professionals: Grassroots Women as Local Experts: A Peruvian Case Study." *Political Geography* 27, no. 2 (2008): 139–59.

Kent, Robert B., and Aldo S. Ricci. "Evolucion y provision de servicios publicos distritales en la Sierra Central del Peru." *Revista Geografica* 99 (1984): 35–55.

Lausent-Herrera, Isabella. "La presencia japonesa en el eje Huanuco-Pucallpa entre 1918 y 1982." *Revista Geografica* 107 (1988): 93–118.

Momsen, Richard P. "The Isconahua Indians: A Study of Change and Diversity in the Peruvian Amazon." *Revista Geografica* 60 (1964): 59–82.

Myers, Sarah K. *Language Shift among Migrants to Lima, Peru*. Chicago: University of Chicago, Department of Geography, Research Paper 147, 1973.

Olson, Elizabeth. "Development, Transnational Religion, and the Power of Ideas in the High Provinces of Cusco, Peru." *Environment and Planning A* 38, no. 5 (2006): 885–902.

Rowe, John H. "The Distribution of Indians and Indian Languages in Peru." *Geographical Review* 37, no. 2 (1947): 202–15.

Skelton, Ronald. "The Evolution of Migration Patterns during Urbanization in Peru." *Geographical Review* 67, no. 3 (1977): 394–411.

Stewart, Norman R. "Migration and Settlement in the Peruvian Montana: The Apurimal Valley." *Geographical Review* 55, no. 2 (1965): 143–57.

White, C. Langdon. "Altitude: Its Role in the Life of Man in the High Peruvian Sierra." *Journal of Geography* 52 (1953): 361–74.

Community and Settlement Studies

Eidt, Robert C. "Pioneer Settlement in Eastern Peru." *Annals of the Association of American Geographers* 52, no. 2 (1962): 255–78.

Maos, Jacob O. "Regional Development and Land Settlement in Southern Peru." *Regional Science Review* 16 (1989): 85–100.

———. "Water Resource Development and Land Settlement in Southern Peru: The Majes Project." *GeoJournal* 11, no. 1 (1985): 69–78.

Salisbury, David S., et al. "Frontera Vivas or Dead Ends? The Impact of Military Settlement Projects in the Amazon Borderlands." *Journal of Latin American Geography* 9, no. 2 (2010): 49–71.

Schellerup, I. "Wayko-Lamas: A Quechuan Community in the Selva Alta of North Peru under Change." *Geografisk Tidsskrift*, Special Issue 1 (1999): 199–208.

Sjoholt, Peter. "Movement of the Colonization Frontier: A Dynamic Regional Development or a Reproduction of Underdevelopment? Some Findings from Peru, 1972–1981." *Norsk Geografisk Tidsskrift* 40, no. 2 (1986): 65–75.

Tourism and Recreation

Bury, Jeffrey. "New Geographies of Tourism in Peru: Nature-Based Tourism and Conservation in the Cordillera Huayhuash." *Tourism Geographies* 10, no. 3 (2008): 312–33.

Desforges, Luke. "State Tourism Institutions and Neo-liberal Development: A Case Study of Peru." *Tourism Geographies* 2, no. 2 (2000): 177–92.

Eiselen, Elizabeth. "A Tourist-Geographer Visits Iquitos, Peru." *Journal of Geography* 55 (1956): 176–82.

Grotzbach, Erwin. "Tourism in the Cordillera Blanca Region, Peru." *Revista Geografica* 133 (2003): 53–72.

Hill, Jennifer, and Ross Hill. "Ecotourism in Amazonian Peru: United Tourism, Conservation, and Community Development." *Geography* 96, no. 2 (2011): 75–85.

O'Hare, Greg, and Hazel Barrett. "The Destination Life Cycle: International Tourism in Peru." *Scottish Geographical Magazine* 113, no. 2 (1997): 66–73.

———. "Regional Inequalities in the Peruvian Tourist Industry." *Geographical Journal* 165, no. 1 (1999): 47–61.

Place Names

Kus, James S. "Changing Names on Peruvian Trucks." *Names* 42, no. 4 (1994): 235–68.

Theses and Dissertations

Alperin, Juan P. "A Spatio-temporal Model for the Evaluation of Education Quality in Peru." Master's thesis, University of Waterloo, Waterloo, Ontario, 2008.

Cole, Christina. "Examining the Spatial Patterns of Dengue Risk Factors and Serological Conversions: A Case Study in Iquitos, Peru." Master's thesis, San Diego State University, San Diego, 2005.

Fernandez-Achenbach, Sandra. "A Perceptual Assessment of Ecotourism Lodges in Peruvian Amazon." Master's thesis, California State University, Northridge, 2009.

Gonzalez, Fernando L. "Fences, Fields, and Fodder: Enclosures in Lari, Valle del Colca, Southern Peru." Master's thesis, University of Wisconsin, Madison, 1996.

Gray, Kenneth. "A Spatial Analysis of Dengue Fever Risk Determinants: A Case Study in Iquitos, Peru." Master's thesis, San Diego State University, San Diego, 2001.

Leahy, Michael. "Spatial Analysis of Education Quality in Peru." Master's thesis, University of Waterloo, Waterloo, Ontario, 2005.

Montoya, Mariana. "How Access, Values, and History Shape the Sustainability of a Social-Ecological Systems: The Case of the Kandozi Indigenous Group of Peru." PhD diss., University of Texas, Austin, 2010.

Pendleton, Jesse L. "Structural Changes of the Peruvian Population 1940–1955." Master's thesis, Clark University, Worcester, Massachusetts, 1961.

Savard, Michael. "Contribution a l'etude de l'integration des paysanneries du 'tiers-monde au systeme alimentaire mondial': l'analyse geographique structurale de l'espace caja marquien (Perou)." Master's thesis, Laval University, Ste. Foy, Quebec, 1990.

Schuh, Janet C. "People and Place in Transition: Pangoa, Upper Central Montana of Peru." PhD diss., University of California, Berkeley, 1977.

Skelton, Ronald. "Migration in a Peasant Society: The Example of Cuzco, Peru." PhD diss., University of Toronto, Toronto, 1974.

Walton, Nyle K. "Human Spatial Organization in an Andean Valley: The Calleon de Huaylas, Peru." PhD diss., University of Georgia, Athens, 1974.

ECONOMIC GEOGRAPHY

General Works

Cardozo, Mario. "Economic Displacement and Local Attitude towards Protected Area Establishment in the Peruvian Amazon." *Geoforum* 42, no. 5 (2011): 603–14.

Cole, John P. "The Economic Possibilities of the Selva Region of Peru." *Tijdschrfit voor Economische and Sociale Geographie* 47, no. 19 (1956): 237–42.

Coomes, Oliver T. "Rain Forest 'Conservation-through-Use'? Chambira Fibre Extraction and Handicraft Production in a Land-Constrained Community, Peruvian Amazon." *Biodiversity and Conservation* 13, no. 2 (2004): 351–60.

Golte, W. "Die Angange der Bewasserungs-wirtschaft imvorspanischen Peru und das Chavin-Problem." *Colloquium Geraficum* 16 (1983): 243–71.

Hardy, Osgood. "Cuzco and Apurimac: A Report on the Commercial and Industrial Outlook of South Central Peru." *Bulletin of the American Geographical Society* 46, no. 7 (1914): 500–512.

Hiraoka, Mario. "Cash Cropping, Wage Labor, and Urbanward Migrations: Changing Subsistence in the Peruvian Amazon." *Studies in Third World Societies* 32 (1985): 199–242.

———. "Riberernos' Changing Economic Patterns in the Peruvian Amazon." *Journal of Cultural Geography* 9, no. 2 (1989): 103–20.

Kent, Robert B., and J. I. Rimapachin. "Rural Public Works Construction in the Andes of Northern Peru." *Third World Planning Review* 16, no. 4 (1994): 358–74.

Morisset, Jean. "The Department of Puno as a Territory to Be Developed in Southern Peru." *Revista Geografica* 79 (1973): 11–40.

Romero, Emilio. *Geografica economica del Peru*. Lima, Peru, 1961.

Takasaki, Yoshito, et al. "Amazonian Peasants, Rain Forest Use, and Income Generation: The Role of Wealth and Geographical Factors." *Society and Natural Resources* 14, no. 4 (2001): 291–308.

———. "Rural Rapid Appraisal in Humid Tropical Forests: An Asset Possession-Based Approach and Validation Methods for Wealth Assessment among Forest Peasant Households." *World Development* 28, no. 1 (2000): 1961–77.

Agriculture and Land Use

Akojima, I. "Cultivated Fields and Water in the Pampa of Nasca and Its Environs, Peru [in Japanese]." *Kigan Chirigaku/Quarterly Journal of Geography* 62 (2010): 223–28.

Amiran, David H. K. "El desierto de Sechura, Peru: Problems of Agricultural Use of Deserts." *Revista Geografica* 72 (1970): 7–12.

Ban, Natalie, and Oliver T. Coomes. "Home Gardens in Amazonian Peru: Diversity and Exchange of Planting Material." *Geographical Review* 94, no. 3 (2004): 348–67.

Barber, Mary L. "National Parks, Conservation, and Agrarian Reform in Peru." *Geographical Review* 70, no. 1 (1980): 1–18.

Bourricaud, F. "Questions agraires du Sud du Perou." *Les Cahiers d'Outre Mer* 13 (1960): 377–87.

Cardick, A. "Native Agriculture in the Highlands of the Peruvian Andes." *National Geographic Research* 3, no. 1 (1987): 22–39.

Chaleard, J. L., and E. Mesclier. "New Actors, New Products, and Expansion of the Land Market in Northern Peru [in French]." *Annales de Geographie* 119, no. 6 (2010): 678–96.

Coomes, Oliver T. "Traditional Peasant Agriculture along a Black Water River of the Peruvian Amazon." *Revista Geografica* 124 (1998): 33–56.

Coomes, Oliver T., et al. "Tropical Forests and Shifting Cultivation: Secondary Forest Fallow Dynamics among Traditional Farmers of the Peruvian Amazon." *Ecological Economics* 32, no. 1 (2000): 109–29.

Delavaud, Collin. "Plantes eruopeennes introduites au Perou." *Les Cahiers d'Outre Mer* 63, nos. 209–210 (2000): 111–28.

Denevan, William M. "Measurement of Abandoned Terracing from Air Photos, Colca Valley, Peru." *Yearbook, Conference of Latin Americanist Geographers* 14 (1988): 20–30.

Dollfus, O. "Quelques donnees sur l'agriculture du Peru." *Les Cahiers d'Outre Mer* 12 (1959): 112–16.

Eidt, Robert C. "Economic Features of Land Opening in the Peruvian Montana." *Professional Geographer* 18 (1966): 147–50.

Feige, Wolfgang. "Irrigation Projects in the Andes Highlands of Peru." *Applied Geography and Development* 34 (1989): 23–45.

Hiraoka, Mario. "Agricultural Systems on the Floodplains of the Peruvian Amazon." In *Fragile Lands of Latin America: Strategies for Sustainable Development*, edited by John O. Browder, 75–101. Boulder, CO: Westview, 1989.

———. "Changing Floodplain Livelihood Patterns in the Peruvian Amazon." *Tsukuba Studies in Human Geography* 9 (1985): 243–75.

———. "Floodplain Farming in the Peruvian Amazon." *Geographical Review of Japan* 58B, no. 1 (1985): 1–23.

———. "Land Use Change in the Amazon Estuary." *Global Environmental Change A* 5, no. 4 (1995): 323–36.

———. "Zonation of Mestizo Riverine Farming Systems in Northeast Peru." *National Geographical Research* 2 (1986): 354–71.

Hodl, Walter, and Jurg Gaschi. "Indian Agriculture as Exemplified by a Secaya Village on the Rio Yubineto in Peru." *Applied Geography and Development* 20 (1982): 20-31.

Holmes, Roland C. *Irrigation in Southern Peru: The Chili Basin*. Chicago: University of Chicago, Department of Geography, Research Paper 212, 1986.

Kariya, Yoshihiko, et al. "Geomorphology and Pastoral-Agricultural Land Use in Cotahuasiand Puica, Souathern Peruvian Andes." *Geographical Review of Japan* 78, no. 12 (2000): 842–52.

Kent, Robert B. "The African Honeybee in Peru: An Insect Invader and Its Impact on Beekeeping." *Applied Geography* 9, no. 4 (1989): 237–57.

Kus, James S. "Chavimochi: A Peruvian Irrigation Project." *Yearbook, Conference of Latin Americanist Geographers* 13 (1987): 19–24.

———. "The Sugar Cane Industry of the Chicoma Valley, Peru." *Revista Geografica* 109 (1989): 57–72.

Milstead, Harley P. "Distribution of Crops in Peru." *Economic Geography* 4, no. 1 (1928): 88–106.

Rosenfeld, Arthur, and Clarence F. Jones. "The Cotton Industry of Peru." *Economic Geography* 3, no. 4 (1927): 507–23.

Salonen, Maria, et al. "Critical Distances: Comparing Measures of Spatial Accessibility in the Riverine Landscapes of Peruvian Amazonia." *Applied Geography* 32, no. 2 (2012): 501–13.

Scott, Geoffrey A. J. "Effects of Shifting Cultivation in the Gran Pajonal, Eastern Peru." *Proceedings of the Association of American Geographers* 6 (1974): 58–61.

———. *Grassland Development in the Gran Pajonal of Eastern Peru*. Honolulu, Hawaii: University of Hawaii, Monograph 1, 1979.

Silvertsen, A., and A. Lundberg. "Farming Practices and Environmental Problems in an Arid Landscape: A Case Study from the Region of Lambayeque, Peru." *Geografiska Annaler* 78B, no. 3 (1996): 147–62.

Smith, Clifford T. "Agriculture and Settlement in Peru." *Geographical Journal* 126, no. 4 (1960): 397–412.

———. "Location and Agrarian Reform: The Peruvian Experience." *Proceedings, Conference of Latin Americanist Geographers* 5 (1974): 141–53.

Sommers, Marjorie S. "Wheat Producing Areas in Peru." *Economic Geography* 25, no. 1 (1949): 13–16.

Tajima, Hisashi. "Agriculture of the Chancay Valley in Peru [in Japanese]." *Annals, Japan Association of Economic Geographers* 10 (1963): 1–10.

Treacy, John M. "Agricultural Terraces in Peru's Colca Valley: Promises and Problems of an Ancient Technology." In *Fragile Lands of Latin America: Strategies for Sustainable Development*, edited by John O. Browder, 209–29. Boulder, CO: Westview, 1989.

———. "Building and Rebuilding Agricultural Terraces in the Colca Valley of Peru." *Yearbook, Conference of Latin Americanist Geographers* 13 (1987): 51–57.

Wernke, Steven A. "A Reduced Landscape: Toward a Multi-causal Understanding of Historic Period Agricultural Deintensification in Highland Peru." *Journal of Latin American Geography* 9, no. 3 (2010): 51–83.

White, Stuart. "Relations of Subsistence to the Vegetation Mosaic of Vilcabamba, Southern Peruvian Andes." *Yearbook, Conference of Latin Americanist Geographers* 11 (1985): 5–12.

Williams, Lyndon S. "Land Use Intensity and Farm Size in Highland Cuzco, Peru." *Journal of Developing Areas* 11, no. 2 (1977): 185–204.

Works, Martha A. "Aguarana Agriculture in Eastern Peru." *Geographical Review* 77, no. 3 (1987): 343–58.

Zimmerer, Karl S. "Labor Shortages and Crop Diversity in the Southern Peruvian Sierra." *Geographical Review* 81, no. 4 (1991): 414–32.

———. "The Loss and Maintenance of Native Crops in Mountain Agriculture." *GeoJournal* 27, no. 1 (1992): 61–72.

———. "Managing Diversity in Potato and Maize Fields of the Peruvian Andes." *Journal of Ethnobiology* 11, no. 1 (1991): 23–49.

———. "Wetland Production and Smallholder Persistence: Agricultural Change in a Highland Peruvian Region." *Annals of the Association of American Geographers* 81, no. 3 (1991): 443–63.

Commerce and Trade

Hays-Mitchell, Maureen. "From Survivor to Entrepreneur: Gendered Dimensions of Microenterprise Development in Peru." *Environment and Planning A* 31, no. 2 (1999): 251–72.

———. "The Ties That Bind: Informal and Formal Sector Linkages in Street Vending: The Case of Peru's Ambulantes." *Environment and Planning A* 25, no. 8 (1993): 1085–102.

Jones, Clarence F. "The Commercial Growth of Peru." *Economic Geography* 3, no. 1 (1927): 23–49.

Manufacturing and Labor Studies

Lattrie, Nina. "Negotiating Femininity: Women and Representation in Emergency Employment in Peru." *Gender, Place, and Culture* 4, no. 2 (1997): 235–52.

———. "State-Based Work Programmes and the Regendering of Work in Peru: Negotiating Femininity in the Provinces." *Environment and Planning A* 31, no. 2 (1999): 229–50.

Rosner, Waltrand. "Migration and the Development of an Industrial District: Footwear Manufacturing in El Porvenir, Trujillo, Peru." *Yearbook, Conference of Latin Americanist Geographers* 23 (1997): 107–18.

White, C. Langdon, and C. Chenkin. "Peru Moves onto the Iron and Steel Map of the Western Hemisphere." *Journal of Inter-American Studies* 1 (1959): 377–86.

192   *Chapter 11*

Energy, Mining, Fishing, and Lumbering

Bury, Jeffrey. "Livelihoods in Transition: Transnational Gold Mining Operations and Local Changes in Cajamarca, Peru." *Geographical Journal* 170, no. 1 (2004): 78–91.

———. "Mining Mountains: Neoliberalism, Land Tenure, Livelihoods, and the New Peruvian Mining Industry in Cajamarca." *Environment and Planning A* 37, no. 2 (2005): 221–40.

Coomes, Oliver T. "Rain Forest 'Conservation through Use'? Chambira Palm Fibre Extraction and Handicraft Production in a Land-Constrained Community, Peruvian Amazon." *Biodiversity and Conservation* 13 (2004): 351–60.

Coomes, Oliver T., and G. J. Burt. "Peasant Charcoal Production in the Peruvian Amazon: Rain Forest Use and Economic Reliance." *Forest Ecology and Management* 140 (2001): 39–50.

Cordova-Aguilar, H. "Firewood Use and the Effect on the Eco-system: A Case Study of the Sierra of Piura, Northwestern Peru." *GeoJournal* 26, no. 3 (1992): 297–310.

Coull, James R. "The Development of the Fishing Industry in Peru." *Geography* 59, no. 4 (1974): 322–32.

Craig, Alan K. "Placer Gold in Eastern Peru: The Great Strike of 1942." *Revista Geografica* 79 (1973): 117–28.

Denevan, William M. "Young Managed Fallows at Brillo Nuevo." In *Swidden-Fallow Agroforestry in the Peruvian Amazon*, edited by William M. Denevan and C. Padoch, 8–46. New York: New York Botanical Gardens, Advances in Economic Botany 5, 1988.

Denevan, William M., et al. "Indigenous Agroforestry in the Peruvian Amazon: Bora Indian Management of Swidden Fallows." *Interciencia* 9 (1984): 346–57.

———. "Indigenous Agroforestry in the Peruvian Amazon: Bora Indian Management of Swidden Fallows." In *Change in the Amazon Basin: Man's Impact on Forests and Rivers*, edited by John Hemming, 137–55. Manchester, England: Manchester University Press, 1985.

Denevan, William M., and C. Padoch, eds. *Swidden-Fallow Agroforestry in the Peruvian Amazon*. New York: New York Botanical Gardens, Advanced Economic Botany 5, 1988.

Echavarria, Fernando. "Cuatificacion de la Deforestation en el Valle del Huallaga, Peru." *Revista Geografica* 114 (1991): 37–53.

Ektvedt, Tone M. "Firewood Consumption amongst Poor Inhabitants in a Semiarid Tropical Forest: A Case Study from Piura, Northern Peru." *Norsk Geografisk Tidsskrift* 65, no. 1 (2011): 28–41.

Fiedler, Reginald H. "The Peruvian Fisheries." *Geographical Review* 34 (1944): 96–119.

Hoy, Don R., and Sten A. Taube. "Power Resources of Peru." *Geographical Review* 43, no. 4 (1963): 580–94.

Hoy, Harri E. "Mahogony Industry of Peru." *Economic Geography* 22, no. 1 (1946): 1–13.

Kallioia, Risto, and Pedro Flores. "Brazil Nut Harvesting in Peruvian Amazonia from the Perspective of Ecosystem Services." *Fennia* 189, no. 2 (2011): 1–13.

Maennling, C. "Iuterne Formen und Folgen von aussen induszierter Entwicklung im peruanischen Amazonasgebiet: Der Goldboom in Madre de Dios." *Zeitschrift fur Wirthschaftsgeographie* 31, nos. 3–4 (1987): 230–41.

Moreau, M. A., and Oliver T. Coomes. "Aquarium Fish Exploitation in Western Amazonia: Conservation Issues in Peru." *Environmental Conservation* 34, no. 1 (2007): 12–22.

Naughton-Torres, Lisa. "Wild Animals in the Garden: Conserving Wildlife in Amazonian Agroecosystems." *Annals of the American Geographical Association* 92, no. 3 (2002): 488–506.

Staver, Charles P., et al. "Land Resources Management and Forest Conservation in Central Amazonian Peru: Regional, Community, and Fari-Level Approaches among Native Peoples." *Mountain Research and Development* 14, no. 2 (1994): 147–57.

White, Stuart. "Cedar and Mahogany Logging in Eastern Peru." *Geographical Review* 68, no. 3 (1978): 394–416.

Young, Kenneth R. "Threats to Biological Diversity Caused by Coca-Cocaine Deforestation in Peru." *Environmental Conservation* 23, no. 1 (1996): 7–15.

Transportation and Communications

Cole, John P. "Ports and Hinterlands in Peru." *Revista Geografica* 49 (1958): 119–28.
MacPhail, Donald. "Peruvian Avenues of Penetration into Amazonia." *Geographical Review* 34, no. 1 (1944): 1–35.
Mikus, Werner. "Regional Development of Transport Systems in the Peruvian Highlands." *Basler Beitrage zur Geographie* 26 (1978).

Theses and Dissertations

Bell, Martha G. "Negotiating the Pottery Exchange Landscape: Handicraft Production, Rural Trade Networks, and Agriculture in Piura, Peru." Master's thesis, University of Wisconsin, Madison, 2007.
Cohalan, Jean-Michel. "River Trading in the Peruvian Amazon: Assessing the Impact of Market Access on Livelihoods." Master's thesis, McGill University, Montreal, Quebec, 2007.
Crook, Carolyn. "Biodiversity Prospects and Agreements: Evaluating Their Economic and Conservation Benefits in Costa Rica and Peru." PhD diss., University of Toronto, Toronto, Ontario, 2001.
Dickey, Knowles B. "The Problems of Peru as an Economically Underdeveloped Country." Master's thesis, Columbia University, New York, 1950.
Drewes, Wolfram D. "The Economic Development of the Western Montana of Central Peru as Related to Transportation." PhD diss., Syracuse University, Syracuse, NY, 1967.
Espeland, Erika. "Monitoring Montane Tropical Forest Land Cover and Land Use Change in the Eastern Peruvian Andes Using Landsat Imagery." Master's thesis, Central Michigan University, Mt. Pleasant, 2009.
Hays-Mitchell, Maureen. "Streetvending in Provincial Peru: A Geographical Analysis of an Informal Economic Activity." PhD diss., Syracuse University, Syracuse, NY, 1990.
Hemley, Matthew. "Frontiers of Capital: Mining, Mobilization, and Resource Governance in Andean Peru." PhD diss., Syracuse University, Syracuse, NY, 2010.
Holmes, Roland C. "Water Use in Southern Peru: The Chili Basin." PhD diss., University of Chicago, Chicago, 1979.
Houlahan, Margaret. "The Limitation of Productive Land in Peru." Master's thesis, Columbia University, New York, 1937.
Kus, James S. "Selected Aspects of Irrigated Agriculture in the Chimu Heartland, Peru." PhD diss., University of California, Los Angeles, 1972.
Lipton, Jennifer K. "Human Dimensions of Conservation, Land Use, and Climate Change in Huascaran National Park, Peru." PhD diss., University of Texas, Austin, 2008.
McCann, Joseph M. "Extraction and Depletion of Fruits and Fibers in Peruvian Amazonia: A Coevolutionary Perspective." Master's thesis, University of Wisconsin, Madison, 1993.
Penn, J. W. "Another Boom for Amazonia? Socioeconomic and Environmental Implications of the New Camu Camu Industry in Peru." PhD diss., University of Florida, Gainesville, 2004.
Savage, Barbara C. "The Changing Fishery of Peru: An Example of Changing Resource Evaluation." Master's thesis, University of California, Berkeley, 1970.
Toth, Julius A. "Economic Geography of the Iquitos Region of Peru." Master's thesis, University of Colorado, Boulder, 1958.
Treacy, John M. "The Fields of Coporaque: Agricultural Terracing and Water Management in the Colca Valley, Arequipa, Peru." PhD diss., University of Wisconsin, Madison, 1989.
Webber, Ellen R. "Cows in the Colca: Household Cattle Raising in Achoma, Peru." Master's thesis, University of Wisconsin, Madison, 1993.
Williams, Lyndon S. "Land Use Intensity and Farm Size: Traditional Agriculture in Cuzco, Peru." PhD diss., University of Kansas, Lawrence, 1973.

HISTORICAL GEOGRAPHY

General Works

Bowman, Isaiah. "A Buried Wall at Cuzco and Its Relation to the Question of a Pre-Inca Race." *American Journal of Science* 34 (1912): 497–509.

Carey, Mark, et al. "Unintended Effects of Technology on Climate Change Adaptation: An Historical Analysis of Water Conflicts Below Andean Glaciers." *Journal of Historical Geography* 38, no. 2 (2012): 181–91.

Chang, Tung Ning. "Diffusion of African Slave Labor to the Coastal Sugarcane Plantations of Colonial Peru." *Geoscience and Man* 24 (1983): 21–30.

Coleman, Katharine. "Provincial Urban Problems: Trujillo, Peru, 1600–1784." In *Social Fabric and Spatial Structure in Colonial Latin America*, edited by David J. Robinson, 369–408. Syracuse, NY: Syracuse University, Department of Geography, 1979

Craig, Alan K. "Exploration of Eastern Peru by the Junta de Vias Fluviales." *Revista Geografica* 90 (1979): 199–212.

Denevan, William M. *The Cultural Ecology, Archeology, and History of Terracing and Terrace Abandonment in the Colca Valley of Southern Peru.* Madison: University of Wisconsin, Department of Geography, Technical Report to the National Science Foundation and National Geographic Society, 1986.

———. "Terrace Abandonment in the Colca Valley, Peru." In *The Ecology and Archaeology of Prehistoric Agricultural Fields in the Central Andes*, edited by William M. Denevan, 1–43. Oxford, England: British Archaelogical Reports, International Series 359, 1987.

Gade, Daniel, and Mairo Excobar. "Village Settlement and the Colonial Legacy in Southern Peru." *Geographical Review* 72, no. 4 (1982): 430–49.

Hardy, Osgood. "Cuzco and Apurimac." *Bulletin of the American Geographical Society* 46, no. 7 (1914): 500–512.

Knapp, Gregory W. "Prehistoric Flood Management on the Peruvian Coast: Reinterpreting the Sunken Fields of Chilca." *American Antiquity* 47 (1982): 144–54.

Kus, James S. "The Chicama-Moche Canal: Failure or Success? An Alternative Explanation for an Incomplete Canal." *American Antiquity* 49, no. 2 (1984): 408–15.

———. "Irrigation and Urbanization in Pre-Hispanic Peru: The Moche Valley." *Yearbook, Association of Pacific Coast Geographers* 36 (1974): 45–56.

LeMoine, G., and J. S. Raymond. "Leishmoniasis and Inca Settlement in the Peruvian Jungle." *Journal of Historical Geography* 13, no. 2 (1987): 113–29.

Park, Chris C. "Water, Resources, and Irrigation in Pre-Hispanic Peru." *Geographical Journal* 149, no. 2 (1983): 153–66.

Parsons, James J., and Norbert Psuty. "Sunken Fields and Prehistoric Subsistence on the Peruvian Coast." *American Antiquity* 40 (1975): 259–82.

Pozorski, Thomas, et al. "Pre-Hispanic Ridged Fields of the Casma Valley, Peru." *Geographical Review* 73, no. 4 (1983): 407–16.

Scott, Heidi V. "The Contested Spaces of the Subterranem: Colonial Governmentality, Mining, and the Nita in Early Spanish Peru." *Journal of Latin American Geography* 11, Special Issue (2012): 7–33.

———. "Contested Territories: Areas of Geographical Knowledge in Early Colonial Peru." *Journal of Historical Geography* 29, no. 2 (2003): 166–88.

———. "Rethinking Landscape and Colonialism in the Context of Early Spanish Peru." *Environment and Planning D* 24, no. 4 (2006): 481–96.

Smith, Clifford T. "Patterns of Urban and Regional Development in Peru on the Eve of the Pacific War." In *Region and Class in Modern Peruvian History*, edited by R. Miller, 77–103. Liverpool, England: University of Liverpool, Institute of Latin American Studies, Monograph 14, 1987.

Theses and Dissertations

Gade, Daniel. "Plant Use and Folk Agriculture in the Vilcanota Valley of Peru: A Cultural-Historical Geography of Plant Resources." PhD diss., University of Wisconsin, Madison, 1967.

Guitierrez, Nicanor F. "400 Anos en la Colonizacion del Nor-Oriente Peruano." Master's thesis, University of Texas, Austin, 1967.

Knapp, Gregory W. "The Sunken Fields of Chilca: Horticulture, Microenvironment, and History in the Peruvian Coastal Desert." Master's thesis, University of Wisconsin, Madison, 1979.

Kus, James S. "A Historical Geography of Irrigated Agriculture in the Chicama Valley, Peru." Master's thesis, Michigan State University, East Lansing, 1967.

PHYSICAL GEOGRAPHY

General Works

Delavaud, Claude C. "Sobre las inundaciones catastroficas del norte del Peru El Nino: Mitos y realidades." *Revista Geografica* 101 (1985): 133–39.

Martinez, Teodoro. "El Peru y los desartres naturales en la historia: De la inquietude cientifica." *Revista Geografica,* 147 (2010): 63–94.

Rodrigues, Efrain. "Consideraciones en Torno de la tierra y el hombre del Peru." *Revista Geografica* 42 (1955): 133–50.

Sears, Alfre. "The Coast Desert of Peru." *Journal of the American Geographical Society of New York* 27, no. 3 (1895): 256–71.

Whitbeck, R. H. "The Coast Waters and the Coast Lands of Peru." *Journal of Geography* 25 (1926): 41–52.

Biogeography

Chrostowski, Marshall, and William M. Denevan. *The Biogeography of a Savanna Landscape: The Gran Pajonal of Eastern Peru.* Montreal, Quebec: McGill University, Savanna Research Series 16, 1970.

Craig, Alan K., and Norbert Psuty. "Paleoecology of Shell Mounds at Oruma, Peru." *Geographical Review* 61 (1971): 125–32.

Dickinson, Joshua C. "The Eucalypt in the Sierra of Southern Peru." *Annals of the Association of American Geographers* 59, no. 2 (1969): 294–307.

Dillon, M., et al. "Floristic Inventory and Biogeographic Analysis of Montane Forests in NW Peru." In *Biodiversity and Conservation of Neotropical Montane Forests,* edited by J. P. Churchill, 251–69. New York: New York Botanical Gardens, 1995.

Ektvedt, Tone M., et al. "Land-Cover Changes during the Past Fifty years in the Semi-arid Tropical Forest Region of Northern Peru." *Erdkunde* 66, no. 1 (2012): 57–76.

El Instituto Nacional de Planification. "Aspectos Geograficas de la selva alta Peruara que sera recarrida par la Corretera Marginal." *Revista Geografica* 67 (1965): 69–82.

Feeley, K. J., et al. "Upslope Migration of Andean Trees." *Journal of Biogeography* 38, no. 4 (2011): 783–91.

Gade, Daniel. *Plants, Man, and the Land in the Vilcanota Valley of Peru.* The Hague, Netherlands: W. Junk BV, Biogeographica 6, 1975.

Galan de Mera, A., et al. "Phytogeographical Sectoring of the Peruvian Coast." *Global Ecology and Biogeography* 6, no. 5 (1997): 349–68.

Graham, Gary L. "Bats versus Birds: Comparisons among Peruvian Vertegrate Faunas along an Elevational Gradient." *Journal of Biogeography* 17, no. 6 (1990): 657–68.

Kristiansen, Thea, et al. "Environment versus Dispersal in the Assembly of Western Amazonian Palm Communities." *Journal of Biogeography* 39, no. 7 (2012): 1318–22.

Moreau, M. A., and Oliver T. Coomes. "Potential Threat of the International Aquarium Fish Trade to Silver Arawana (*Osteoglossum bicirrhosum*) in the Peruvian Amazon." *Oryx* 40, no. 2 (2006): 1–9.

Oka, Shuichi, and Hajine Ogawa. "The Distribution of Lomas Vegetation and Its Climatic Environments along the Pacific Coast of Peru." *Geographical Reports of Tokyo Metropolitan University* 19 (1984): 113–26.

Ouhakka, Maarit, et al. "River Types and Site Evolution and Successful Vegetation Patterns in Peruvian Amazonia." *Journal of Biogeography* 19, no. 6 (1992): 651–66.

Oyama, Shuichi. "Ecology and Wildlife Conservation of Vicuna in Peruvian Andes." *Geographical Reports of Tokyo Metropolitan University* 41 (2006): 27–44.

Penn, J. W. "The Domestication of Camu Camu (*Myrciaria dubia*): A Tree Planning Programme in the Peruvian Amazon." *Forest, Trees, and Livelihoods* 16, no. 1 (2006): 85–101.

Reese, Carl A., and Kam Bui Liu. "Pollen Dispersal and Deposition on the Quelecaya Ice Cap, Peru." *Physical Geography* 23, no. 1 (2002): 44–58.

Richter, Michael, and M. Ise. "Monitoring Plant Development after El Nino 1997–98 in Northwestern Peru." *Erdkunde* 59, no. 2 (2005): 136–55.

Roberts, Jennifer, et al. "Rapid Diversification of Colouration among Populations of a Poison Frog Isolated on Sky Peninsulas in the Central Cordilleras of Peru." *Journal of Biogeography* 34, no. 3 (2007): 417–26.

Rodriguez, R., et al. "El Nino Events Recorded in Dry-Forest Species of the Lowlands of Northwest Peru." *Dendrochonologia* 22 (2005): 181–86.

Sabelli, Andrea. "A New Solution to a Persistent Problem: Addressing Tropical Deforestation with Carbon Forest Offset Projects." *Journal of Latin American Geography* 10, no. 1 (2011): 109–30.

Scott, Geoffrey A. J. "The Role of Fire in the Creation and Maintenance of Savanna in the Montana of Peru." *Journal of Biogeography* 4, no. 2 (1977): 143–68.

Silva, D. "Observations on the Diversity and Distribution of the Spiders of Peruvian Montane Forests." In *Biogeografia, Ecologia, y Conservacion del Bosque Montano en el Peru*, edited by Kenneth R. Young and N. Valencia, 31–37. Lima, Peru: Universidad Nacional Major de San Marcos, Memoria de Museo de Historia Natural UNMSM 21, 1992.

Valencia, B. G., et al. "From Ice Age to Modern: A Record of Landscape Change in an Andean Cloud Forest." *Journal of Biogeography* 37, no. 9 (2010): 1637–47.

Weberbauer, A. "Phytogeography of the Peruvian Andes." In *Flora of Peru*, edited by J. F. Macbride, 1:18–31. Chicago: Field Museum of Natual History, 1932.

Young, Kenneth R. "Biogeography of the Montane Forest Zone of the Eastern Slopes of Peru." In *Biogeografia, Ecologia, y Conservacion del Bosque Montanos el Peru*, edited by Kenneth R. Young and N. Valencia, 119–40. Lima, Peru: Unviersidad Nacional Mayor de San Marcos, Memoria del Museo de Historia Natural, UNMSM 21, 1992.

———. "National Park Protection in Relation to the Ecological Zonation of a Neighboring Community: An Example from Northern Peru." *Mountain Research and Development* 14, no. 3 (1993): 267–80.

———. "Threats to Biological Diversity Caused by Coca/Cocaine Deforestation in Peru." *Environmental Conservation* 23 (1996): 7–15.

———. "Tropical Timberlines: Changes in Forest Structure and Regeneration between Two Peruvian Timberline Margins." *Arctic and Alpine Research* 25, no. 3 (1993): 167–74.

Young, Kenneth R., and Blanca Leon. "Curvature of Woody Plants of a Timberline Montana Forest in Peru." *Physical Geography* 11, no. 1 (1990): 66–74.

———. "Distributions and Conservation of Peru's Montane Forests: Interactions between the Biota and Human Society." In *Tropical Montane Cloud Forests*, edited by L. S. Hamilton et al., 363–76. New York: Springer-Verlag, 1995.

Zimmerer, Karl S. "Managing Diversity in Potato and Maize Fields of the Peruvian Andes." *Journal of Ethnobiology* 11, no. 1 (1991): 23–49.

———. "The Regional Biogeography of Native Potato Cultivars in Highland Peru." *Journal of Biogeography* 18, no. 2 (1991): 165–78.

Climatology and Weather

Coker, R. E. "Ocean Temperature off the Coast of Peru." *Geographical Review* 5 (1918): 127–35.
Dornbusch, Uwe. "Current Large-Scale Climate Conditions in Southern Peru and Their Influence on Snowline Altitudes." *Erdkunde* 52, no. 1 (1998): 41–53.
Fontugne, Michel, et al. "El Nino Variability in the Coastal Desert of Southern Peru during the Mid-Holocene." *Quaternary Research* 52, no. 2 (1999): 171–79.
Tapley, Thomas, and Peter R. Waylen. "A Mixture Model of Annual Precipitation in Peru." *Professional Geographer* 41, no. 1 (1989): 62–70.
Wilson, Lucy L. "Climate and Man in Peru: The Coast, Parts I–II." *Bulletin of the Geographical Society of Philadelphia* 8, no. 3 (1910): 1–19; 8, no. 4 (1910): 27–45, 79–97, 153–71.

Geomorphology, Landforms, and Volcanism

Abizaio, Christian. "An Anthropogenic Meander Cutoff along the Ucayali River, Peruvian Amazon." *Geographical Review* 95, no. 1 (2005): 122–35.
Bowman, Isaiah. "The Canon of the Urubamba." *Bulletin of the American Geographical Society* 44 (1912): 881–97.
Grolier, M. J., et al. *The Desert Land Forms of Peru.* Washington, DC: GPO, 1974.
Hastenrath, S. L. "The Borchan Dunes of Southern Peru Revisited." *Zeitschrift fur Geomorphologie* 31, no. 2 (1987): 167–78.
———. "The Borchans of the Arequipa Region, Southern Peru." *Zeitschrift fur Geomorphologie* 11, no. 3 (1967): 300–331.
Inbar, Moske, and Carlos Llerena Pinto. "Efectos de la erupcion del volcan Sabancaya, Peru-1990. Evaluacion preliminar." *Investigaciones Geograficas* 30 (1995): 79–96.
Kaszowski, L. "Altitudinal Alteration of the Morphological Systems in the Rio Cheeras Basin of the Peruvian Andes." *Geographia Polonica* 60 (1992): 25–32.
Roosevelt, Theodore. "The Andes of Southern Peru." *Geographical Review* 3 (1917): 317–22.
Vilimek, V. "Neotectonic Activities in the Fault Zone of the Cordillera Blanca Mountains." *Acta Universitatis Carolinae, Geographica* 33, no. 1 (1998): 59–69.
Wells, Lisa. "Holocene History of the El Nino Phenomenon as Recorded in Flood Sediments of Northern Coastal Peru." *Geology* 18 (1990): 1134–37.
———. "Holocene Landscape Change on the Santa Delta, Peru: Impact on Archeological Site Disturbances." *The Holocene* 2, no. 3 (1992): 193–204.
———. "The Santa Beach Ridge Complex: Sea-Level and Progradational History of an Open Gravel Coast in Central Peru." *Journal of Coastal Research* 12, no. 1 (1996): 1–17.

Hydrology and Glaciology

Caviedes, Cesar N, and Peter R. Waylen. "El Nino and Annual Floods in Coastal Peru." In *Catastrophic Flooding,* edited by L. Mayer and D. Nash, 57–78. Boston: Allen & Unwin, 1987.
Clapperton, Chalmers M. "The Pleistocene Moraine Stages of West-Central Peru." *Journal of Glaciology* 11, no. 62 (1972): 255–64.
Craig, Alan K., and I. Shimoda. "El Nino Flood Deposits at Batan Grande, Northern Peru." *Geoarcheology* 1 (1986): 29–38.
Gellers, Joshua. "Here Comes the Rain Again: Flooding and Disaster Mitigation in Peru. A Case Study from the 1997–1998 El Nino." *Florida Geographer* 36 (2005): 99–106.
George, Christian. "Twentieth Century Glacier Fluctuations in the Tropical Cordillera Blanca, Peru." *Arctic, Antarctic, and Alpine Research* 36, no. 1 (2004): 100–107.
Hubbard, B., et al. "Impact of a Rock Avalanche on a Moraine-Dammed Proglacial Lake: Laguna Safuna Alta, Cordillera Blanca, Peru." *Earth Surface Processes and Landforms* 30, no. 10 (2005): 1251–64.

Kaser, G., et al. "The Impact of Glaciers on the Runoff and the Reconstruction of Mass Balance History from Hydrological Data in the Tropical Cordillera Blanca, Peru." *Journal of Hydrology* 282, nos. 1–4 (2003): 130–44.

———. "On the Mass Balance of Low Latitude Glaciers with Particular Consideration of the Peruvian Cordillera Blanca." *Geografiska Annaler* 81A, no. 4 (1999): 643–52.

Leuthart, Clara A., et al. "Reaction of a Low-Gradient Tropical Rainforest Stream in the Amazon Basin of Peru." *Physical Geography* 19, no. 2 (1998): 147–61.

Mark, Bryan, et al. "Climate Change and Tropical Andean Glacier Recession: Evaluating Hydrologic Changes and Livelihood Vulnerability in the Cordillera Blanca, Peru." *Annals of the Association of American Geographers* 100, no. 4 (2010): 794–805.

Mercer, John H., and O. Palacios. "Radiocarbon Dating of the Late Glaciation in Peru." *Geology* 5 (1977): 600–604.

Murphy, Robert C. "Notes on the Findings of the 'William Scaresby' in the Peru Coastal Current." *Geographical Review* 27 (1937): 295–300.

Thompson, L. G., et al. "Late Glacial Stage and Holocene Tropical Ice Core Records from Huascaran, Peru." *Science* 269 (1995): 46–50.

Waylen, Peter R., and Cesar N. Caviedes. "El Nino and Annual Floods on the North Peruvian Littoral." *Journal of Hydrology* 89 (1986): 141–56.

Wright, H. E. "Late Pleistocene Glaciation and Climate around the Junin Plain, Central Peruvian Highlands." *Geografiska Annaler* 65A, nos. 1–2 (1983): 35–44.

Soils

Andres, Nuria, et al. "Ground Thermal Conditions at Chachani Volcano, Southern Peru." *Geografiska Annaler* 93A, no. 3 (2011): 151–62.

Scott, Geoffrey A. J. "Soil Profile Changes Resulting from the Conversion of Forest to Grassland in the Montana of Peru." *Great Plains–Rocky Mountain Geographical Journal* 4 (1975): 124–30.

Theses and Dissertations

Alvarez, Nora. "Deforestation in the Southeastern Peruvian Amazon: Linking Remote Sensing Analysis to Local Views of Landscape Change." Master's thesis, University of Wisconsin, Madison, 2001.

Bumbaco, Karin. "Seasonal and Diurnal Variations in Hydrometeorological Variables in an Andean Mountain Environment and Comparison to the NCEP/NCAR Reanalysis Data." Master's thesis, Ohio State University, Columbus, 2008.

Cohen, Perry. "Geographical Aspects of the Peruvian Coastal Current." Master's thesis, Ohio State University, Columbus, 1950.

Conas, C. M. "Hydrologic Regime and Downstream Movement of Catfish Larvae in the Madre de Dios River, Southeastern Peru." Master's thesis, University of Florida, Gainesville, 2007.

Echavarria, Fernando. "Remote Sensing Systems for Monitoring and Quantifying Tropical Deforestation in the Huallaga River Valley of Peru." Master's thesis, Florida Atlantic University, Boca Raton, 1989.

Gonzalez, Fernando L. "Socio-environmental Dynamics of Biological Invasions: The Case of Kikuyu Grass (*Pennisetum clandestinum*) in Cotahuasi Valley, Southern Peru." PhD diss., University of Wisconsin, Madison, 2009.

Li, Xu Ming. "Geochemistry of Lavas from Minor Centers in Southern Peru." Master's thesis, Indiana State University, Terre Haute, 1996.

Nickl, Elsa. "Telecommunications and the Climate of the Peruvian Andes." Master's thesis, University of Delaware, Newark, 2007.

Pomora, Lazarus Y. "Biogeography of Upland Bird Communities in the Peruvian Amazon." PhD diss., University of Texas, Austin, 2009.

Scott, Geoffrey A. J. "Grassland Creation in a Montane Tropical Rainforest and Its Effects on Soil-Vegetation Nutrient Pools and Nutrient Cycles: A Case Study in the Gran Pajonal of Eastern Peru." PhD diss., University of Hawaii, Manoa, 1974.

Seimon, Anton. "Climatic Variability and Environmental Responses in an Andean Alpine Watershed, Cordillera Vilcanota, Peru." PhD diss., University of Colorado, Boulder, 2004.

Van Sledright, Malinda. "Documenting Aguaje Palm Swamp Area and Conditions in the Peruvian Amazon: Lessons for Conservation." Master's thesis, Western Michigan University, Kalamazoo, 2008.

POLITICAL GEOGRAPHY

General Works

Auty, R. M. "Populist Policies and Accelerating Economic Weaknesses in Peru." In *Sustaining Development in Mineral Economies: The Resource Curse Thesis*, edited by R. M. Auty, 91–109. London: Routledge, 1993.

Bury, Jeffrey, and Adam Kolff. "Livelihoods, Mining, and Peasant Protests in the Peruvian Andes." *Journal of Latin American Geography* 1, no. 1 (2002): 3–16.

Coomes, Oliver T. "State Credit Programs and the Peasantry under Populist Regimes: Lessons from the APRA Experience in the Peruvian Amazon." *World Development* 24, no. 8 (1996): 1333–46.

Degg, Martin R., and David K. Chester. "Seismic and Volcanic Hazards in Peru: Changing Attitudes to Disaster Mitigation." *Geographical Journal* 171, no. 2 (2005): 125–45.

Gorman, S. M. "Geopolitics and Peruvian Foreign Policy." *Inter-American Economic Affairs* 36, no. 2 (1982): 233–46.

Haarstad, H. O., and Arnt Floysand. "Globalization and the Power of Rescaled Narratives: A Case of Opposition to Mining in Tambogrande, Peru." *Political Geography* 26, no. 3 (2007): 289–308.

Hiraoka, Mario. "Aquatic and Land Fauna Management among the Floodplain Riberenos of the Peruvian Amazon." In *The Fragile Tropics of Latin America: Sustainable Management of Changing Environments*, edited by Toshie Nishizawa and J. Vitto, 201–25. Tokyo, Japan: UN University Press, 1995.

Kent, Robert B. "Beekeeping Regions, Technical Assistance and Development Policy in Peru." *Yearbook, Conference of Latin Americanist Geographers* 12 (1986): 22–31.

———. "Geographical Dimensions of the Shining Path Insurgency in Peru." *Geographical Review* 83, no. 4 (1993): 441–54.

Kent, Robert B., and Aldo S. Ricci. "Evolucion y provision de servicios publicos distritales en la Sierra Central de Peru." *Revista Geografica* 99 (1984): 35–55.

Kent, Robert B., and Jesus C. Rimarachin. "Rural Public Works Construction in the Andes of Northern Peru: The Role of Community Participation." *Third World Planning Review* 16, no. 4 (1994): 357–74.

Laurie, Nina. "From Work to Welfare: The Response of the Peruvian State to the Feminization of Emerging Work." *Political Geography* 16, no. 8 (1997): 691–714.

Naughton-Treves, Lisa. "Deforestation and Carbon Emissions at Tropical Frontiers: A Case Study from the Peruvian Amazon." *World Development* 32 (2004): 173–90.

Roucek, Joseph S. "Peru in Geopolitics." *Contemporary Review* 204 (1963): 310–15; 205 (1964): 24–31.

Zimmerer, Karl S., and Martha G. Bell. "An Early Framework of National Land Use and Geovisualization: Policy Attributes and Application of Pulgar Vidal's State-Indigenous Vision of Peru (1941–Present)." *Land Use Policy* 30, no. 1 (2013): 305–16.

Theses and Dissertations

Chaves, A. B. "Public Policy and Spatial Variation in Land Use and Land Cover in the Southeastern Peruvian Amazon." PhD diss., University of Florida, Gainesville, 2009.

Flores-Carrasco, Cecilia. "Planning for Development in Peru, 1960–1985: From Growth Poles to Microregions." Master's thesis, Syracuse University, New York, 1990.

Gianella, Natalia. "Harvests of the War: Peasant Communities after Political Violence in Ayacucho, Peru." Master's thesis, Syracuse University, New York, 1990.

Gonzales, R. M. "The Political Ecology of Scallop (*Argopecten purpuratus*) Use and Management in the Piseo-Paracas Region." PhD diss., University of Hawaii, Manoa, 2008.

Pruett, Timothy S. "The Lost Valley of Peru? A Geographic Analysis of Illicit Coca Production and Terrorism in the Upper Huallaga Valley." Master's thesis, Western Kentucky University, Bowling Green, 2000.

URBAN GEOGRAPHY

General Works

Dollfus, O. "Lima, quelques aspects de la capitale du Perou en 1958." *Les Cahiers d'Outre Mer* 11 (1958): 258–71.

Gade, Daniel. "Regional Isolation of Ayacucho, a City in the Peruvian Andes." *Yearbook, Association of Pacific Coast Geographers* 29 (1967): 111–20.

Goluchowska, Katarzyna. "La camplejidad de la ciudad intermedia andina en el Peru hacia un modelo ambiental." *Revista Geografica* 132 (2002): 5–13.

Kent, Robert B. "Peru." In *Latin American Urbanization: Historical Profiles of Major Cities*, edited by Gerald M. Greenfield, 446–67. Westport, CT: Greenwood, 1994.

———. "Spatial and Temporal Variations of Sound in an Andean City: Cajamarca, Peru." *GeoJournal* 33, no. 4 (1994): 453–58.

Leonard, J. B. "Lima." *Cities* 17, no. 6 (2000): 433–45.

Morisset, Jean. "Urban Interrelationships and Regional Planning in the Development of Puno, Southern Peru." *Cahiers de Geographie du Quebec* 20, no. 49 (1976): 93–126.

Rosner, Waltrand. "La ciudad de Chiclayo (Peru) frente al reto de la sostenibilidad: Crecimiento urbano y problemas ambientales de cena metropolis regional." *Revista Geografica* 130 (2001): 131–62.

Urban Social Geography

Avellaneda, Pau G. "Mobility, Poverty, and Social Exclusion in Lima." *Anales de Geografia de la Universidad Complutense* 28, no. 2 (2008): 9–35.

Brisseau-Loaiza, J. "Le role du camion dans les relations ville-campagne dans la region du Cuzco (Perou)." *Les Cahiers d'Outre Mer* 25 (1972): 27–56.

Cowder, Stella. "Migration and Urbanization in Peru." In *Spatial Aspects of Development*, edited by Brian K. Hoyle, 209–30. London: Wiley, 1974.

Fernandez-Maldonado, Ana Maria. "Expanding Networks for the Urban Poor: Water and Telecommunications Services in Lima, Peru." *Geoforum* 39, no. 6 (2008): 1884–96.

Ioris, Antonio A. R. "The Persistent Water Problems of Lima, Peru: Neoliberalism, Institutional Failures, and Social Inequalities." *Singapore Journal of Tropical Geography* 33, no. 3 (2012): 335–50.

Mesa, Jose L. "The Metropolitan Area Concept and Urban Population Growth: The Example of Lima, Peru." *Proceedings of the Association of American Geographers* 6 (1974): 117–20.

Neilson, Nora G., and Martyn Bowden. "Bound to the Barridas: Migration to the Peripheral Squatter Settlements of Lima." *Proceedings, New England–St. Lawrence Valley Division, Association of American Geographers* 2 (1972): 28–38.

Peters, Paul A., and Emily H. Skop. "Socio-spatial Segregation in Metropolitan Lima, Peru." *Journal of Latin American Geography* 6, no. 1 (2007): 149–71.

Ploger, Jorg. *Die abgeschotteten Nachborschaften in Lima (Peru). Eine Analyse sozialraulicher kontroll massnahmen im Kontext zunehmender Unsicherheiten.* Kiel, Germany. Kieler Geographische Schriften, Band 112, 2006.

Morphology and Neighborhoods

Chambers, B. "The Barricadas of Lima: Slums of Hope or Despair? Problems or Solutions." *Geography* 90 (2000): 200–224.

Deler, J. P. "Croissance acceleree et formes de sous-developpement urbain a Lima." *Les Cahiers d'Outre Mer*, 23 (1970): 73–94.

Ploger, Jorg. "Gated Barriadas: Responses to Urban Insecurity in Marginal Settlements in Lima, Peru." *Singapore Journal of Tropical Geography* 33, no. 2 (2012): 212–25.

Urban Economic Geography

Hays-Mitchell, Maureen. "Streetvending in Peruvian Cities: The Spatio-temporal Behavior of Ambulantes." *Professional Geographer* 46, no. 4 (1994): 425–38.

Urban Environments

Ioris, Antonio A. R. "The Geography of Multiple Scarcities: Urban Development and Water Problems in Lima, Peru." *Geoforum* 43, no. 3 (2012): 612–22.

———. "The Neoliberalization of Water in Lima, Peru." *Political Geography* 31, no. 5 (2012): 266–78.

Molina, Jose M. "Lima, un clima de desierto litoral." *Anales de Geografía de la Universidad Complutense* 19 (1999): 25–45.

Pol, M. H., et al. "Biodiversity Conservation Implications of Landscaping Change in an Urbanizing Desert of Southwestern Peru." *Urban Ecosystems* 8 (2005): 313–34.

Theses and Dissertations

Fisher, Winnifred V. "The Geographical Structure of Lima, Peru." Master's thesis, Rutgers University, New Brunswick, NJ, 1962.

Mesa, Jose L. "The Metropolitan Area Concept and Its Development in Lima, Peru." Master's thesis, Western Michigan University, Kalamazoo, 1966.

Myers, Sarah K. "Language Shifts among Migrants to Lima, Peru." PhD diss., University of Chicago, Chicago, 1971.

Neilson, Nora G. "Bound for the Bariadas Marginales: A Study of Migrations to the Peripheral Squatter Settlements of Lima, Peru." Master's thesis, Clark University, Worcester, MA, 1970.

Sinclair, Joseph T. "Lima, Peru: A Study in Urban Geography." PhD diss., University of Michigan, Ann Arbor, 1959.

Williams, Lyndon S. "The Suburban Bariadas of Lima: Squatter Settlements as a Type of Peripheral Urban Growth in Peru." Master's thesis, California State University, Long Beach, 1969.

Chapter Twelve

Surinam

GENERAL WORKS

Books, Monographs, and Texts

Nystrom, John W. *Surinam: A Geographical Study.* New York: Netherlands Information Bureau, 1942.

Articles and Book Chapters

De Bruijne, G. A. "Surinam in Regional Geography: An Alternative to Preston James' Latin America." *Geografisch Tijdschrift* 5, no. 3 (1971): 228–31.
———. "Surinam and the Netherlands Antilles: Their Place in the World." *Geografisch Tidschrift* 5, no. 4 (1971): 517–24.
Tirtha, Ranjit, and Cornelium Loester. "Surinam." In *Focus on South America*, edited by Alice Taylor, 94–105. New York: Praeger, 1973.

Theses and Dissertations

Fisher, Albert L. "Surinam's Development Possibilities: A Geographical Study." PhD diss., Johns Hopkins University, Baltimore, 1954.
Nystrom, John W. "Surinam: A Geographic Study." PhD diss., Clark University, Worcester, Massachusetts, 1942.

CULTURAL AND SOCIAL GEOGRAPHY

General Works

Zonneveld, J. I. S. "Luchtfoto-geographie in Suriname." *De West-Indische Gids* 33 (1952): 35–48.

ECONOMIC GEOGRAPHY

General Works

Hanrath, J. J. "The Economic-Geographical Structure of Surinam: A Character Sketch." *Revista Geografica* 48 (1958): 60–65.

PHYSICAL GEOGRAPHY

General Works

Maguire, Bassett. "Notes on the Geology and Geography of Tafelberg, Surinam." *Geographical Review* 35, no. 4 (1945): 563–79.

Geomorphology, Landforms, and Volcanism

Riezebos, H. T. "Geomorphology and Savannization in the Upper Sipaliwini River Basin (Southern Surinam)." *Zeitschrift fur Geomorphologie* 28, no. 3 (1984): 265–84.

POLITICAL GEOGRAPHY

General Works

De Blij, Harm J. "Cultural Pluralism and the Political Geography of Decolonization: The Case of Surinam." *Pennsylvania Geographer* 8, no. 1 (1970): 1–11.

URBAN GEOGRAPHY

General Works

De Bruijne, G. A., and A. Schalkwijk. "Paramaibo: Its Characteristics as a Caribbean City." *Beitrage zur Geographischen Regionalforschung in Lateinamerika* 10 (1997): 93–103.
Verrest, Hebe. "Paramaibo." *Cities* 27, no. 1 (2010): 50–60.

Chapter Thirteen

Uruguay

GENERAL WORKS

Articles and Book Chapters

Crist, Raymond E., and Edward P. Leahy. "Uruguay." In *Focus on South America*, edited by Alice Taylor, 217–27. New York: Praeger, 1973.
Kirby, John B. "Uruguay and New Zealand, Paths to Progress." *Revista Geografica* 107 (1988): 119–50.
Lemert, Ben F. "Uruguay and the Uruguayans." *Journal of Geography* 33, no. 8 (1934): 289–303.

CULTURAL AND SOCIAL GEOGRAPHY

General Works

Pietri Livy, Anne-Lise. "Disparites culturelles et organization de l'espace en Uruguay." *Geographie et Cultures* 14 (1995): 43–66.

Ethnic, Social, and Population Geography

Tata, Robert J. "Uruguay: Population Geography of a Troubled Welfare State." *Journal of Geography* 76, no. 2 (1977): 46–50.

ECONOMIC GEOGRAPHY

General Works

Deffontaines, Pierre. "Routes du betail et types de foires en Uruguay." *Les Cahiers d'Outre Mer* 4 (1951): 93–100.

Agriculture and Land Use

Fitzgibbon, Russel H. "Uruguay's Agricultural Problems." *Economic Geography* 29 (1953): 251–62.
Griffin, Ernst C. "The Agricultural Land Use Regions in Uruguay." *Revista Geografica* 76 (1972): 121–52.
———. "Causal Factors Influencing Agricultural Land Use Patterns in Uruguay." *Revista Geografica* 80 (1974): 13–34.
———. "Testing the Von Thunen Theory in Uruguay." *Geographical Review* 63, no. 4 (1973): 500–516.

Commerce and Trade

Jones, Clarence F. "The Trade of Uruguay." *Economic Geography* 3, no. 4 (1927): 361–81.

Energy, Mining, Fishing, and Lumbering

Chehataroff, Jorge. "Fuentes, Produccion, Consumo de Energia en el Uruguay." *Revista Geografica* 63 (1965): 83–108.
Lehtinen, A. "Lessons from Fray Bentos, Forest Industry, Overseas Investments and Discursive Regulation." *Fennia* 186, no. 2 (2008): 69–82.

Theses and Dissertations

Griffin, Ernst C. "Agricultural Land Use in Uruguay." PhD diss., Michigan State University, East Lansing, 1972.

HISTORICAL GEOGRAPHY

General Works

Gautreau, Pierre. "Rethinking the Dynamics of Woody Vegetation in Uruguayan Compos, 1800–2000." *Journal of Historical Geography* 36, no. 2 (2010): 194–204.
Kleinpenning, J. M. G. *Peopling the Purple Land: A Historical Geography of Rural Uruguay, 1500–1915*. Amsterdam, Netherlands: CEDLA, 1995.

PHYSICAL GEOGRAPHY

General Works

Chebataroff, Jorge. "Las regions naturales de Rio Grande del Sur y de la Republica Oriental del Uruguary." *Revista Geografica*, 31–36 (1951–1952): 59–95.

Geomorphology, Landforms, and Volcanism

Fernandez, Gabriela, et al. "Caracterizacion geomorfologica de la cuenca del banado de Farropos-Rio Negro, Uruguay." *Revista Geografica* 148 (2010): 23–42.
Munka, Carolina, et al. "Long-Term Variation in Rainfall Erosivity in Uruguay: A Preliminary Fourier Approach." *GeoJournal* 70, no. 4 (2007): 257–63.

Solari, Alberto B. "La Riqueza Hidrografica de la Republica Oriental del Uruguay." *Revista Geografica* 88 (1978): 221–28.

Hydrology and Glaciology

Mari, Ema. "Obras hidraulicas e hidroelectricas del Uruguay." *Revista Geografica*, 9–10 (1949–50): 133–45.

POLITICAL GEOGRAPHY

General Works

Kleinpenning, J. M. G. "Uruguary: The Rise and Fall of a Welfare State Seen against a Background of Dependency Theory." *Revista Geografica* 93 (1981): 101–17.

Theses and Dissertations

Buchert, Beverly. "The Tupamaros: Anomalies of Guerrilla War." PhD diss., University of Kansas, Lawrence, 1979.

URBAN GEOGRAPHY

General Works

Collin-Delavaud, A. "L'Uruguay, exemple d'urbanisation originale en pays d'elevage." *Les Cahiers d'Outre Mer* 25 (1972): 361–89.
Ferrari, Maria S. G. "Montevideo." *Cities* 23, no. 5 (2006): 382–99.
Martins, Gunter, et al. *Beitrage zur Stadtgeographie von Montevideo*. Marburg-Lahn, Germany: Marburger Geographische Schriften, Heft 108, 1987.
Sargent, Charles S. "Uruguay." In *Latin American Urbanization: Historical Profiles of Major Cities*, edited by Gerald M. Greenfield, 468–85. Westport, CT: Greenwood, 1994.

Urban Economic Geography

Snyder, David E. "Commercial Passenger Lineages and the Metropolitan Nodality of Montevideo." *Economic Geography* 38 (1962): 95–112.

Theses and Dissertations

Snyder, David E. "Urban Places and Passenger Transportation in Uruguay." PhD diss., Northwestern University, Evanston, IL, 1959.

Chapter Fourteen

Venezuela

GENERAL WORKS

Atlases and Graphic Presentations

Mohn, M. "La cartographe au Venezuela." *Les Cahiers d'Outre Mer* 21 (1968): 424–30.

Books, Monographs, and Texts

Alexander, Charles. *The Geography of Margarita and Adjacent Islands.* Berkeley: University of California Press, 1958.
Crist, Raymond E. *Venezuela.* Garden City, NY: Doubleday, 1964.
Crist, Raymond E., and Edward P. Leahy. *Venezuela: Search for a Middle Ground.* New York: Van Nostrand, 1969.
Crooker, Richard A. *Venezuela.* Broomall, PA: Chelsea, 2005.

Articles and Book Chapters

Anthony, H. E., H. A. Gleason, and Raye R. Platt. "The Pacaraima-Venezuela Expedition." *Geographical Review* 21 (1931): 353–62.
De Booy, Theodore. "An Exploration of the Sierra de Perija, Venezuela." *Geographical Review* 6 (1918): 385–410.
———. "The Western Maracaibo Lowland, Venezuela." *Geographical Review* 6 (1918): 481–500.
Burchfield, William. "Our Tropical Outpost in Venezuela." *Journal of Geography* 44, no. 5 (1945): 192–201.
Crist, Raymond E., and Edward P. Leahy. "Venezuela." In *Focus on South America*, edited by Alice Taylor, 67–79. New York: Praeger, 1973.
Eden, Michael J. "Scientific Exploration in Venezuelan Amazonas." *Geographical Journal* 137, no. 2 (1971): 149–56.
Hitchcock, Charles B. "The Orinoco-Ventuari Region, Venezuela." *Geographical Review* 37, no. 4 (1947): 525–66.
Holdridge, Desmond. "Exploration between the Rio Branco and the Sierra Parima." *Geographical Review* 23 (1933): 372–93.

———. "Notes on an Exploratory Journey in Southeastern Venezuela." *Geographical Review* 21 (1931): 373–78.
Martinez, Mario. "El suroeste de Venezuela: Espacios de integracion fronteriza." *Anales de Geografia de la Universidad Complutense* 18 (1998): 139–58.
Miller, Leo E. "The Land of the Maquiritares." *Geographical Review* 3 (1917): 356–74.
———. "Up the Orinoco to the Land of the Maquiritores." *Geographical Review* 3 (1917): 258–77.
Price, Marie D. "The Venezuela Andes and the Geographical Imagination." *Geographical Review* 86, no. 3 (1996): 334–55.
Sunal, Cynthia S., et al. "Using the Five Themes of Geography to Teach about Venezuela." *The Social Studies* 96, no. 4 (1995): 169–74.
Tate, G. H. H. "Auyontepui: Notes on the Phelps Venezuelan." *Geographical Review* 28, no. 4 (1938): 452–74.
Tate, G. H. H., and Charles B. Hitchcock. "The Cerro Duida Region of Venezuela." *Geographical Review* 20, no. 1 (1930): 31–52.
White, C. Langdon. "Sleepy Orinoco Valley Comes to Life." *Journal of Geography* 55, no. 3 (1956): 111–20.

Theses and Dissertations

Alexander, Charles. "The Geography of Margarita and Adjacent Islands, Venezuela." PhD diss., University of California, Berkeley, 1955.

CULTURAL AND SOCIAL GEOGRAPHY

General Works

Acevedo, Miguel, et al. "Models of Natural and Human Dynamics in Forest Landscapes: Cross-Site and Cross-Cultural Synthesis." *Geoforum* 39, no. 2 (2008): 846–66.
Crist, Raymond E. "Development and Agrarian Land Reform in Venezuela's Pioneer Zone: Social Progress along the Llanos-Andes Border in a Half Century of Political Advance." *American Journal of Economics and Sociology* 43 (1984): 149–58.
———. "Life on the Llanos of Venezuela." *Bulletin of the Geographical Society of Philadelphia* 35, no. 2 (1937): 13–25.
Rojas, Temistocles. "Metodologia para la evalacion de areas sensibles en Venezuela." *Revista Geografica* 102 (1985): 29–41.

Ethnic, Social, and Population Geography

Betancourt, Jose. "Estimating Interstate Internal Migration from Place-of-Birth Data." *Revista Geografia* 88 (1976): 61–78.
Brown, Lawrence A., and Andrew Gretz. "Development-Related Contextual Effects and Individual Attributes in Third World Migration Processes: A Venezuelan Example." *Demography* 24 (1987): 497–516.
Brown, Lawrence A., and J. E. Kondras. "Migration, Human Resource Transfers and Development Contexts: A Logit Analysis of Venezuela Data." *Geographical Analysis* 19, no. 3 (1987): 243–63.
Brown, Lawrence A., and Victoria A. Lawson. "Polarization Reversal, Migration Related Shifts in Human Resource Profiles, and Spatial Growth Policies: A Venezuelan Study." *International Regional Science Review* 12, no. 2 (1989): 165–88.
Crist, Raymond E. "The Bases of Social Instability in Venezuela." *American Journal of Economics and Sociology* 1, no. 1 (1941): 37–44.

———. "Ethnogeography: Pile-Dwellers and Coconut Culture in the Laguna de Sinamain, Venezuela." *Journal of the Washington Academy of Science* 48 (1958): 380–86.

Delavaud, Anne. "Papel geografico de las misiones religiosas contemparaneas en el delta del Orinoco y en la gran savanna en Venzuela." *Revista Geografica* 94 (1981): 53–65.

Denevan, William M., and K. Schwerin. "Adaptive Strategies in Karinya Subsistence, Venezuelan Llanos." *Anthropologia* 50 (1978): 3–91.

Eastwood, David A. "Motivation and Intra-Rural Migration: A Case Study in Merida State, Venezuela." *Journal of Tropical Geography* 49 (1979): 1–10.

———. "Reality or Delusion: Migrant Perception of Level of Living and Opportunity in Venezuela, 1961–1971." *Journal of Developing Areas* 17, no. 4 (1983): 491–98.

———. "Venezuela: The 1980 Census Shows Continued Rapid Population Growth." *Geography* 68, no. 4 (1983): 345–47.

Jones, Richard C. "Latent Migration Potential between a Depressed Region and Alternative Destinations: A Venezuelan Case Study." *Proceedings of the American Association of Geographers* 7 (1975): 104–9.

———. "Myth Maps and Migration in Venezuela." *Economic Geography* 54, no. 1 (1978): 75–91.

———. "El potencial latent migratorio entre una region de primida y destinos alternatives: Un ejempro venezolana." *Revista Geografica, Venezuela,* (1976): 127–37.

Jones, Richard C., and Lawrence A. Brown. "Cross-National Tests of a Third World Development-Migration Paradigm: With Particular Attention to Venezuela." *Socio-economic Planning Sciences* 19 (1985): 357–61.

Lawson, Victoria A., and Lawrence A. Brown. "Structural Tension, Migration, and Development: A Case Study of Venezuela." *Professional Geographer* 39, no. 2 (1987): 179–88.

Levy, Mildred B., and W. J. Wadycki. "Lifetime versus One-Year Migration in Venezuela." *Journal of Regional Science* 12, no. 3 (1972): 407–15.

Miller, Willard. "Population Growth and Agricultural Development in the Western Llanos of Venezuela." *Revista Geografica* 69 (1968): 7–28.

Osorio, Emilio. "Presentacion geodemografica contemporanea de Venezuela." *Revista Geografica* 102 (1985): 43–54.

Ravuri, Evelyn D. "Life-Time and Recent Migration to Bolivar State, Venezuela, 1990: The Effect of the Guayana Program on Migration." *Journal of Latin American Geography* 1, no. 1 (2002): 69–81.

Sletto, Bjorn. "Indigenous People Don't Have Boundaries: Rebordering, Fire Management, and Productions of Authenticities in Indigenous Landscapes." *Cultural Geographies,* 16, no. 2 (2009): 253–77.

Community and Settlement Studies

Camacho, O. O. "Venezuela's National Colonization Program: The Tovar Colony, a German Agricultural Settlement." *Journal of Historical Geography* 10, no. 3 (1984): 279–90.

Eidt, Robert C. *Agrarian Reform and the Growth of New Rural Settlements in Venezuela.* Milwaukee: University of Wisconsin–Milwaukee, Center for Latin America, Publication 55, 1975.

———. "Agricultural Reform and the Growth of New Rural Settlements in Venezuela." *Erdkunde* 29 (1975): 118–33.

Platt, Robert S. "Pattern of Occupance in the Maracaibo Basin." *Annals of the Association of American Geographers* 24 (1934): 157–73.

Samudio, Edda A. "The Dissolutions of Indian Community Lands in the Venezuelan Andes: The Case of La Mesa." *Yearbook, Conference of Latin American Geographers* 23 (1997): 17–26.

Tourism and Recreation

Sota, Jesus. "Resort Development in the Venezuelan Central Littoral: The Case of Naiguta." *Geographical Bulletin* 4 (1972): 50–59.

Theses and Dissertations

Betancourt, Jose. "Those Who Remain: One Aspect of the Process of Migration in a Rapidly Changing Country: Venezuela." PhD diss., University of Iowa, Iowa City, 1976.

Enberg, Dennis P. "Distance and the Motive for Migration: Movement to Agricultural Settlements in Venezuela." PhD diss., University of North Carolina, Chapel Hill, 1975.

De Ramirez, Arminda U. "Causes and Consequences of Internal Migration in Zulia State, Venezuela." PhD diss., University of Florida, Gainesville, 1978.

Salazar, Deborah. "Through Sickness and in Health: A Tropical Ethnoecology of Traditional Medicine among the Peman Indians of the Venezuelan Gran Sabana." PhD diss., University of Texas, Austin, 1995.

ECONOMIC GEOGRAPHY

General Works

Crist, Raymond E. "Westward Thrust the Pioneer Zone in Venezuela: A Half Century of Economic Development along the Llanos-Andes Border." *American Journal of Economics and Sociology* 42 (1983): 451–62.

Jones, Richard C. "Regional Income Inequalities and Government Investment in Venezuela." *Journal of Developing Areas* 16, no. 3 (1982): 373–90.

Martinson, Tom L. "Oil Profits, Farm to Market Roads, and Corn and Coffee Production in Venezuela." *Proceedings, Indiana Academy of the Social Science* 9 (1974): 75–83.

Walters, Rudy F. "Economic Backwardness in the Venezuelan Andes: A Study of the Traditional Sector of the Dual Economy." *Pacific Viewpoint* 8, no. 1 (1967): 17–67.

Agriculture and Land Use

Anderson, Thomas D. "Subsystems of Conuco Agriculture in the Basin of Lake Valencia, Venezuela: A Classification and Description." *Ohio Geographer* 4 (1976): 23–34.

Angeliaume-Descamps, A., and J. Oballos. "Le maraichage intensif irrigue dans les hautes vallees andines venezueliennes: Quelle remise en question?" *Les Cahiers d'Outre Mer* 62 (2009): 439–68.

Behrens, C. A., et al. "A Regional Analysis of Bari Land Use Intensification and Its Impact on Landscape Heterogeneity." *Human Ecology* 22, no. 3 (1994): 279–316.

Crist, Raymond E. "Development and Agrarian Land Reform in Venezuela's Pioneer Zone." *American Journal of Economics and Sociology* 43, no. 2 (1984): 149–58.

———. "Land Tenure Problems in Venezuela." *American Journal of Economics and Sociology* 32, no. 1 (1942): 143–54.

———. "Wheat Raising in the Venezuelan Andes." *Scientific Monthly* 56 (1943): 332–38.

Crist, Raymond E., and Carlos E. Chardon. "Changing Patterns of Land Use in the Valencia Lake Basin of Venezuela." *Geographical Review* 31, no. 3 (1941): 430–43.

Denevan, William M., and Roland Bergman. "Karinya Indian Swamp Cultivation in the Venezuelan Llano." *Yearbook, Association of Pacific Coast Geographers* 37 (1975): 23–37.

Eastwood, David A. "The Meriden Lowlands of Venezuela: A Waster of Agricultural Potential." *Geography* 64, no. 3 (1979): 182–87.

Eden, Michael J. "Irrigation Systems and the Development of Peasant Agriculture in Venezuela." *Tijdschrift voor Economische en Sociale Geographie* 65, no. 1 (1974): 48–54.

Eidt, Robert C. "Agricultural Reform and the Growth of New Rural Settlements in Venezuela." *Erdkunde* 29 (1975): 118–33.

Harris, David. "The Ecology of Swidden Cultivation in the Upper Orinoco Rain Forest, Venezuela." *Geographical Review* 61, no. 3 (1971): 475–95.

James, Preston E. "The Possibilities of Cattle Production in Venezuela." *Bulletin of the Geographical Society of Philadelphia* 22, no. 2 (1924): 9–20.

Newsome, Tracey, et al. "Land Contamination Adjacent to Roadways in Trujillo, Venezuela." *Professional Geographer* 49, no. 3 (1997): 331–41.

Powers, W. L. "Soil Development and Land Use in Northern Venezuela." *Geographical Review* 35, no. 2 (1945): 273–85.

Rojas Lopex, Jose. "La emergencia de nuevos sistemas de produccion y la profundizacion de las desigualdades agroespaciales en el campo andino venezolano." *Revista Geografica* 102 (1985): 107–14.

Rudolph, William E. "Agricultural Possibilities in Northwestern Venezuela." *Geographical Review* 34, no. 1 (1944): 36–56.

Sarmiento, L., et al. "Ecological Bases, Sustainability, and Current Trends in Traditional Agriculture in the Venezuelan High Andes." *Mountain Research and Development* 13, no. 2 (1993): 167–76.

Sebastiani, M. "La agricultura des de la perspective de los sistemas y regions agricolas: une linea de investigacion en Venezuela." *Revista Geografica Venezolara* 32 (1991): 65–86.

Vargas, Luis. "Manual de geografia agrarian: Para el uso de los tecnicas de conservacion de suelos y aguas." *Revista Geografica* 57 (1962): 81–109.

White, C. Langdon. "Cattle Raising: A Way of Life in the Venezuelan Llanos." *Scientific Monthly* 83 (1956): 122–29.

Commerce and Trade

Chaves, Luis F. "The Organization of Venezuelan Space According to the Banking Function." *Revista Geografica* 67 (1967): 43–68.

Manufacturing and Labor Studies

Auty, R. M. "Resource-Based Industrialization and Country Size: Venezuela and Trinidad and Tobago." *Geoforum* 17, nos. 3–4 (1986): 326–28.

Brown, Lawrence A., et al. "Location, Social Categories, and Individual Labor Market Experiences in Developing Economies: The Venezuelan Case." *International Regional Science Review* 12, no. 1 (1989): 1–28.

Karlsson, Weine. *Manufacturing in Venezuela: Inequalities and Government Development and Location.* Stockholm, Sweden: Almquist and Wiksell International, 1975.

Miller, Elbert E. "The Guayana Region, Venezuela: A Study in Industrial and Urban Development." *Yearbook, Association of Pacific Coast Geographers* 27 (1965): 77–88.

Miller, Eugene W. "Petroleum in the Economy of Venezuela." *Economic Geography* 16, no. 2 (1940): 204–10.

Energy, Mining, Fishing, and Lumbering

Corfield, George S. "Recent Activities in Venezuela's Petroleum Industry." *Economic Geography* 24 (1948): 114–18.

Muller, B. "Gold- und Diamantenabbau in Sudost-Venezuela." *Zeitschrift fur Wirtschaftsgeographie* 43, nos. 3–4 (1999): 229–44.

Sequera de Segnini, Isbelia. "Reflexiones sobre la incidencia ecologica de la explotacion petrolera en Venezuela." *Revista Geografica* 102 (1985): 101–5.

Silva Aristegueta, Jose L., et al. "Sistema de informacion geografica para el analisis especial de la pesqueria ortesanal asentada en la peninsula de Araya, Estado Sucre-Venezuela." *Revista Geografica de American Central* 45 (2010): 149–77.

Vila, Marco. "La electrificacion de los grandes paisajes geograficas Venezolanos." *Revista Geografica* 46 (1957): 1–12.

White, C. Langdon. "Venezuela Moves onto the World's Iron Ore Map." *California Geographer* 1 (1960): 1–6.

Transportation and Communications

Marchand, Bernard. "Deformation of a Transportation Surface." *Annals of the Association of American Geographers* 63, no. 4 (1973): 507–21.

Theses and Dissertations

Daum, Mary. "Land Amalgamation in Government Colonies in the Aroa Valley and Varinas Piedmont Region of Venezuela." PhD diss., University of Wisconsin, Madison, 1977.

Denny, Evan. "Economic Development: A Case Study of the Caroni River Region, Venezuela." Master's thesis, University of Washington, Seattle, 1965.

Heyman, Arthur. "Physical Parameters in the Development of Peasant Horticulture in the Highland Guyana Region, Venezuela." PhD diss., Columbia University, New York, 1967.

Hoffman, Shirley D. "Subsistence in Transition: Indigenous Agriculture in Amazonia, Venezuela." PhD diss., University of California, Berkeley, 1993.

Jaspe-Alvarez, Jose. "Estudio de la Distribucion especial del sistema cooperative de Ferias de Consamo Familiar (FCF) y de su papelen et abastecimiento alimentario de la region Centrooccidental de Venezuela." Master's thesis, Laval University, Ste. Foy, Quebec, 1991.

Jones, Richard C. "Descriptive and Policy Linear Programing Models for Region Crop Allocation: The Western Llanos of Venezuela." PhD diss., Ohio State University, Columbus, 1973.

Kingsbury, Nancy. "Increasing Pressure on Decreasing Resources: A Case Study of Peman Amerindian Shifting Cultivation in the Gran Sabana, Venezuela." PhD diss., York University, North York, Ontario, 1999.

Miller, Willard. "The Economic Geography of the Petroleum Industry of Venezuela." Master's thesis, University of Nebraska, Lincoln, 1939.

Minkel, Clarence W. "The Industrial Development of the Valencia Basin, Venezuela." PhD diss., Syracuse University, NY, 1960.

Ravuri, Evelyn D. "Reassessing the Growth Pole Phenomenon: Migration and Economic Development in Venezuela, 1950–1990." PhD diss., University of Cincinnati, Cincinnati, OH, 2001.

Villaroel, Jesus. "A Study of Waste Water Treatments System (Oxidation Ditch) and Its Potential Use in a Tropical Area (Venezuela)." Master's thesis, Johns Hopkins University, Baltimore, 1970.

HISTORICAL GEOGRAPHY

General Works

Denevan, William M., and A. Zucchi. "Ridged Fields Excavations in the Central Orinoco Llanos, Venezuela." In *Advances in Andean Archeology*, edited by D. L. Browman, 235–46. The Hague, Netherlands: Mouton, 1978.

diPolo, Mario, and M. M. Suarez. "History, Patterns, and Migration: A Case Study in the Venezuelan Andes." *Human Organization* 33, no. 2 (1974): 183–95.

Grace, Pedro. "Desarticulacion de las paisajes regionales venezolanos en el siglo XIX." *Revista Geografica* 102 (1985): 115–21.

Lombardi, John V. *People and Places in Colonial Venezuela*. Bloomington: Indiana University Press, 1976.

Price, Marie D. "Hands for the Coffee: Migrants and Western Venezuela's Coffee Economy, 1870–1930." *Journal of Historical Geography* 20, no. 1 (1994): 62–80.

Samudio, Edda A., and David J. Robinson. *Jesuit Estates of the College of Merida, Venezuela, 1629–1767.* Syracuse, NY: Syracuse University, Department of Geography, Discussion Paper 99, 1989.

Schuller, Rudolph. "The Date of Oviedo's Map of the Maracaibo Region." *Geographical Review* 3 (1917): 294–302.

PHYSICAL GEOGRAPHY

General Works

Perez, Francisco L. "Needle-Ice Activity and the Distribution of Stem-Resette Species in a Venezuelan Paramo." *Arctic and Alpine Research* 19, no. 2 (1987): 135–53.

Biogeography

Anderson, Robert P. "Real versus Artefactual Absences in Species Distribution: Tests for *Oryzomys albigularis* (Rodentia: Muridae) in Venezuela." *Journal of Biogeography* 30, no. 4 (2003): 591–606.

Bidulph, J., and Martin Kellman. "Fuels and Fire at Savanna-Gallery Forest Boundaries in Southeastern Venezuela." *Journal of Tropical Ecology* 14 (1998): 445–61.

Burnadt, Charles, and Rafael Campias. "Colonization, Extinction, and Species Numbers of Vascular Plants for the Island Gran Roque, Venezuela." *Journal of Biogeography* 13, no. 6 (1986): 541–50.

Eden, Michael J. "Paleoclimatic Influences and the Development of Savanna in Southern Venezuela." *Journal of Biogeography* 1, no. 2 (1974): 95–104.

Foldats, E., and E. Rutkis. "Ecological Studies of Chaparro and Manteco in Venezuela." *Journal of Biogeography* 2, no. 3 (1975): 159–78.

Huber, Otto. "Herbaceous Ecosystems on the Guayana Shield: A Regional Overview." *Journal of Biogeography* 33, no. 3 (2006): 464–75.

Kingsbury, Nancy, and Martin Kellman. "Root Mat Depths and Surface Soil Chemistry in Southeastern Venezuela." *Journal of Tropical Ecology* 13 (1997): 475–79.

Leon-Vargas, Y., et al. "Micro-climate, Light Adaptation, and Desiccation Tolerance of Epiphytic Bryophytes in Two Venezuelan Cloud Forests." *Journal of Biogeography* 33, no. 5 (2006): 901–13.

Medina, Ernesto, and Norma Motta. "Metabolism and Distribution of Grasses in Tropical Flooded Savannas in Venezuela." *Journal of Tropical Ecology* 6, no. 1 (1990): 77–89.

Monasterio, M., and G. Sarmiento. "Phenological Strategies of Plant Species in the Tropical Savanna and Semi-deciduous Forest of the Venezuelan Llanos." *Journal of Biogeography* 3, no. 4 (1976): 325–36.

Murphy, Robert C. "Bird Islands of Venezuela." *Geographical Review* 42, no. 4 (1952): 551–61.

Pannier, F. "Mangroves Impacted by Human-Induced Disturbances: A Case Study of the Orinoco Delta Mangrove Ecosystem." *Environmental Management* 3, no. 3 (1979): 205–16.

Perez, Francisco L. "The Ecological Impact of Cattle on Caulescent Andean Rosettes in a High Venezuelan Paramo." *Mountain Research and Development* 12, no. 1 (1992): 29–46.

———. "Ecology and Morphology of Globular Mosses of Grimmia longirostris in the Paramo de Piedras Blancas, Venezuelan Andes." *Arctic and Alpine Research* 23, no. 2 (1991): 133–48.

———. "Vagrant Cryptogames in a Paramo of the High Venezuelan Andes." *Flora* 189 (1994): 263–76.

Rull, Valenti. "Holocene Global Warming and the Origin of the Neotropical Gran Sabana in the Venezuelan Guayana." *Journal of Biogeography* 34, no. 2 (2007): 279–88.

————. "Successional Patterns of the Gran Sabana (Southeastern Venezuela) Vegetation during the Last 5000 Years and Its Responses to Climatic Fluctuations and Fire." *Journal of Biogeography* 19, no. 3 (1992): 329–38.

Salgado-Labouriou, Maria, et al. "Paleoecologic Analysis of a Late-Quaternary Terrace from Macubaji, Venezuelan Andes." *Journal of Biogeography* 4, no. 4 (1977): 313–26.

Sarmiento, G. "Patterns of Specific and Phenological Diversity in the Grass Community of the Venezuelan Tropical Savannas." *Journal of Biogeography* 10, no. 5 (1983): 373–91.

Sarmiento, G., and M. Pinillos. "Patterns and Processes in a Seasonally Flooded Tropical Plain: The Apure Llanos, Venezuela." *Journal of Biogeography* 28, no. 8 (2001): 985–96.

Silva, J. F., et al. "Increase in the Woody Component of Seasonal Savannas under Different Fire Regimes in Calabozo, Venezuela." *Journal of Biogeography* 28, no. 8 (2001): 977–84.

Williams, Llewelyn. "The Caura Valley and Its Forests." *Geographical Review* 31, no. 3 (1941): 414–29.

Climatology and Weather

Diaz, Jose M. "Caracterizacion climatica de Venezuela." *Revista Geografica* 102 (1985): 7–15.

Martelo, Maria, et al. "Projecto piloto de regionalizacion de parametros climaticas en una zona al Nor-oeste de Venezuela en el contexto de sistema de informacion computnizada del ambiente (SICA)." *Revista Geografica* 116 (1992): 17–29.

Marytin, Carlos E., et al. "Potential Effects of Global Climatic Change on the Phenology and Yield of Maize in Venezuela." *Clim atic Change* 29, no. 2 (1995): 189–212.

Pulwartz, Roger, et al. "Annual and Seasonal Patterns of Rainfall Variability over Venezuela." *Erdkunde* 46, nos. 3–4 (1992): 273–89

Geomorphology, Landforms, and Volcanism

Alexander, Charles. "A Comparative Study of Modern and Ancient Beach Morphologies: Insights into the Paleoclimatic of Margarita Island, Venezuela." *Journal of Geology* 90 (1982): 663–78.

Almeida, Yajaida. "Estudio preliminary sobre la variabilidad del nivel de Mar en las costas de Venezuela." *Revista Geografica* 115 (1992): 5–26.

Briceno, H. O., and C. Schubert. "Geomorphology of the Gran Sabana, Guayana Shield, Southeastern Venezuela." *Geomorphology* 3, no. 2 (1990): 125–42.

Camacho, Ronna, et al. "Caracterizacion geomorfologica de las dunas longtitudinales del Istmo de Medanos, estado Falcon, Venezuela." *Investigaciones Geograficas* 76 (2011): 7–19.

Chaves, Luis F. "Consideraciones sobre la fisiografia del Orienta Venezolano." *Revista Geografica* 54 (1961): 35–46.

Doerr, S. H. "Karst-Like Landforms and Hydrology in Quartziles of the Venezuelan Guyana Shield: Pseudokarst or 'Real' Karst." *Zeitschrift fur Geomorphologie* 43, no. 1 (1999): 1–18.

Ellenberg, L. "Coastal Types of Venezuela: An Application of Coastal Classification." *Zeitschrift fur Geomorphologie* 22, no. 4 (1978): 439–56.

————. *Morphologie Venezolandischer Kusten.* Berlin, Germany: Berliner Geographische Studien, Band 5, 1979.

Gonzalez, Luis. "Evolucion geomorfologica de la planicie lodosa de la Macolla, Peninsula de Paraguana, Estado Falcon, Venezuela." *Investigaciones Geografica* 62 (2007): 7–30.

Hitchcock, Charles B. "The Sierra de Perija, Venezuela." *Geographical Review* 44, no. 1 (1954): 1–28.

Maguire, Bassett. "Cerro de la Neblinda, Amazonas, Venezuela: A Newly Discovered Sandstone Mountain." *Geographical Review* 45, no. 1 (1955): 27–51.

Mahaney, W. C., et al. "Nye Channels (Flutinjaps) on the Humbolt Massif, Northern Venezuelan Andes." *Zeitschrift fur Geomorphologie* 49, no. 2 (2005): 253–64.

Perez, Francisco L. "Downslope Stone Transport by Needle Ice in a High Andean Area (Venezuela)." *Revue de Geomorphologie Dynamique* 36 (1987): 33–51.

———. "Miniature Sorted Strips in the Paramo de Piedras Blancas (Venezuela, Andes)." In *Periglacial Geomorphology,* edited by John C. Dixon and Athol D. Abrahams, 125–57. New York: Wiley, 1992.
———. "The Movement of Debris on a High Andean Talus." *Zeitschrift fur Geomorphologie* 32, no. 1 (1988): 77–100.
Villavicencio, Jose. "Analisis geomorfologico de la Peninsula de Paraguana." *Revista Geografica* 102 (1985): 17–27.

Soils

Perez, Francisco L. "Cryptogamic Soil Buds in the Equatorial Andes of Venezuela." *Permafrost and Periglacial Processes* 7, no. 3 (1996): 229–56.
———. "The Effects of Giant Andean Rosettes on Surface Soils along a High Paramo Toposequence." *GeoJournal* 40, no. 3 (1996): 283–98.
———. "A High-Andean Toposequence; The Geoecology of Coalescent Paramo Rosettes." *Mountain Research and Development* 15, no. 2 (1995): 133–52.
———. "The Influence of Organic Matter Addition by Caulescent Andean Rosettes on Surficial Soil Properties." *Geoderma* 54 (1991): 151–71.
———. "Matrix Granulometry of Catastropic Debris Flows (December 1999) in Central Coastal Venezuela." *Catena* 45, no. 3 (2001): 163–84.
———. "Particle Sorting due to Off-Road Vehicle Traffic in a High Andean Paramo." *Catena* 18, nos. 3–4 (1991): 239–54.
———. "Striated Soil in an Andean Paramo of Venezuela." *Arctic and Alpine Research* 16, no. 3 (1984): 277–89.
———. "Surficial Talus Movement in an Andean Paramo of Venezuela." *Geografiska Annaler* 67A, nos. 3–4 (1985): 221–38.
Schubert, C. "Geomorphology and Glacier Retreat in the Pico Bolivar Area, Sierra Nevada de Merida, Venezuela." *Zeitschrift fur Gletscherkunde und Glaziolgeologie* 8 (1972): 271–84.
———. "Glaciation and Periglacial Morphology in the Northwestern Venezuelan Andes." *Eiszeitalter un Gegenwart* 26 (1975): 196–211.

Theses and Dissertations

Bayard, C. W. "Ecological Footprint of Energy Development in Eastern Venezuela's Heavy Oil Belt." PhD diss., University of Florida, Gainesville, 2008.
Carlsson, E. "Understanding Influences on Harvesting Species of the Genes Heteropsis and Basket Production by Indigenous Ye'kwana of the Orinoco Basin, Venezuela." Master's thesis, University of Florida, Gainesville, 2012.
Lahey, James F. "Dry Littoral of Northern Venezuela." Master's thesis, University of Wisconsin, Madison, 1948.

POLITICAL GEOGRAPHY

General Works

Austin, Joseph. "Venezuela's Territorial Claims." *Bulletin of the Geographical Society of Philadelphia* 2, no. 1 (1896): 1–19.
Bressan, Pedro A. "La regionalizacion en Venezuela: macroadministrativo." *Revista Geografica* 102 (1985): 123–32.
Chaves, Luis F. "Division politico-territorial y ordenacion del territorio: El caso del Estado Merida." *Revista Geografica* 102 (1985): 143–57.
Cordero, Elias. "Aspectos espaciales del VII Plan de la Nacion." *Revista Geografica* 102 (1985): 133–41.

Crist, Raymond E. "Along the Llanos-Andes Border in Venezuela: Then and Now." *Geographical Review* 46, no. 2 (1956): 187–208.
———. "Along the Llanos-Andes Border in Zamora, Venezuela." *Geographical Review* 22, no. 3 (1932): 411–22.
Jaspe-Alvarez, Jose, and Paul-Yves Denis. "Estudio de la distribucion especial del sistema cooperative de 'Ferias de Consumo Fanciliar (FCF) de su Papel en el abas tecimiento alimentario en la region centro-occidental de Venezuela." *Revista Geografica* 114 (1991): 5–36.
Jones, Richard C. "Regional Income Inequalities and Government Investment in Venezuela." *Journal of Developing Areas* 16 (1982): 373–89.
Olivo, Beatriz. "Hacia el desarrollo integral de las islas venezolans en el Caribe." *Revista Geografica* 102 (1985): 169–75.
Rivero Santos, Angelo A. "Neighborhood Associations in Venezuela: Los Vecinas Voice Their Dissent." *Yearbook, Conference of Latin Americanist Geographers* 21 (1995): 1–12.
Roucek, Joseph S. "Venezuela in Geopolitics." *Contemporary Review* 203, no. 1 (1963): 84–87; 203, no. 2 (1963): 126–32.
Segnini, Isbelia, et al. "Algunos aspectos del conflicto urbano-rural, casos espacificas en Venezuela." *Revista Geografica* 91–92 (1980): 69–88.
Sequera de Segnini, Isbelia, et al. "La ocupacion de los espacios fronterizos como medio para reafirmar la soberania territorial: Case espacios colindontes con la Guayana Esequiba." *Revista Geografica* 102 (1985): 159–62.
Sletto, Bjorn. "Conservation Planning, Boundary-Making, and Border Terrains: The Desire for Forest and Order in the Gran Sabana, Venezuela." *Geoforum* 42, no. 2 (2011): 197–210.
Tarazona, Angel, and Andre Gide. "Conservacion y administracion del ambiente en Venezuela." *Revista Geografica* 102 (1985): 163–67.

URBAN GEOGRAPHY

General Works

Abreu, E., and Y. Verhasselt. "Quelques aspects geographiques du developpement de Caracas." *Les Cahiers d'Outre Mer* 34 (1981): 180–88.
Bendrat, T. A. "Ciudad Bolivar." *Journal of Geography* 8 (1910): 218–22.
Chaves, Luis F. "Changes in Settlement Patterns as a Result of Urbanization in Latin America: The Case of Venezuela." *Geografia Polonica* 39 (1978): 189–98.
Crist, Raymond E. "Merida, Venezuela: From Isolation to Integration." *Scientific Monthly* 55 (1942): 114–31.
Diaz, Keissy. "Los estudios geograficas sobre la ciudad de vida en Venezuela." *Revista Geografica* 102 (1985): 55–71.
Garcia-Sanchez, Pedro J. "La forma privative de l'urbanite-emprise securitaire et homogenization socio-spatiale a Caracas." *L'Espace Geographique* 33, no. 2 (2004): 114–30.
Gilbert, Alan, and Patsy Healy. *The Political Economy of Land: Urban Development in an Oil Economy*. Aldershot, England: Gower, 1985.
Grau, Pedro. "Ciudad venezolana medio ambiente en el siglo XIX." *Anales de Geografia de la Universidad Complutense* 15 (1995): 247–56.
Grillet, Rodolfo H. "Ciudad Guayana: Un polo de desarrollo o un enclave regional?" *Revista Geografica* 102 (1985): 93–99.
Hernandez, Eduardo. "Relaciones ciudad-campo en America Latina: El caso de Venezuela." *Revista Geografica de America Central* 15–16 (1983): 163–74.
Irazabal, Clara. "A Planned City Comes of Age: Rethinking Ciudad Guayana Today." *Journal of Latin American Geography* 3, no. 1 (2004): 22–51.
Jones, Emrys. "Aspects of Urbanisation in Venezuela." *Ekistics* 18 (1964): 420–25.
Lombardi, John V. "The Rise of Caracas as a Primate City." In *Social Fabric and Spatial Structure in Colonial Latin America*, edited by David J. Robinson, 433–72. Syracuse, NY: Syracuse University, Department of Geography, 1979.

Lynch, Edward. "Propositions for Planning New Towns in Venezuela." *Journal of Developing Areas* 7, no. 4 (1973): 549–70.

Martinez-Tirado, Nestor. "Incidencias del proceso de urbanization en Venezuela." *Revista Geografica* 102 (1985): 73–80.

Morris, Arthur S. "Urban Growth Patterns in Latin America with Illustrations from Caracas." *Urban Studies* 15, no. 3 (1978): 299–312.

Penfold, Anthony H. "Ciudad Guyana: Planning A New City in Venezuela." *Town Planning Review* 36 (1966): 225–48.

Robinson, David J. "The City as Centre of Change in Modern Venezuela." In *Cities in a Changing Latin America: Two Studies of Urban Growth in the Development of Mexico and Venezuela*, edited by David Fox and David J. Robinson, 23–48. London: Latin American Publications Fund, 1969.

Robinson, David J., and Michael Swann. "Geographical Interpretations of the Hispanic-American City: A Case Study of Caracas in the Late Eighteenth Century." *Proceedings, Conference of Latin Americanist Geographers* 5 (1974): 1–15.

Snyder, David E. "Ciudad Guyana: A Planned Metropolis on the Orinoco." *Inter-American Economic Affairs* 5, no. 3 (1965): 405–12.

Turner, A., and J. Smulian. "New Cities in Venezuela." *Town Planning Review* 42 (1971): 3–27.

Urban Social Geography

Jones, Richard C., and G. Zannaras. "Perceived versus Objective Urban Opportunities and the Migration of Venezuelan Youths." *Annals of Regional Science* 10 (1976): 83–97.

———. "The Role of Awareness Space in Urban Residential Preferences: A Case Study of Venezuelan Youth." *Annals of Regional Science* 12 (1978): 36–52.

Morphology and Neighborhoods

Brisseau-Loaiza, J. "Les 'Barrios' de Petare, faubourges populaires d'une banlieu de Caracas." *Les Cahiers d'Outre Mer* 16 (1963): 5–42.

Chaves, Luis F. "La estructura de la ciudad de el Vigia: Desarrollo structural de un centro poblado venezolano en tierras de colonizacion reciente y en posicion nodal en e contacto de regions diversas." *Revista Geograficas* 63 (1965): 51–66.

Mitchell, Jeffrey. "Neoliberal Economic Policies and the Changing Geography of Office Development in Caracas, Venezuela." *Papers of the Applied Geography Conferences* 20 (1997): 101–9.

Penfold, Anthony H. "Caracas: Urban Growth and Transportation." *Town Planning Review* 41 (1970): 103–20.

Tata, Robert J., and Maria Campbell. "La variabilidad de los barrios de Caracas." *Revista Geografica* 102 (1985): 81–92.

Urban Economic Geography

Marchand, B. "Les ranchos de Caracas, contribucion a l'etude des bidon villes." *Les Cahiers d'Outre Mer* 19 (1966): 105–43.

Theses and Dissertations

Mitchell, Jeffrey. "From Oil Rents to Urban Rents: The Inconstant Regulation of Private Investment in the Built Environment in Caracas." PhD diss., Clark University, Worcester, MA, 1996.

Author Index

About the Author

Thomas A. Rumney, now retired, was professor of geography at Plattsburgh State University, New York. He is the author of *The Study of Agricultural Geography* (2005), *Canadian Geography: A Scholarly Bibliography* (2010), and *Caribbean Geography: A Scholarly Bibliography* (2012), and *The Geography of Central America and Mexico: A Scholarly Guide and Bibliography* (2013), all published by Scarecrow Press.

CPSIA information can be obtained
at www.ICGtesting.com
Printed in the USA
LVHW091801081219
639821LV00008B/132/P